AF358484

Colloidal Quantum Dots for Nanophotonic Devices

Colloidal Quantum Dots for Nanophotonic Devices

Guest Editors

Menglu Chen
Qun Hao

Basel • Beijing • Wuhan • Barcelona • Belgrade • Novi Sad • Cluj • Manchester

Guest Editors

Menglu Chen
School of Optics and Phonics
Beijing Institute of Technology
Beijing
China

Qun Hao
School of Physics
Changchun University of
Science And Technology
Changchun
China

Editorial Office
MDPI AG
Grosspeteranlage 5
4052 Basel, Switzerland

This is a reprint of the Special Issue, published open access by the journal *Materials* (ISSN 1996-1944), freely accessible at: www.mdpi.com/journal/materials/special_issues/CQD_Materials.

For citation purposes, cite each article independently as indicated on the article page online and using the guide below:

Lastname, A.A.; Lastname, B.B. Article Title. *Journal Name* **Year**, *Volume Number*, Page Range.

ISBN 978-3-7258-2622-3 (Hbk)
ISBN 978-3-7258-2621-6 (PDF)
https://doi.org/10.3390/books978-3-7258-2621-6

© 2024 by the authors. Articles in this book are Open Access and distributed under the Creative Commons Attribution (CC BY) license. The book as a whole is distributed by MDPI under the terms and conditions of the Creative Commons Attribution-NonCommercial-NoDerivs (CC BY-NC-ND) license (https://creativecommons.org/licenses/by-nc-nd/4.0/).

Contents

About the Editors

Menglu Chen

Prof. Dr. Menglu Chen obtained a B.S. from the School of Gifted Young, University of Science and Technology of China, in 2015, and a Ph.D. supervised by Prof. Philippe Guyot-Sionnest from the Physics Department, University of Chicago, in 2020. She is now a full professor at the School of Optics and Photonics, Beijing Institute of Technology. She achieves honors like the IAAM Scientist Medal (2022), Young Elite Scientists by the China Association for Science and Technology (2021), Excellent Young Scholars overseas (2021), Beijing Nova Program (2021), and Yodh Prize (University of Chicago, 2019). Her research mainly focuses on colloidal quantum dots and their application to photodetection, especially at infrared wavelengths. She has published more than 40 papers, including *Nature Materials, Light: Science & Applications, ACS Nano*, etc.

Qun Hao

Professor Dr. Qun Hao was born in 1968. She received her Ph.D. degree in optical instruments from Tsinghua University, Beijing, China, in 1998. She is a visiting professor at the Case Western Reserve University, Cleveland, OH, USA (2011) and a guest researcher at the University of Tokyo, Tokyo, Japan (1999–2001). She is an executive director of the Chinese Optical Society, the director of the China Instrument and Control Society, the director of the Optics Council in the China Ordnance Society, the director of the Photoelectric Technology Council in the Chinese Optical Society, and the director of the Chinese Society for Measurement. She became a full professor in 2003. She is currently the president of Changchun University of Science and Technology, China. Her research interests include infrared materials and photodetection, measurement technology and intelligent instruments, imaging systems and technology, and precision measurement for bionics studies. She has authored more than 150 *SCI* papers, including *Nature Photonics, Light: Science and Applications, Science Advances*, etc.

 *materials*

Editorial

Colloidal Quantum Dots for Nanophotonic Devices

Menglu Chen [1] and Qun Hao [1,2,*]

1   School of Optics and Photonics, Beijing Institute of Technology, Beijing 100081, China; menglu@bit.edu.cn
2   Physics Department, Changchun University of Science and Technology, Changchun 130022, China
*   Correspondence: qhao@bit.edu.cn

check for
updates

Citation: Chen, M.; Hao, Q. Colloidal Quantum Dots for Nanophotonic Devices. *Materials* **2024**, *17*, 2471.
https://doi.org/10.3390/ma17112471

Received: 7 May 2024
Revised: 10 May 2024
Accepted: 14 May 2024
Published: 21 May 2024

Copyright: © 2024 by the authors. Licensee MDPI, Basel, Switzerland. This article is an open access article distributed under the terms and conditions of the Creative Commons Attribution (CC BY) license (https://creativecommons.org/licenses/by/4.0/).

Colloidal quantum dots (CQDs) have unique advantages in the wide tunability of visible-to-infrared emission wavelength and low-cost solution processibility [1–5]. Therefore, they have become an important class of materials with great potential for applications in fields such as biological medicine [6,7], optoelectronics [8–12], and quantum information [13]. The performance of CQD-based photovoltaic and light-emitting devices has become competitive with other state-of-the-art materials [14–16]. Narrow-band semiconductor CQDs also hold unique promise for infrared technologies [17,18]. Thus, new and in-depth insights into CQD growth, chemical transformations, and physical properties would benefit not only the purely fundamental side but also commercialization. This Special Issue (SI), "Colloidal Quantum Dots for Nanophotonic Devices", presents recent and CQD-related information ranging from CQD material chemistry and characterization to processing and device fabrication. This SI contains ten articles, including seven research articles and three review articles. This editorial aims to summarize the publications included in this SI.

CQDs have the advantages of a broad spectral tuning range, low preparation cost, and compatibility with silicon-based readout integrated circuits via solution processing [1,5,8,19–21]. As a result, CQD photodetectors have become a research hotspot in recent years. Lead chalcogenide CQDs and mercury chalcogenide CQDs, as the mainstream materials of CQDs, have excellent infrared detection performance and have become the most ideal materials for infrared photodetectors [22–25]. Zhao et al. [26] and Hao et al. [27] summarized the recent development of infrared photodetectors based on lead chalcogenide CQDs and mercury chalcogenide CQDs, respectively.

In addition to showing superior performance in the detector field, CQDs also have obvious advantages in spectral filtering [28–30]. This indicates that CQDs have great potential for applications in microspectrometers. Qiu et al. [31] presented the advances of micro spectrometers based on material nanoarchitectonics, which pointed to the direction for researchers to study novel low-dimensional materials in this field. Qiao et al. [32] reported that CQDs can be introduced as a sacrificial layer when polishing single-crystal silicon carbide (SiC) using pulsed ion beam sputtering to improve surface quality. This provides a new idea for achieving high-precision fabrication of ultra-smooth single-crystal SiC surfaces.

Low-dimensional materials have a wide range of applications in various fields, such as photovoltaic devices, of which well-designed heterojunctions can make it possible to improve the performance of the devices [33]. In the study by Wang et al. [34], $CH_3NH_3PbI_3/Au/Mg_{0.2}Zn_{0.8}O$ heterojunction self-powered photodetectors with a high responsivity of 0.58 A/W were demonstrated, where $CH_3NH_3PbI_3$ and $Mg_{0.2}Zn_{0.8}O$ acted as the p-type and n-type layer, respectively. This work provides new concepts for the study of perovskite photodetectors with low dark current and high detectivity. Sun et al. [35] reported flexible CZTSSe/ZnO solar cells by optimizing ZnO buffer layers, achieving the maximum power conversion efficiency of 5.0%.

To improve the detection performance of photodetectors, optical structures and photosensitive materials need to be efficiently combined to enhance light absorption. In

Materials **2024**, *17*, 2471

photoelectric devices, a metal microstructure can be engineered to squeeze light into the sub-diffractive region, and then, the plasmon exciton resonance phenomenon boosts light absorption [36]. Lin et al. [37] investigated a performance-enhanced GaAs nanowire photodetector by introducing Au nanoparticles prepared by thermal evaporation. Assisted by the coupling of electron gas in the Au nanoparticles to the excitation light, the photocurrent and responsivity of this proposed photodetector were raised. Pierini et al. [38] designed a phototransistor that combined two resonances by utilizing a lithium-ion glass gating of HgTe nanocrystal film, realizing a high responsivity.

In addition, the performance and reliability of optoelectronics can also be improved by studying and optimizing the optical material characteristics. Zhao et al. [39] enhanced the homogeneity of large-scale nanorods by controlling the solution flow and tuning the electric field distribution, thus improving their light utilization efficiency. The optimized ZnO NRs have promising applications in solar cells and collector systems. Bai et al. [40] investigated the spatial shifts of the reflected light beam on hexagonal boron nitride (hBN)/alpha-molybdenum (α-MoO$_3$) trioxide structure. They successfully enhanced hBN in-plane anisotropy by twisting. This study provides theoretical guidance for novel nanophotonic devices and optical encoders.

Conflicts of Interest: The authors declare no conflict of interest.

References

1. Talapin, D.V.; Lee, J.-S.; Kovalenko, M.V.; Shevchenko, E.V. Prospects of Colloidal Nanocrystals for Electronic and Optoelectronic Applications. *Chem. Rev.* **2010**, *110*, 389–458. [CrossRef] [PubMed]
2. Smith, A.M.; Nie, S. Semiconductor Nanocrystals: Structure, Properties, and Band Gap Engineering. *Acc. Chem. Res.* **2010**, *43*, 190–200. [CrossRef] [PubMed]
3. Efros, A.L.; Brus, L.E. Nanocrystal Quantum Dots: From Discovery to Modern Development. *ACS Nano* **2021**, *15*, 6192–6210. [CrossRef] [PubMed]
4. Kovalenko, M.V.; Manna, L.; Cabot, A.; Hens, Z.; Talapin, D.V.; Kagan, C.R.; Klimov, V.I.; Rogach, A.L.; Reiss, P.; Milliron, D.J.; et al. Prospects of Nanoscience with Nanocrystals. *ACS Nano* **2015**, *9*, 1012–1057. [CrossRef] [PubMed]
5. Lu, H.; Carroll, G.M.; Neale, N.R.; Beard, M.C. Infrared Quantum Dots: Progress, Challenges, and Opportunities. *ACS Nano* **2019**, *13*, 939–953. [CrossRef] [PubMed]
6. Medintz, I.L.; Uyeda, H.T.; Goldman, E.R.; Mattoussi, H. Quantum Dot Bioconjugates for Imaging, Labelling and Sensing. *Nat. Mater.* **2005**, *4*, 435–446. [CrossRef] [PubMed]
7. Chattopadhyay, P.K.; Price, D.A.; Harper, T.F.; Betts, M.R.; Yu, J.; Gostick, E.; Perfetto, S.P.; Goepfert, P.; Koup, R.A.; De Rosa, S.C.; et al. Quantum Dot Semiconductor Nanocrystals for Immunophenotyping by Polychromatic Flow Cytometry. *Nat. Med.* **2006**, *12*, 972–977. [CrossRef]
8. Konstantatos, G.; Howard, I.; Fischer, A.; Hoogland, S.; Clifford, J.; Klem, E.; Levina, L.; Sargent, E.H. Ultrasensitive Solution-Cast Quantum Dot Photodetectors. *Nature* **2006**, *442*, 180–183. [CrossRef] [PubMed]
9. Keuleyan, S.; Lhuillier, E.; Brajuskovic, V.; Guyot-Sionnnest, P. Mid-infrared HgTe Colloidal Quantum Dot Photodetectors. *Nat. Photonics* **2011**, *5*, 489–493. [CrossRef]
10. Caillas, A.; Guyot-Sionnest, P. Uncooled High Detectivity Mid-Infrared Photoconductor Using HgTe Quantum Dots and Nanoantennas. *ACS Nano* **2024**, *18*, 8952–8960. [CrossRef]
11. Xue, X.; Chen, M.; Luo, Y.; Qin, T.; Tang, X.; Hao, Q. High-Operating-Temperature Mid-Infrared Photodetectors via Quantum Dot Gradient Homojunction. *Light Sci. Appl.* **2023**, *12*, 2. [CrossRef] [PubMed]
12. Xue, X.; Hao, Q.; Chen, M. Very Long Wave Infrared Quantum Dot Photodetector up to 18 µm. *Light Sci. Appl.* **2024**, *13*, 89. [CrossRef] [PubMed]
13. Hanschke, L.; Fischer, K.A.; Appel, S.; Lukin, D.; Wierzbowski, J.; Sun, S.; Trivedi, R.; Vučković, J.; Finley, J.J.; Müller, K. Quantum Dot Single-Photon Sources with Ultra-Low Multi-Photon Probability. *NPJ Quantum Inf.* **2018**, *4*, 43. [CrossRef]
14. Yuan, M.; Liu, M.; Sargent, E. Colloidal Quantum Dot Solids for Solution-Processed Solar Cells. *Nat. Energy* **2016**, *1*, 16016. [CrossRef]
15. Gong, X.; Yang, Z.; Walters, G.; Comin, R.; Ning, Z.; Beauregard, E.; Adinolfi, V.; Voznyy, O.; Sargent, E.H. Highly Efficient Quantum Dot Near-Infrared Light-Emitting Diodes. *Nat. Photonics* **2016**, *10*, 253–257. [CrossRef]
16. Pradhan, S.; Di Stasio, F.; Bi, Y.; Gupta, S.; Christodoulou, S.; Stavrinadis, A.; Konstantatos, G. High-Efficiency Colloidal Quantum Dot Infrared Light-Emitting Diodes via Engineering at the Supra-Nanocrystalline Level. *Nat. Nanotechnol.* **2019**, *14*, 72–79. [CrossRef] [PubMed]
17. Kovalenko, M.V.; Kaufmann, E.; Pachinger, D.; Roither, J.; Huber, M.; Stangl, J.; Hesser, G.; Schäffler, F.; Heiss, W. Colloidal HgTe Nanocrystals with Widely Tunable Narrow Band Gap Energies: From Telecommunications to Molecular Vibrations. *J. Am. Chem. Soc.* **2006**, *128*, 3516–3517. [CrossRef] [PubMed]

18. Song, H.; Eom, S.Y.; Kim, G.; Jung, Y.S.; Choi, D.; Kumar, G.S.; Lee, J.H.; Kang, H.S.; Ban, J.; Seo, G.W.; et al. Narrow Bandgap Silver Mercury Telluride Alloy Semiconductor Nanocrystal for Self-Powered Midwavelength-Infrared Photodiode. *Commun. Mater.* **2024**, *5*, 60. [CrossRef]
19. Chu, A.; Martinez, B.; Ferré, S.; Noguier, V.; Gréboval, C.; Livache, C.; Qu, J.; Prado, Y.; Casaretto, N.; Goubet, N.; et al. HgTe Nanocrystals for SWIR Detection and Their Integration up to the Focal Plane Array. *ACS Appl. Mater. Interfaces* **2019**, *11*, 33116–33123. [CrossRef]
20. Greboval, C.; Darson, D.; Parahyba, V.; Alchaar, R.; Abadie, C.; Noguier, V.; Ferre, S.; Izquierdo, E.; Khalili, A.; Prado, Y.; et al. Photoconductive Focal Plane Array Based on HgTe Quantum Dots for Fast and Cost-Effective Short-Wave Infrared Imaging. *Nanoscale* **2022**, *14*, 9359–9368. [CrossRef]
21. Liu, J.; Liu, P.; Chen, D.; Shi, T.; Qu, X.; Chen, L.; Wu, T.; Ke, J.; Xiong, K.; Li, M.; et al. A Near-Infrared Colloidal Quantum Dot Imager with Monolithically Integrated Readout Circuitry. *Nat. Electron.* **2022**, *5*, 443–451. [CrossRef]
22. Pietryga, J.M.; Schaller, R.D.; Werder, D.; Stewart, M.H.; Klimov, V.I.; Hollingsworth, J.A. Pushing the Band Gap Envelope: Mid-Infrared Emitting Colloidal PbSe Quantum Dots. *J. Am. Chem. Soc.* **2004**, *126*, 11752–11753. [CrossRef] [PubMed]
23. Ackerman, M.M.; Tang, X.; Guyot-Sionnest, P. Fast and Sensitive Colloidal Quantum Dot Mid-Wave Infrared Photodetectors. *ACS Nano* **2018**, *12*, 7264–7271. [CrossRef] [PubMed]
24. Chen, M.; Hao, Q.; Luo, Y.; Tang, X. Mid-Infrared Intraband Photodetector via High Carrier Mobility HgSe Colloidal Quantum Dots. *ACS Nano* **2022**, *16*, 11027–11035. [CrossRef]
25. Deng, Y.; Pang, C.; Kheradmand, E.; Leemans, J.; Bai, J.; Minjauw, M.; Liu, J.; Molkens, K.; Beeckman, J.; Detavernier, C.; et al. Short-Wave Infrared Colloidal QD Photodetector with Nanosecond Response Times Enabled by Ultrathin Absorber Layers. *Adv. Mater.* **2024**, 2402002. [CrossRef] [PubMed]
26. Zhao, X.; Ma, H.; Cai, H.; Wei, Z.; Bi, Y.; Tang, X.; Qin, T. Lead Chalcogenide Colloidal Quantum Dots for Infrared Photodetectors. *Materials* **2023**, *16*, 5790. [CrossRef]
27. Hao, Q.; Ma, H.; Xing, X.; Tang, X.; Wei, Z.; Zhao, X.; Chen, M. Mercury Chalcogenide Colloidal Quantum Dots for Infrared Photodetectors. *Materials* **2023**, *16*, 7321. [CrossRef]
28. Bao, J.; Bawendi, M. A Colloidal Quantum Dot Spectrometer. *Nature* **2015**, *523*, 67–70. [CrossRef] [PubMed]
29. Zhu, X.; Bian, L.; Fu, H.; Wang, L.; Zou, B.; Dai, Q.; Zhang, J.; Zhong, H. Broadband Perovskite Quantum Dot Spectrometer Beyond Human Visual Resolution. *Light Sci. Appl.* **2020**, *9*, 73. [CrossRef]
30. Li, H.; Bian, L.; Gu, K.; Fu, H.; Yang, G.; Zhong, H.; Zhang, J. A Near-Infrared Miniature Quantum Dot Spectrometer. *Adv. Opt. Mater.* **2021**, *9*, 2100376. [CrossRef]
31. Qiu, Y.; Zhou, X.; Tang, X.; Hao, Q.; Chen, M. Micro Spectrometers Based on Materials Nanoarchitectonics. *Materials* **2023**, *16*, 2253. [CrossRef]
32. Qiao, D.; Shi, F.; Tian, Y.; Zhang, W.; Xie, L.; Guo, S.; Song, C.; Tie, G. Ultra-Smooth Polishing of Single-Crystal Silicon Carbide by Pulsed-Ion-Beam Sputtering of Quantum-Dot Sacrificial Layers. *Materials* **2024**, *17*, 157. [CrossRef]
33. Furchi, M.M.; Pospischil, A.; Libisch, F.; Burgdörfer, J.; Mueller, T. Photovoltaic Effect in an Electrically Tunable van der Waals Heterojunction. *Nano Lett.* **2014**, *14*, 4785–4791. [CrossRef] [PubMed]
34. Wang, M.; Zhao, M.; Jiang, D. $CH_3NH_3PbI_3$/Au/$Mg_{0.2}Zn_{0.8}O$ Heterojunction Self-Powered Photodetectors with Suppressed Dark Current and Enhanced Detectivity. *Materials* **2023**, *16*, 4330. [CrossRef]
35. Sun, Q.; Tang, J.; Zhang, C.; Li, Y.; Xie, W.; Deng, H.; Zheng, Q.; Wu, J.; Cheng, S. Efficient Environmentally Friendly Flexible CZTSSe/ZnO Solar Cells by Optimizing ZnO Buffer Layers. *Materials* **2023**, *16*, 2869. [CrossRef] [PubMed]
36. Schuller, J.A.; Barnard, E.S.; Cai, W.; Jun, Y.C.; White, J.S.; Brongersma, M.L. Plasmonics for Extreme Light Concentration and Manipulation. *Nat. Mater.* **2010**, *9*, 193–204. [CrossRef] [PubMed]
37. Lin, F.; Cui, J.; Zhang, Z.; Wei, Z.; Hou, X.; Meng, B.; Liu, Y.; Tang, J.; Li, K.; Liao, L.; et al. GaAs Nanowire Photodetectors Based on Au Nanoparticles Modification. *Materials* **2023**, *16*, 1735. [CrossRef]
38. Pierini, S.; Abadie, C.; Dang, T.H.; Khalili, A.; Zhang, H.; Cavallo, M.; Prado, Y.; Gallas, B.; Ithurria, S.; Sauvage, S.; et al. Lithium-Ion Glass Gating of HgTe Nanocrystal Film with Designed Light-Matter Coupling. *Materials* **2023**, *16*, 2335. [CrossRef]
39. Zhao, Y.; Li, K.; Hu, Y.; Hou, X.; Lin, F.; Tang, J.; Tang, X.; Xing, X.; Zhao, X.; Zhu, H.; et al. The Effect of the Solution Flow and Electrical Field on the Homogeneity of Large-Scale Electrodeposited ZnO Nanorods. *Materials* **2024**, *17*, 1241. [CrossRef]
40. Bai, S.; Li, Y.; Cui, X.; Fu, S.; Zhou, S.; Wang, X.; Zhang, Q. Spatial Shifts of Reflected Light Beam on Hexagonal Boron Nitride/Alpha-Molybdenum Trioxide Structure. *Materials* **2024**, *17*, 1625. [CrossRef]

Disclaimer/Publisher's Note: The statements, opinions and data contained in all publications are solely those of the individual author(s) and contributor(s) and not of MDPI and/or the editor(s). MDPI and/or the editor(s) disclaim responsibility for any injury to people or property resulting from any ideas, methods, instructions or products referred to in the content.

materials

Review

Mercury Chalcogenide Colloidal Quantum Dots for Infrared Photodetectors

Qun Hao [1,2,†], Haifei Ma [2,†], Xida Xing [1], Xin Tang [1,2], Zhipeng Wei [1], Xue Zhao [2,*] and Menglu Chen [1,2,*]

[1] Physics Department, Changchun University of Science and Technology, Changchun 130022, China; qhao@bit.edu.cn (Q.H.); xingxida@163.com (X.X.); xintang@bit.edu.cn (X.T.); weizp@cust.edu.cn (Z.W.)
[2] School of Optics and Photonics, Beijing Institute of Technology, Beijing 100081, China; 3220220493@bit.edu.cn
* Correspondence: 3220210544@bit.edu.cn (X.Z.); menglu@bit.edu.cn (M.C.)
† These authors contributed equally to this work.

Abstract: In recent years, mercury chalcogenide colloidal quantum dots (CQDs) have attracted widespread research interest due to their unique electronic structure and optical properties. Mercury chalcogenide CQDs demonstrate an exceptionally broad spectrum and tunable light response across the short-wave to long-wave infrared spectrum. Photodetectors based on mercury chalcogenide CQDs have attracted considerable attention due to their advantages, including solution processability, low manufacturing costs, and excellent compatibility with silicon substrates, which offers significant potential for applications in infrared detection and imaging. However, practical applications of mercury-chalcogenide-CQD-based photodetectors encounter several challenges, including material stability, morphology control, surface modification, and passivation issues. These challenges act as bottlenecks in further advancing the technology. This review article delves into three types of materials, providing detailed insights into the synthesis methods, control of physical properties, and device engineering aspects of mercury-chalcogenide-CQD-based infrared photodetectors. This systematic review aids researchers in gaining a better understanding of the current state of research and provides clear directions for future investigations.

Keywords: colloidal quantum dots; infrared photodetector; mercury chalcogenide

Citation: Hao, Q.; Ma, H.; Xing, X.; Tang, X.; Wei, Z.; Zhao, X.; Chen, M. Mercury Chalcogenide Colloidal Quantum Dots for Infrared Photodetectors. *Materials* **2023**, *16*, 7321. https://doi.org/10.3390/ma16237321

Academic Editor: Heesun Yang

Received: 24 October 2023
Revised: 21 November 2023
Accepted: 22 November 2023
Published: 24 November 2023

Copyright: © 2023 by the authors. Licensee MDPI, Basel, Switzerland. This article is an open access article distributed under the terms and conditions of the Creative Commons Attribution (CC BY) license (https:// creativecommons.org/licenses/by/ 4.0/).

1. Introduction

Extensive applications abound for infrared (IR) detection technology, spanning telecommunications, chemical spectroscopy, gas sensing, biomedical applications, military night vision, and autonomous driving [1–11]. However, the current commercial IR photodetectors are primarily based on epitaxially grown semiconductors such as InGaAs and HgCdTe [12–20]. Although these detectors offer advantages such as high sensitivity and good stability, they suffer from drawbacks including high manufacturing costs and incompatibility with silicon-based readout integrated circuits and are hindered by intricate epitaxial growth processes, which hinder further advancements in IR detection technology [21–24]. Building upon this, novel infrared detectors such as the quantum well IR photodetector [25–30], type-II superlattice [31–35], infrared detectors based on two-dimensional materials [36–40], and quantum dot IR photodetector [41–48] have begun to be extensively investigated.

Colloidal quantum dots (CQDs), as alternatives to traditional epitaxial-grown semiconductors, have emerged with unique advantages such as the ease of processing solutions, cost-effectiveness, scalability, and the ability to adjust size [49–53]. These advantages present a promising outlook for low-cost, large-scale, high-resolution, and small-pixel IR detectors. Over the past decade, CQDs have found widespread applications in various fields, including solar cells [54–56], spectroscopy [57], phototransistors [58–60], focal plane array (FPA) imagers [61], lasers [62–64], and light-emitting diodes [65,66]. Among them, mercury chalcogenide CQDs, due to their large Bohr radius and wide tunable size range, theoretically cover the critical IR spectral bands, positioning them as ideal materials for IR detectors that have the potential for superior detection performance [67,68]. Mercury chalcogenide CQDs encompass materials including HgTe, HgSe, and HgS CQDs. However, quantum dot photodetectors still face challenges, including material stability, surface modification, and difficulties in integrating with FPA [69].

This paper provides an overview of the recent progress on mercury chalcogenide CQD IR detectors and discusses the future directions for mercury chalcogenide CQDs in IR detection technology. Compared with other review articles [70,71], our discussion takes a detailed approach to three different materials and covers aspects such as material synthesis, optimization, and device fabrication. Additionally, we provide an in-depth analysis of the infrared detector structure based on HgTe CQDs, considering material synthesis, optimization, and device preparation. This comprehensive research approach makes an important contribution to the understanding of these materials in the context of current infrared detection technologies.

2. Research on HgTe CQDs
2.1. Synthesis of HgTe CQDs

Zero-gap HgTe quantum dots exhibit extensive potential applications owing to their elevated absorption coefficients, adjustable bandgaps spanning the entire infrared range, and distinctive optoelectronic characteristics [72–74].

In the late 1990s, Rogach et al. drew inspiration from their previous research on CdTe nanoparticles to propose an aqueous colloidal growth technique to synthesize HgTe CQDs [75–77]. They utilized 1-thioglycerol (1-mercaptopropane-2,3-diol) as the size-regulating capping agent. In the presence of thioglycerol as a ligand, mercury perchlorate was dissolved in water and, simultaneously, H_2Te was bubbled into this solution as a source of Te^{2-}. The inclusion of H_2Te increased the complexity of the synthetic setup for aqueous synthesis. As a result, alternative methods for growth in organic media were subsequently developed.

In 2011, Keuleyan et al. successfully extended the absorption edge of HgTe CQDs to over 5 µm by improving the synthesis technology. They conducted a comprehensive discussion on the synthesis process of HgTe CQDs [78] (Figure 1a–c) and studied the optical properties [79], photoluminescence properties [80], and transport properties [81,82] of HgTe CQD films. These detailed studies provided a solid theoretical foundation for the design of HgTe CQD infrared detectors. They also first reported HgTe CQD photodetectors with a room-temperature photoresponse beyond 5 µm [83]. In the same year, Liu et al. discussed the electronic structure of HgTe CQDs [84] (Figure 1d–f). Employing liquid gating, they characterized the electrochemical response of HgTe CQD films, investigating the feasibility of achieving n-type and p-type charging and exploring the distinctions in electron and hole transport properties. They demonstrated electrochemical modulation of n-type and p-type carriers in HgTe CQD films. The results showed that although the carrier mobility of both n-type and p-type HgTe CQD films could reach 0.1 cm^2/Vs, the p-type films demonstrated an increased photocurrent, whereas the n-type films displayed a more pronounced magnetoresistance effect.

In order to ensure the stable growth and good colloidal dispersion of CQDs, long-chain organic molecules are used as ligands during the synthesis stage of CQDs [85]. However, due to the influence of their own volume, substantial steric hindrance arises from the presence of elongated organic molecules with extensive chain lengths. Therefore, ligand exchange is often required during the preparation of CQDs to increase the coupling between CQDs via exchanging long-chain ligands for short-chain ligands in order to improve the carrier mobility of CQDs. Amine ligands exhibit low stability, making them prone to detachment during the centrifugal purification stage and thereby causing the aggregation of CQDs. On the other hand, thiol ligands form robust bonds with the Hg sites on the surface of HgTe CQDs, posing challenges for subsequent surface modifications [71].

Therefore, the selection of ligands and the improvement of the ligand-exchange technique are of paramount importance.

In order to find suitable ligands, Lhuillier et al. explored the inorganic ligand exchange route in 2013, where they used arsenic sulfide (As_2S_3) to perform ligand exchange with HgTe CQD films [86]. The mobility of the As_2S_3-treated film was approximately 10^{-2} cm^2/Vs, which was 100 times higher than that of the EDT-treated film. This study showed that the exchange and encapsulation of inorganic ligands could improve the electrical and optical properties of CQD films. In addition, using As_2S_3 solution to encapsulate CQDs could slow down the oxidation rate of CQD films in air. In order to enhance the stability of CQDs, Keuleyan et al. synthesized HgTe CQDs with high monodispersity by diluting the TOP (Te precursor with oleylamine), with a size up to 15 nm, corresponding to the room-temperature absorption edge at 5 μm [87] (Figure 1g–i). Diluted Te precursors could be fully mixed with $HgCl_2$ solution to ensure a more uniform mixture before significant growth of CQDs occurred while maintaining a low enough temperature to sufficiently prevent additional nucleation. This made the size distribution of HgTe CQDs much improved. In 2017, Shen et al. mitigated the tendency of HgTe CQD aggregation by employing a more reactive tellurium source and an excess of mercury precursors. In addition, the size of the CQDs was adjustable from 4.8 to 11.5 nm [88] (Figure 2a). At the same time, the CQDs could be stably dispersed without mercaptan, which was conducive to further ligand exchange and the nuclear/shell structure growth of CQDs to improve the device performance.

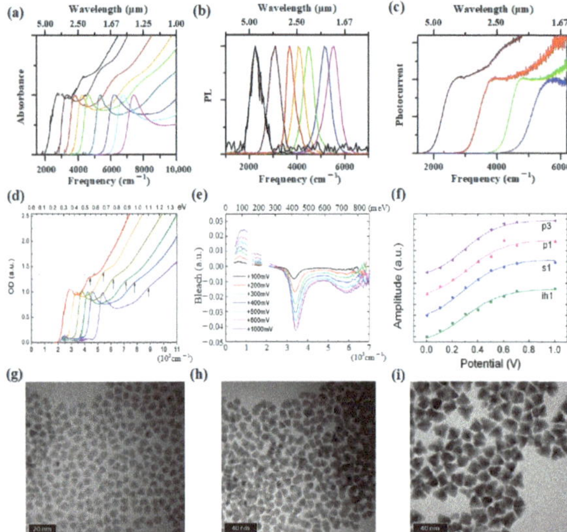

Figure 1. HgTe CQD Characterization. (**a**) Absorption spectra of HgTe CQDs of different sizes in C_2Cl_4 are presented, with the C-H absorbance from the ligands subtracted for clarity. (**b**) Photoluminescence (PL) spectra of HgTe CQDs of different sizes in C_2Cl_4. (**c**) Photoconduction spectra are presented for films composed of colloidal HgTe nanoparticles with varying sizes. The films exhibit typical optical densities of approximately 0.1 at the exciton and have an approximate thickness of 100 nm [78]. Copyright 2011, *J Am Chem Soc*. (**d**) The absorption spectra of HgTe CQDs of different sizes in tetrachloroethylene (TCE) solution at room temperature are depicted. A second feature is emphasized with vertical arrows. The region around the C-H stretch (2900 cm^{-1}) has been excluded and substituted with dashed lines. (**e**) Differential spectra of a HgTe film recorded under positive potentials at 210 K is presented. Spectral regions where the electrolyte absorbs more than 70% have been excluded from the data. (**f**) The amplitude of the Gaussians corresponding to each transition as

a function of the applied potential [84]. Copyright 2011, *The Journal of Physical Chemistry C*. (**g–i**) Transmission electron microscopy (TEM) images of particles with sizes of (**g**) 5.5, (**h**) 9.1, and (**i**) 15.5 nm [87]. Copyright 2014, *ACS Nano*.

For the improvement of ligand exchange technique, Chen et al. proposed a novel manufacturing process for HgTe CQD films in 2019 [89] (Figure 2d–f). They replaced the bulky oleylamine ligands with a combination of $HgCl_2$, 2-mercaptoethanol, n-butylamine, and n-butylammonium chloride. This substitution resulted in the formation of stable nanoinks in N, N-dimethylformamide (DMF) and achieved high electron mobility in HgTe CQD films without compromising the discrete nature of the electronic states. During the ligand exchange process, the doping state of HgTe CQDs was controlled by adjusting the concentration of $HgCl_2$. The optimized HgTe CQD films exhibited a high mobility up to $1 \text{ cm}^2/\text{Vs}$. To further improve the mobility, in 2020 Lan et al. demonstrated that CQD solids can achieve high electron mobility without compromising the discrete nature of the electronic states [90]. They increased the mobility up to $8 \text{ cm}^2/\text{Vs}$. Subsequently, Xue et al. further optimized the ligand exchange technique by introducing a mixed-phase ligand exchange method. This approach segregates mobility enhancement, doping control, and Fermi level adjustment into separate steps, providing precise control over the transport properties of CQDs [91] (Figure 2b).

Researchers are constantly exploring methods to ensure the stable growth and good colloidal dispersion of HgTe CQDs. In 2022, Liu et al. introduced an innovative method for synthesizing uniform mid-infrared HgTe CQDs and optimized the synthetic process of the mercury precursor solution [92]. They used HgI_2 as the mercury source and broke the strong Hg-I bond by changing the reaction temperature. Finally, they achieved the successful in situ passivation of HgTe CQDs with I^-, enabling tunable sizes ranging from 8 to 15 nm.

In addition, Yang et al. studied a ligand engineering method that could produce well-separated HgTe CQDs [93] (Figure 2c). Their strategy initially involved using strongly bound alkylthiol ligands to synthesize well-dispersed HgTe cores. This was succeeded by a secondary growth process and a final step of ligand modification to augment the colloidal stability. Using this method, the HgTe CQDs could be highly monodisperse in a large size range of 4.2~15.0 nm. The edge absorption was tunable in the wide infrared region of 1.7~6.3 μm, completely covering short-wave and mid-wave infrared regions. Moreover, electron mobility reached a record high of $18.4 \text{ cm}^2/\text{Vs}$.

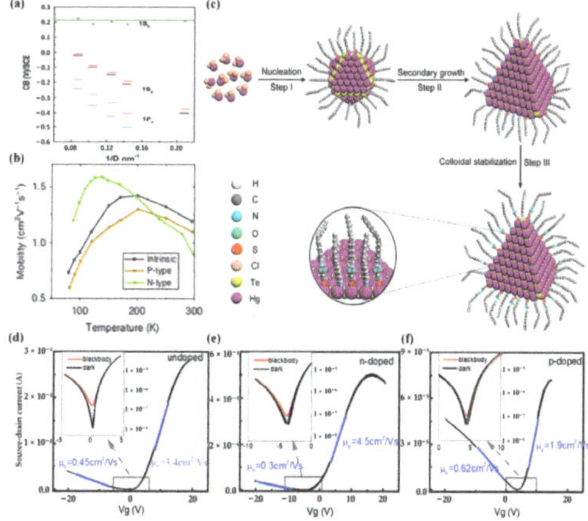

Figure 2. Ligand exchange on HgTe CQDs. (**a**) Energies of the states in HgTe synthesized using (trimethylsilyl)telluride (TMSTe) are plotted against size. The black and blue dashed lines represent

the 1S$_e$ and 1P$_e$ states derived from electrochemistry, respectively. The red and magenta dashed lines depict the 1S$_e$ and 1P$_e$ states corrected by the charging energy. The green dots indicate the estimated conduction band minimum obtained from optical interband absorption [88]. Copyright 2017, *J Phys Chem Lett*. (**b**) Temperature-dependent mobility [91]. Copyright 2023, *Light Sci Appl*. (**c**) Visual representation of the controlled growth of HgTe CQDs through ligand engineering [93]. Copyright 2022, *Nano Lett*. (**d–f**) FET transfer characteristics at 80 K for near-intrinsic, n-doped, and p-doped HgTe/hybrid ligand films, respectively. Insets display enlarged regions near the conductance minima highlighting the photocurrent (red) and dark current (black). The blue lines represent the slopes utilized for calculating hole and electron mobility [89]. Copyright 2019, *ACS Photonics*.

2.2. HgTe CQD Photodetectors

HgTe CQDs have received considerable attention in optoelectronic devices such as photoconductive devices, phototransistors, photovoltaic devices, and infrared focal plane array detectors due to their excellent electronic and optical properties, as well as controllable synthesis.

Photoconductive devices are fabricated by directly depositing CQD films onto a substrate or dielectric layer with interdigitated electrodes. The typical device structure is shown in Figure 3a [71].

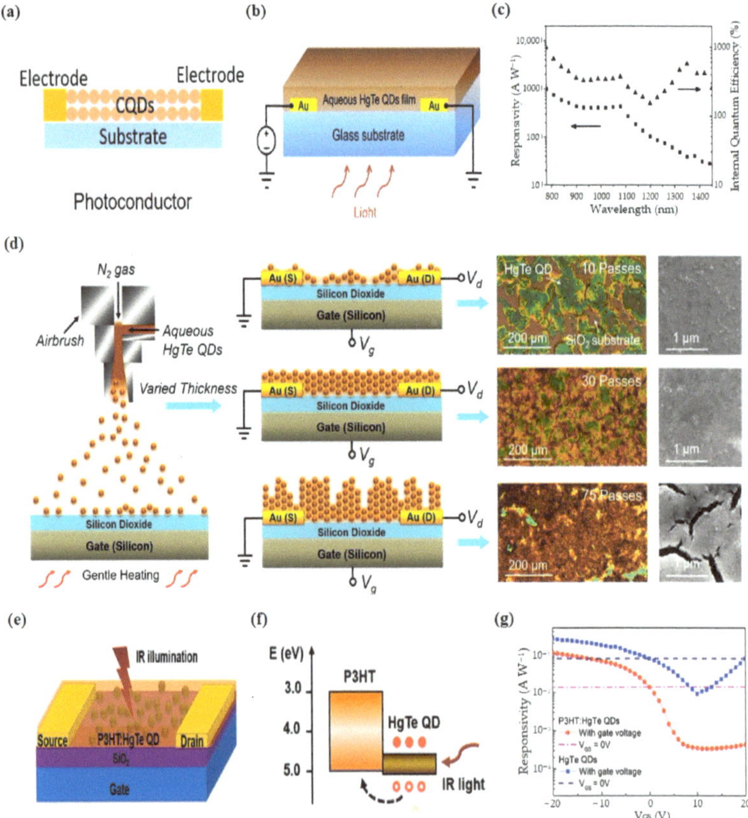

Figure 3. HgTe CQD photoconductive devices and phototransistors. (**a**) The typical device structure of photoconductive devices [71]. Copyright 2023, *Nanoscale*. (**b**) The structure of the photoconductors. (**c**) Responsivity and internal quantum efficiency as a function of wavelength at 295 K with the bias = 60 V, The triangle represents internal quantum efficiency and the square represents responsivity [94].

Copyright 2014, *Advanced Functional Materials*. (**d**) Schematics illustrating the spray-coating setup and the device structures of phototransistors based on aqueous HgTe quantum dots (QDs) with varying QD film thicknesses achieved through multiple spray passes. The corresponding optical microscopy and SEM images of the QDs films are presented on the right [95]. Copyright 2017, *ACS Nano*. (**e**) Illustrations showing the device architecture of the P3HT:HgTe QD hybrid phototransistor. (**f**) Schematic diagram of the energy band alignment between P3HT and HgTe QDs. (**g**) Gate-voltage-dependent responses of a phototransistor based on HgTe CQDs and a phototransistor based on a P3HT:HgTe CQD hybrid. The responsivity at VGS = 0 V is used for comparison. The illumination level is 1550 nm, 22 mW cm^{-2}. All data were acquired at room temperature [96]. Copyright 2020, *Adv Sci (Weinh)*.

Keuleyan et al. prepared the first HgTe CQD photoconductor in 2011, which exhibited a photoresponse extending beyond 5 μm at room temperature. A detectivity of 2×10^9 Jones at 130 K for a sample with a 5 μm cut-off was reported, which generated substantial research interest in HgTe CQDs as an affordable material system for IR photodetection [83]. In 2014, Chen et al. demonstrated a single-layer photoconductor structure based on aqueous HgTe CQDs with an ultrafast time response of up to 2 μs at 1600 nm [94] (Figure 3b,c). The device was fabricated using a simple spray-coating process and showed excellent stability in ambient conditions. The results exhibited that the gain and temporal response of aqueous HgTe-CQD-based photoconductors could be modulated by controlling the size and surface chemical properties of the CQDs. This provided a feasible method for optimizing photodetectors with optional sensitivity and operation bandwidth.

Using methods such as ligand exchange can enhance the material's intrinsic properties, leading to an improvement in the detector's performance. In 2019, Chen et al. introduced the novel approach of mixed ligand exchange. Compared with the previous "solid-state ligand exchange" CQDs using ethylene glycol, the new process increased the electron and hole mobility by a factor of 100, the response rate by a factor of 100, and the detectivity by a factor of 10. At 80 K, the detectivity reached 4.5×10^{10} Jones with a cut-off wavelength of 5 μm [89]. In a comparison of mid-infrared photodetectors, those fabricated using HgTe CQDs with novel hybrid mercaptoethanol-HgCl$_2$ (ME-HgCl$_2$) ligands were contrasted with those employing the conventional 1,2-ethanedithiol (EDT) surface treatment, and the authors found that the responsivity of the ME-HgCl$_2$ ligand device increased by a factor of 50, reaching 0.23 A/W at 80 K [90]. After Liu et al. successfully synthesized I$^-$ in situ passivated HgTe CQDs in 2022, they prepared a photoconductive device. By reducing the temperature, the device demonstrated exceptional detection capabilities under 2000 nm light; the noise current density of the photoconductive device was notably minimized to about 10^{-11} A/Hz$^{1/2}$ at 130 K with the frequency of 1 Hz [92].

Phototransistors are another commonly used photodetector structure. Fabrication of typical phototransistors involves applying a coating of CQDs onto a silicon wafer that is affixed with a layer of silicon dioxide and patterned metal electrodes.

In 2013, Lhuillier et al. used the inorganic ligand As$_2$S$_3$ to passivate the HgTe CQDs via solid-film ligand exchange. They achieved a responsivity greater than 100 mA/W and a detectivity of 3.5×10^{10} Jones in transistor samples with a 3.5 μm cut-off at 230 K [86].

In order to improve the detectivity of the HgTe CQD photodetectors towards longer wavelengths, in 2017, to overcome this performance bottleneck, Chen et al. employed a synergistic approach combining synthetic chemistry and device engineering [95] (Figure 3d). Initially, they devised a fully automated synthesis method using an aprotic solvent and gas injection. This allowed for scalable fabrication of large-sized HgTe CQDs with high quality, showcasing a record-high photoluminescence quantum yield of 17% at 2080 nm. Subsequently, they achieved a HgTe-CQD-based phototransistor with a high specific detectivity of up to 2×10^{10} Jones in the 2000 nm wavelength range at room temperature utilizing a layer-by-layer spray-coating technique for the CQD layer. The achieved performance level was on par with that of commercially available room-temperature-operated, epitaxially grown photodetectors. In 2020, Dong et al. demonstrated a P3HT:HgTe QD hybrid phototransistor [96] (Figure 3e–g). By meticulously controlling the co-blend stirring and ligand exchange processes, the nanoscale morphology and charge transport of the hybrid layer were fine-tuned, leading to an optimized uniform phase distribution. Consequently, the device attained a specific detectivity exceeding 1×10^{11} Jones and a response time of less than 1.5 µs under room-temperature operation at 2400 nm.

In addition, photovoltaic devices feature a vertical multilayer geometric structure. Typically, CQD films serve as the photoactive layer sandwiched between the electron transport layer and the hole transport layer.

In 2015, Sionnest et al. pioneered the development of the initial HgTe CQD photovoltaic devices functioning in the mid-infrared, which achieved a specific detectivity of 4.2×10^{10} Jones and microsecond response times with a cut-off wavelength of 5.25 µm at 90 K [97]. This work has greatly stimulated the development of CQDs with fast, low-cost, and sensitive thermal infrared detection.

To realize high-performance optoelectronic devices, in 2018 Ackerman et al. demonstrated improvements in the sensitivity and efficiency of photovoltaic MWIR detectors by developing a p–n junction with enhanced light collection [98] (Figure 4a,b). They introduced a method of solid-state cation exchange, whereas the interface potential was chemically modified, leading to the improvement of the external quantum efficiency at room temperature by an order of magnitude. A sensitivity of 1×10^9 Jones was achieved with a cut-off wavelength between 4 and 5 µm at 230 K. The responsivity was improved to 1.3 A/W at 85 K and the detectivity was improved to 3.3×10^{11} Jones with a cut-off wavelength of 5 µm. In 2022, Yang et al. developed a new p–i–n photodiode from the traditional p–i device structure [99] (Figure 4c,d). Bismuth sulfide (Bi_2S_3) films were adopted as the electron transport layer, which were beneficial to absorber deposition, superior charge extraction, and suppressed interfacial loss. The Bi_2S_3-based photodiodes showed a specific detectivity up to 1×10^{11} Jones at room temperature. Additionally, the Bi_2S_3-based photodiodes achieved a dark current density as low as 1.6×10^{-5} A/cm^2 at -400 mV at room temperature.

The use of metal halide ligands can improve the photovoltaic performance [100] and enhance the photoluminescence of CQDs [101]. In 2020, Ackerman et al. introduced $HgCl_2$ treatment to each layer of HgTe CQDs prior to the crosslinking with EDT/HCl [102]. The treatment with $HgCl_2$ improves the R_0A (the shunt resistance area product) by an order of magnitude and also slightly improves the responsiveness at room temperature. HgTe CQD photodiodes could achieve external quantum efficiencies of more than 50% at room temperature, detectivities of up to 1×10^{11} Jones at 2.2 µm, and response times at the microsecond level.

Achieving superior light detection often requires cooling down CQD infrared detectors due to the thermal carriers generated by a narrow mid-infrared energy gap. As a result, increasing the operating temperature of the detectors becomes crucial. In 2023, Xue et al. demonstrated a high-operating-temperature mid-infrared photodetector with a HgTe CQD gradient homojunction [91] (Figure 4e,f). The detector attained background-limited performance, exhibiting a specific detectivity of up to 2.7×10^{11} Jones at 80 K on 4.2 µm, exceeding 10^{11} Jones up to 200 K, surpassing 10^{10} Jones up to 280 K, and reaching 7.6×10^9 Jones at 300 K for a wavelength of 3.5 µm. The external quantum efficiency also exceeded 77%, with a responsivity of 2.7 A/W at zero bias. Compared with single-spectrum detection, multispectral detection offers improved target recognition and enables the precise determination of the target's infrared characteristics. Tang et al. used two different-sized HgTe CQD photodiodes stacked together to form a photodetector with dual-band detection in 2019 [103] (Figure 4g,h). By establishing stable spatial doping, a vertical stack of two rectifying junctions in a back-to-back diode configuration was created. The new device architecture allowed a bias-switchable spectral response between the short-wave infrared (SWIR) and the MWIR. This covered two critical atmospheric windows for infrared imaging. By manipulating the bias polarity and magnitude, the detector demonstrated rapid switching between short-wave infrared and mid-wave infrared at low temperatures, achieving modulation frequencies of up to 100 kHz and specific detectivities greater than 1×10^{10} Jones. Subsequently, Zhao et al. introduced an innovative dual-band detector utilizing CdTe and HgTe CQDs [104] (Figure 4i,j). This dual-band device, designed with precision, offered the capability to switch between the visible and the SWIR modes through adjustments in bias polarity and magnitude. The response peaks of the device were at 700 and 2100 nm in the visible and SWIR modes, respectively. Remarkably, this dual-band device exhibited exceptional performance metrics and featured a remarkably low noise current of approximately 10^{-13} A/Hz$^{1/2}$ coupled with impressively high responsivity values of 0.5 A/W (VIS) and 1.1 A/W (SWIR). Moreover, it achieved a remarkable detectivity exceeding 10^{11} Jones in both the visible and SWIR modes, even when operating at room temperature.

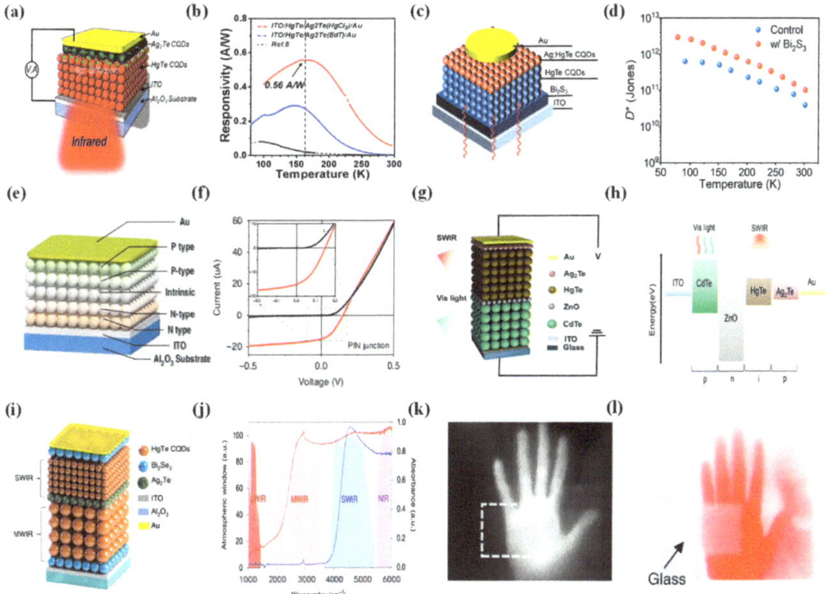

Figure 4. Photovoltaic devices using HgTe CQDs. (**a**) The common structures of photovoltaic infrared detectors. (**b**) The red circle represents the responsivity (T) of the HgTe CQDs MWIR detectors treated with HgCl$_2$, whereas the black arrow represents the responsivity (T) of the HgTe PV MWIR detectors [100]. Copyright 2018, *ACS Nano*. (**c**) The device structure of HgTe CQD photodetectors with an added electron transport layer (ETL). (**d**) The blue line represents the relationship between the temperature and detectivity performances for the device without ETL, whereas the red line represents the relationship between the temperature and detectivity performances for the device based on Bi$_2$S$_3$ [101]. Copyright 2023, *ACS Photonics*. (**e**) The diagram illustrates the structure of a homojunction photodiode with high-mobility CQDs. (**f**) The I-V curve depicts a high-mobility PIN gradient homojunction device, with the inset focusing on the near-zero bias region at 80 K [93]. Copyright 2023, *Light Sci Appl*. (**g**) Structure diagram of the dual-band infrared detector. (**h**) Energy band structure diagram of the dual-band infrared detector [103]. Copyright 2023, *Journal of Materials Chemistry C*. (**i**) Depiction of a dual-band CQD imaging device structure with a bias voltage applied between the indium tin oxide (ITO) and the grounded Au contact. (**j**) The left axis represents the atmospheric transmission window, and the right axis indicates the optical absorption of SWIR and MWIR HgTe CQDs utilized in the fabrication of the dual-band device. The red line indicates HgTe CQDs with absorption characteristics in the MWIR region. The blue line indicates HgTe CQDs with absorption characteristics in the SWIR region (**k**) Using the detector to obtain SWIR images of a hand

placed behind glass. The dotted line indicates the position of the glass. (**l**) Using the detector to obtain MWIR images of a hand placed behind glass [102]. Copyright 2019, *Nature Photonics*.

To further enhance the performance of devices based on HgTe CQDs, the design of the device structure plays a pivotal role. The incorporation of structures such as plasmonic arrays and resonant cavities can effectively serve the purpose of improving detector performance.

In 2018, Tang et al. introduced a HgTe CQD photovoltaic detector that integrated HgTe CQDs and plasmonic structures and was aimed at enhancing the light absorption of the thin HgTe CQD layer [105] (Figure 5a–c). This integration resulted in a significant 2- to 3-fold improvement in responsivity at 5 μm, reaching an impressive 1.62 A/W. Moreover, this enhanced performance spanned a wide range of operating temperatures, from 295 to 85 K, achieving a detectivity of 4×10^{11} Jones at cryogenic temperatures. At an acquisition rate of 1 kHz for a 200 μm pixel, the noise equivalent temperature difference was measured at just 14 mK. Integrating a photodetector with a resonant cavity can enhance the light collection efficiency and spectral selectivity, thereby improving the overall performance of the detector. Luo et al. demonstrated the integration of a HgTe CQD dual-band infrared photodetector with a Fabry–Perot resonant cavity [106] (Figure 5d–f). This integrated photodetector exhibited an impressive responsivity of 1.1 A/W in the SWIR and 1.6 A/W in the MWIR. Notably, the detectivity was significantly enhanced, increasing by a factor of 2 and reaching up to 2×10^{11} Jones.

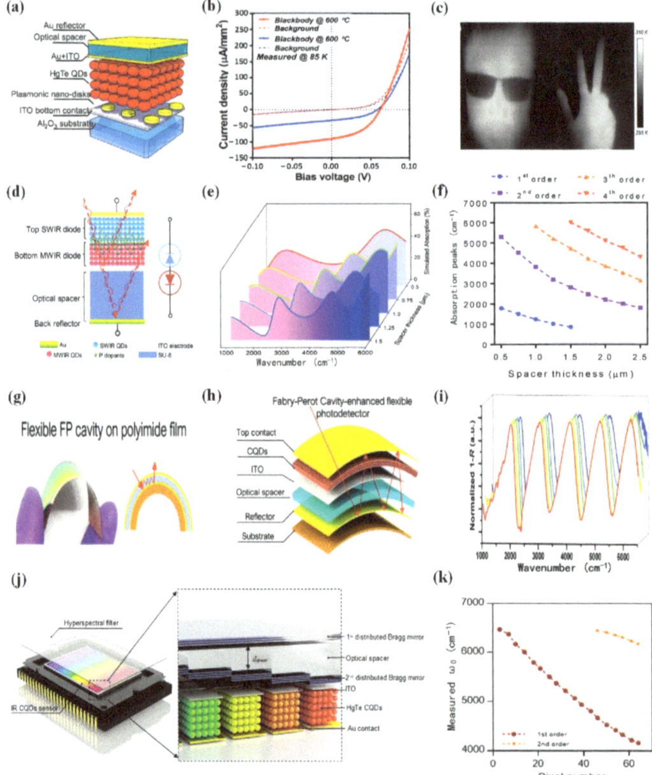

Figure 5. Microstructure HgTe CQD photodetection devices. (**a**) An illustration of the detector with an interference structure and plasmonic disk array based on HgTe CQDs. (**b**) The blue line represents

the IV curve for the reference detector, whereas the red line depicts the IV curve for the detector based on 350 nm plasmonic disks. (**c**) When the MWIR HgTe-CQD-based detectors operated at 90 K, thermal images were captured [105]. Copyright 2018, *ACS Nano*. (**d**) The figure depicts the structure of a dual-band infrared detector based on CQDs. The equivalent circuit of the dual-band infrared photodiode is presented on the right. (**e,f**) Simulating the absorption for different thicknesses of the HgTe layer [106]. Copyright 2022, *Journal of Materials Chemistry C*. (**g**) The figure illustrates a flexible polyimide substrate Fabry–Perot cavity. (**h**) Schematic diagram of a HgTe CQD detector enhanced by a Fabry–Perot cavity. (**i**) The figure demonstrates measurements of "1—reflectance R" of the flexible Fabry–Perot cavity under different bending radii [107]. Copyright 2019, *Small*. (**j**) This image displays the internal structure of a CQD hyperspectral sensor. The sensor comprises a CQD sensor and hyperspectral filter. The optical spacers are positioned between two distributed Bragg reflectors and the CQD sensor array can be directly produced on the filter array. (**k**) The curve in the figure illustrates the function of peak responsivity (ω_0) with respect to the pixel index [108]. Copyright 2019, *Laser & Photonics Reviews*.

Flexible detectors and hyperspectral sensors have been demonstrated in a series of innovative applications. In 2019, Tang et al. successfully fabricated flexible HgTe CQD photovoltaic detectors and proposed a method to further enhance the light absorption in detectors by integrating a Fabry–Perot resonator cavity [107] (Figure 5g–i). Using this method, they prepared flexible infrared detectors with mechanical flexibility and high detectivity. Integrated short-wave IR detectors on flexible substrates had a peak detectivity of 7.5×10^{10} Jones at 2.2 μm at room temperature. In the same year, Tang et al. proposed the design of a CQD hyperspectral sensor by integrating a HgTe CQD photovoltaic sensor with a distributed Bragg mirror filter array [108] (Figure 5g,k). By directly integrating the CQD sensors with a distributed Bragg mirror filter array, 64 narrowband channels with full width at half maxima down to $\approx$30 cm^{-1} could be realized. The experiment effectively demonstrated high-resolution spectral measurements, achieving a resolving power of up to 180. Furthermore, it successfully acquired a hyperspectral image cube within the short-wave infrared range and benefitted from a swift response time of approximately $\approx$120 ns and an impressive sensitivity surpassing >10^{10} Jones that was exhibited by the CQD sensors.

FPA detectors can offer high spatial resolution. CQDs, in contrast to traditional bulk semiconductor materials in the infrared range, have significant advantages in FPA processing. They eliminate the need for complex molecular beam epitaxy processes and flip-chip bonding techniques, substantially reducing the manufacturing cost of infrared FPA detectors.

In 2016, Sionnest et al. presented the inaugural CQD FPA infrared detector designed for mid-infrared imaging. This was achieved by integrating HgTe CQDs with a silicon readout integrated circuit (ROIC) [109]. They showcased a straightforward fabrication approach for an economical HgTe CQD FPA thermal camera, achieving a noise equivalent differential temperature of 1.02 mK at 5 μm and capturing images at a rate of 120 frames per second.

Currently, most efforts on FPA research have focused on the device structure, which requires a multilayer deposition. The optimization of the design appears particularly crucial. Gréboval et al. achieved a photoconductive FPA based on HgTe CQDs [110] (Figure 6a–c). They demonstrated an FPA with a 15 μm pixel pitch presenting an external quantum efficiency of 4–5% (15% internal quantum efficiency) for a cut-off around 1.8 μm with Peltier cooling alone. HgTe-CQD-based FPAs operating in the photoconductive mode are prone to exhibiting larger dark currents. Subsequently, Alchaar et al. designed an FPA based on HgTe CQDs with photovoltaic operation [111]. They designed a diode stack compatible with a readout integrated circuit whose back-end processing was optimized to ensure compatibility with a complete diode stack deposition. The diode design was also optimized to generate a Fabry–Pérot cavity in which 50% of the light was effectively absorbed at the band edge. High-resolution images were achieved through the utilization of the optimized structure.

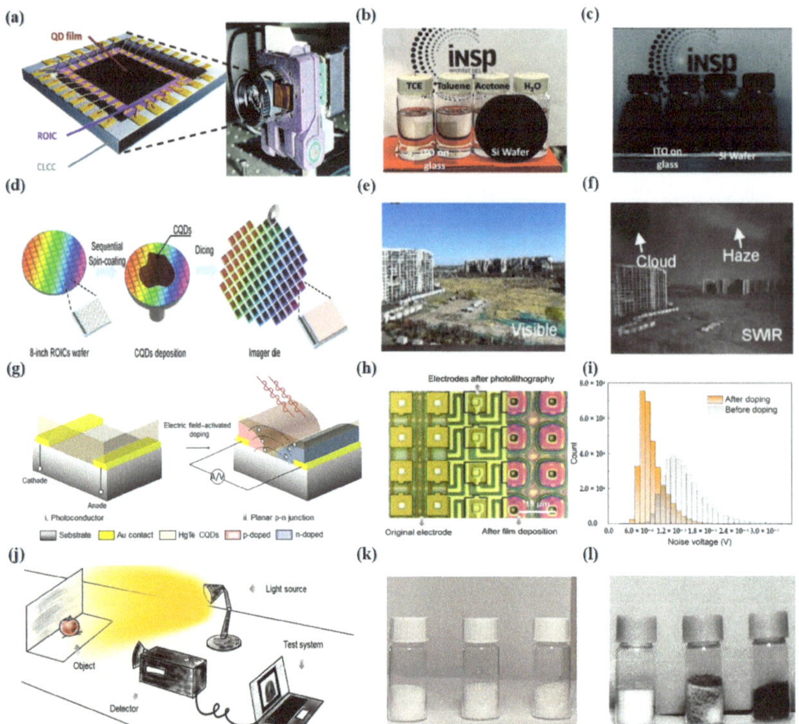

Figure 6. HgTe CQD focal plane arrays. (**a**) The figure shows the HgTe-CQD-based SWIR camera and the camera captures pictures with an M1614-SW objective. (**b**) A visible image captured using a smartphone camera depicting a scene with four vials arranged from left to right: tetrachloroethylene, toluene, acetone, and water. In front of the vials there is an ITO-covered glass slide and a two-inch silicon wafer. (**c**) Using an FPA based on HgTe CQDs to image the same thing as in (**b**) to obtain an SWIR imager [110]. Copyright 2022, *Nanoscale*. (**d**) The image provides a simple demonstration of CMOS-compatible manufacturing of CQDs with an ROIC. The wafer has an 8-inch diameter. Initially, the ROIC wafer undergoes a cleaning process, followed by spin-coating the CQD solution onto the wafer. Once the CQD thickness reaches the desired value, the 8-inch wafer is then cut into individual imaging sensor chips. (**e**) A visible image captured using a smartphone camera depicting a scene with sky and buildings. (**f**) Using an imaging sensor chip to image the same scene as in (**e**) to obtain an SWIR imager [112]. Copyright 2023, *ACS Photonics*. (**g**) Through in situ electric field activation doping, researchers successfully transitioned the working mode of the FPA detector from a photoducting mode to a planar pn photovoltaic mode, as schematically illustrated in the figure. (**h**) This image contains optical microscopic imaging of the substrate in three states. The leftmost depicts the original electrode, the middle part shows the electrode after photolithography, and the rightmost displays the electrode after deposition of quantum dots on the photolithographed electrode. (**i**) A comparison of the noise in the FPA imager before and after doping is presented in histogram form. (**j**) Illustration depicting the imaging device. (**k**) A visible image captured using a smartphone camera depicting a scene with three vials containing from left to right: salt, a mixture of salt and sugar, and sugar. (**l**) Using the FPA imager to image the same scene as in (**k**) to obtain an SWIR imager [69]. Copyright 2023, *Science Advances*.

In 2023, Zhang et al. achieved a complementary metal oxide semiconductor (CMOS) compatible infrared device with trapping-mode photodetectors, in which the minority of carriers were trapped by a vertical built-in potential, contributing to both decreased dark current density and improved quantum efficiency [112] (Figure 6d–f). These trapping-mode CMOS imagers exhibited excellent photoresponse non-uniformity down to 4%, a dead pixel rate down to 0%, external quantum efficiency up to 175%, and detectivity up to 2×10^{11} Jones for SWIR (cut-off wavelength = 2.5 μm) at 300 K and 8×10^{10} Jones for MWIR (cut-off wavelength = 5.5 μm) at 80 K.

Subsequently, Qin et al. discovered through their research that uneven and uncontrollable doping methods, along with complex device configurations, limited the performance of FPA imagers in the photovoltaic mode [69] (Figure 6g–l). Consequently, they proposed a controllable in situ electric field activation doping method to construct lateral p-n junctions for simple planar-structured photodetectors based on SWIR HgTe CQDs. A flat p-n junction FPA imager with 640×512 pixels was successfully manufactured.

We summarize the performance of detectors based on HgTe CQDs in Table 1.

Table 1. The summary of photodetectors based on HgTe CQDs.

Year	Photoactive Material	Detection Range (nm)	Detectivity (Jones)	Responsivity (A/W)	Rise Decay Time	Ref.
			Photodetectors based on HgTe CQDs			
2011	HgTe CQDs	5000	2×10^9	--	--	[83]
2013	HgTe CQDs/As$_2$S$_3$	3500	3.5×10^{10}	0.1	--	[86]
2014	HgTe CQDs	1600	--	--	2 μs	[94]
2015	HgTe CQDs	5250	4.2×10^{10}	--	--	[97]
2016	HgTe CQDs	3600	2×10^{10}	--	--	[109]
2017	HgTe CQDs	2000	2×10^{10}	0.4	12.6 μs	[95]
2018	HgTe CQDs	4000/4500	1×10^{10} @4000 nm 4×10^{10} @4500 nm	1.62 @4500 nm	--	[105]
2018	HgTe CQDs/Ag$_2$Te	5000	3.3×10^{11}	1.3	--	[98]
2019	HgTe CQDs	5000	5.4×10^{10}	--	--	[89]
2019	HgTe CQDs	2200	7.5×10^{10}	--	--	[107]
2019	HgTe CQDs	2500/4500	1×10^{10}	--	--	[103]
2020	HgTe CQDs	4000	5.4×10^{10}	0.23	--	[90]
2020	HgTe CQDs/P3HT	2400	$>1 \times 10^{11}$	--	<1.5 μs	[96]
2022	HgTe CQDs	2350/4000	2×10^{11}	1.1 @2350 nm 1.6 @4000 nm	--	[106]
2023	HgTe CQDs	3500/4200	7.6×10^9 @3500 nm @300 K 2.7×10^{11} @4200 nm @80 K $>1 \times 10^{11}$ @4200 nm @200 K $>1 \times 10^{10}$ @4200 nm @280 K	2.7	--	[91]
2023	HgTe CQDs/Bi$_2$S$_3$	2200	1×10^{11}	--	8 μs	[99]
2023	HgTe CQDs/CdTe CQDs	700/2100	1×10^{11} @700 nm 4.5×10^{11} @2100 nm	0.5 @700 nm 1.1 @2100 nm	--	[104]
2023	HgTe CQDs	2500/5500	2×10^{11} @250 nm 8×10^{10} @5500 nm	--	--	[112]

3. Research on HgSe CQDs

3.1. Synthesis of HgSe CQDs

HgSe CQDs are naturally n-doped in ambient conditions. They typically present several carriers per nanocrystal and consequently exhibit interband and intraband absorption in the mid-infrared range [113–117].

In 2014, Deng et al. conducted a study on the photoconductivity of doped HgSe CQDs with intraband transitions [118] (Figure 7a–c). HgSe CQDs also exhibited intraband photoluminescence. Compared with the interband transitions of CQDs, intraband transitions provided a selective spectral detection, which opens up the possibility for more materials to be applied to infrared applications.

The intraband carrier lifetime is vital in device applications, which are mainly limited by nonradiative processes and a low photoluminescence quantum yield. A core/shell nanocrystal structure can significantly enhance photoluminescence [119–122] and improve the chemical and thermal stability of materials. The shell can passivate surface states of the core, reducing non-radiative recombination pathways. Based on this, in 2015, Deng et al. synthesized HgSe/CdS core/shell CQDs and investigated the changes in their optical absorption and emission [123]. The intraband photoluminescence of HgSe/CdS films was higher than for HgSe films. However, the shell greatly improved the film's resistance. After annealing at 200 °C, HgSe/CdS films maintained narrow intraband emission and higher laser power at a wavelength of 5 μm. In 2018, Shen et al. synthesized HgSe/CdSe core/shell CQDs, achieving the brightest emission at 5 μm, with a quantum yield of approximately 10^{-3} [124] (Figure 7d).

In 2021, Kamath et al. conducted an in-depth study of the core–shell structure [125]. They reported the synthesis and spectral analysis of thick-shell n-type HgSe/CdS core/shell CQDs. Under the milder conditions, HgSe/CdS CQDs with thick shells were synthesized via a two-step growth process utilizing highly reactive single-source precursors. It was found that the intraband lifetime and photoluminescence quantum yield increased with an increase in shell thickness. The maximum photoluminescence efficiency and longest intraband lifetime were achieved at 2000 cm^{-1}. This is the brightest solution phase mid-infrared chromophore reported to date, achieving an internal photoluminescence quantum yield of 2%. Additionally, this photoluminescence had an intraband lifetime of over 10 ns. Later in 2023, Shen et al. presented electroluminescence at 5 μm by utilizing the intraband transition between the $1S_e$ and $1P_e$ states located within the conduction band of core–shell HgSe–CdSe CQDs [126].

To boost the performances, i.e., reducing dark current and improving the time response of the HgSe CQDs, Martinez et al. determined the electronic spectrum on an absolute energy scale for HgSe CQDs with different sizes and different ligands in 2017 [127] (Figure 7e–h). Subsequently, they introduced a technique involving the grafting of functionalized polyoxometalates (POMs) onto the surface of HgSe CQDs. This approach resulted in a substantial tuning of the carrier density (approximately five electrons per nanoparticle) and conduction properties simultaneously [128]. This method showed great promise for achieving significant tuning of the carrier density in degenerately doped semiconductor nanoparticles.

In order to further improve the mobility of CQDs, Chen et al. proposed a method of hybrid ligand exchange (HgSe/hybrid) to prepare the HgSe CQD film and then compared it with the solid-state ligand exchange of ethanedithiol (HgSe/EDT) [129]. In contrast, the mobility of HgSe/hybrid films showed a 100-fold increase in mobility, reaching 1 cm^2/Vs for particles with a diameter of 7.5 nm. Subsequently, they introduced a room-temperature mixed-phase ligand exchange method that allowed for air-stable doping between the $1S_e$ and $1P_e$ states of high-mobility HgSe CQDs that could be adjusted [130] (Figure 7i). Moreover, Chen et al. explored how the size distribution of intraband HgSe CQD films impacts transport and photodetection [131]. The results indicated that enhancing the uniformity of HgSe CQD sizes leads to a significant increase in mobility that correlates with the energy distribution resulting from this size consistency.

3.2. HgSe CQD Photodetectors

As HgSe CQD synthesis technology has gradually matured, researchers have delved into the study of photodetectors based on HgSe CQDs. In 2017, Tang et al. prepared a kind of plasmonic nanodisk array narrowband photodetector by using HgSe CQD films as narrowband light absorption sensing materials [132] (Figure 8a,b). The responsivity at the center wavelength of the narrowband detector was improved by integration with the plasmonic nanodisk array. This was the first time that plasmonic nanodisk arrays had been integrated into MIR detectors based on HgSe CQDs. The results showed that the responsivity at the center wavelength increases significantly and that the full width at half maximum (FWHM) values of the photodetectors decreases. This work was of great significance for the development of filter-free narrowband infrared detectors and cameras operating at room temperature.

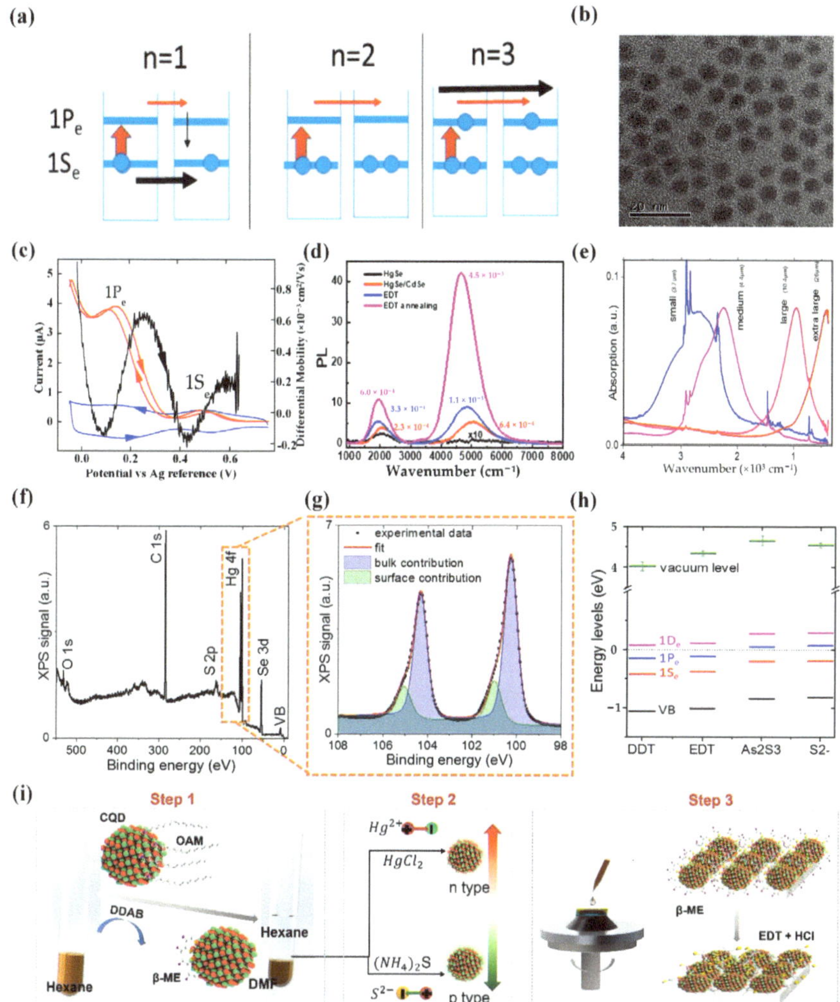

Figure 7. HgSe CQD characterization. (**a**) Band/intraband doping states of QDs. The red horizontal arrows represent photocurrent, and the black arrows represent dark current. The dark current is minimized when $n = 2$. (**b**) TEM image displaying HgSe CQDs. The particles exhibit an average diameter of 6.2 nm, and the standard deviation is 0.76 nm. (**c**) The blue line represents the Faradaic current, the red line represents the conduction current at a bias of 10 mV, and the black line represents the differential mobility of the HgSe CQD film on a platinum interdigitated electrode. The scan rate is 80 mV/s, and the electrolyte is tetrabutylammonium hexafluorophosphate in acetonitrile. The peaks at 0.5 V and 0.15 V in the reversible reduction/oxidation correspond to electron injection into the 1Se and 1Pe states, respectively. The conduction minimum at approximately 0.35 V signifies the filling of the 1Se state. The second minimum near 0 V indicates a conductivity gap between the 1Pe state and higher states. The arrows denote the scan direction [118]. Copyright 2014, *ACS Nano*. (**d**) PL spectra measured in air for HgSe/CdSe films subjected to different treatments [124]. Copyright 2017, *Chemistry of Materials*. (**e**) Absorption spectra of HgSe CQDs with different sizes. (**f**) Measurement of XPS signals for HgSe CQD film with an incident photon energy of 600 eV. (**g**) Photoemission spectrum of Hg 4f core-level emission in medium HgSe CQDs capped with As_2S_3. (**h**) Electronic spectrum representation

(valence band in black, $1S_e$ state in red, $1P_e$ state in blue, $1D_e$ state in pink, and vacuum level in green) for medium-sized HgSe CQDs capped with four different ligands. The energy scale is referenced to the Fermi level and error bars indicate variations in the work function determined through different methods [127]. Copyright 2017, *ACS Appl Mater Interfaces*. (i) Illustration of the mixed-phase ligand exchange process. Step 1 involves liquid-phase ligand exchange. Step 2 includes doping modification using additional salts. Step 3 consists of solid-phase ligand exchange [130]. Copyright 2022, *ACS Nano*.

Modifying HgSe CQDs hinges significantly on the selection of appropriate ligands. Notably, hybrid ligand exchange has made substantial advancements in achieving tunable doping and high mobility in interband CQDs. However, research on intraband CQD detectors is still scarce. Therefore, Chen et al. developed a high-performance photodetector in 2022 [130] (Figure 8e–i). This photodetector could serve as an intraband infrared camera for thermal imaging, as well as a CO_2 gas sensor with a range from 0.25 to 2000 ppm. A relatively high carrier mobility (1 cm^2/Vs) in HgSe intraband CQD solids was obtained by utilizing a room-temperature mixed-phase ligand exchange method. Furthermore, the high mobility and controllable doping proved beneficial for a mid-infrared photodetector utilizing the $1S_e$ to $1P_e$ transition, with a 1000-fold improvement in response speed, which was several μs, a 55-fold increase in responsivity, which was 77 mA/W, and a 10-fold increase in detectivity, which was above 1.7×10^9 Jones at 80 K. In 2022, Sokolova et al. compared the effects of four different types of ligands (EDT, BeSH, S^{2-}, and SCN) on the sensitivity of infrared photodetectors based on HgSe CQDs [133]. The maximum sensitivity of the photodetectors was obtained when SCN was used for the ligands. The obtained responsivity and detectivity of the photodetectors were 0.5 A/W and 3.1×10^7 Jones, respectively.

By mixing with other materials, it is possible to overcome the inherent flaws of the material itself while preserving its original advantages, thus enhancing the performance of the detector [134]. In 2019, Livache et al. proposed a design for a CQD infrared photodetector metamaterial from a mixture of HgSe and HgTe CQDs [135] (Figure 8c,d). At the same time, they integrated this material into a photodiode and achieved a detectivity of 1.5×10^9 Jones at 80 K, which was two folds higher than HgSe CQDs operating at the same temperature and wavelength. In 2022, Khalili et al. proposed a dye-sensitization method to overcome the limitations (high dark current, slow response, low activation energy) observed from internal materials. They used a mixture of HgSe and HgTe as a dye-sensitized infrared sensor [136] (Figure 8j,k). This hybrid material maintained the internal absorption of HgSe CQDs while reducing the dark current, increasing activation energy, and fixing time response. They also studied carrier dynamics in the material using infrared transient spectroscopy and measured the coupling between the two materials. On this basis, they proposed a strategy to improve the optical detection performance of hybrid materials by coupling internal absorption. The mixed material was coupled with a guided mode resonator. The detectivity reached 10^9 Jones at 80 K and the responsivity reached 3 mA/W with a weak temperature dependence.

The design of a core–shell structure can enhance the performance of a photodetector by improving the intrinsic properties of the material itself. In 2023, Shen et al. designed a core–shell structure core–shell HgSe–CdSe CQDs [126] (Figure 8l,m). They applied this material in a photodiode; the device exhibited an external quantum efficiency (EQE) of 4.5% at 2 A/cm^2, with a power efficiency of 0.05%. This 4.5% EQE was comparable with that of commercial epitaxial cascade quantum well light-emitting diodes.

Based on the above work, it can be seen that effective ligand exchange, mixing with other materials, and the design of the core/shell structure are all strategies for improving device performance.

We summarize the performance of detectors based on HgSe CQDs in Table 2.

Table 2. The summary of photodetectors based on HgSe CQDs.

Year	Photoactive Material	Detection Range (nm)	Detectivity (Jones)	Responsivity (A/W)	Rise Decay Time	Ref.
		Photodetectors based on HgSe CQDs				
2014	HgSe CQDs	5000	8.5×10^8	5×10^{-4}	--	[118]
2017	HgSe CQDs	4200/6400/ 7200/9000	--	0.145 @4200 nm 0.092 @6400 nm 0.088 @7200 nm 0.086 @9000 nm	--	[132]
2019	HgSe CQDs/HgTe CQDs	4400	1.5×10^9	--	<500 ns	[125]
2022	HgSe CQDs	4000	1.7×10^9	0.077	--	[131]
2022	HgSe CQDs	--	3.1×10^7	0.5	--	[134]
2022	HgSe CQDs/HgTe CQDs	5000	1×10^9	0.003	<200 ns	[136]

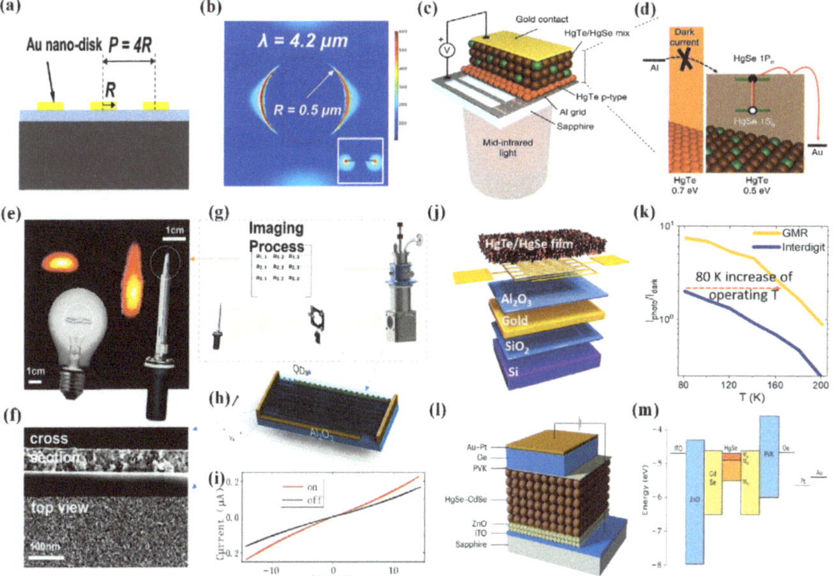

Figure 8. HgSe CQD photodetector. (**a**) The illustration shows a visual depiction of plasmonic disk arrays situated on a SiO_2/Si substrate. (**b**) The graphic illustrates the simulated enhancement ratio corresponding to the resonance wavelength. The inset provides a detailed cross-sectional view for a more comprehensive understanding [132]. Copyright 2017, *Journal of Materials Chemistry C*. (**c**) The device structure. Lighting is supplied from the device's rear, with the sapphire substrate enabling a 70% transmission of mid-infrared light. (**d**) The band diagram illustrates the alignment for the diode structure. HgTe 6k functions as a unipolar barrier, effectively filtering the injection of dark current into the active HgSe/HgTe 4k layer [135]. Copyright 2019, *Nat Commun*. (**e**) Infrared thermal image obtained with a HgSe intraband CQD photodetector. (**f**) SEM of top view and cross section of a photoconductor based on HgSe CQDs. (**g**) Diagram of the imaging scanning device. (**h**) Schematic diagram of a photoconductor based on HgSe CQDs. (**i**) The current of the HgSe-CQD-based photoconductor is depicted by the black line in the absence of blackbody radiation, whereas the red line represents the current in the presence of blackbody radiation. The effective surface area is approximately 0.3 mm × 0.16 mm [130]. Copyright 2022, *ACS Nano*. (**j**) The diagram illustrates the GMR (giant magnetoresistance) device integrated with an intraband absorbing film composed of the

HgSe/HgTe nanocrystal mixture. (**k**) The variation in the intraband photocurrent (induced by QCL illumination at 4.4 µm) to dark current ratio is depicted as a function of temperature. This comparison is made between reference interdigitated electrodes (without resonance) and GMR electrodes [136]. Copyright 2022, *ACS Photonics*. (**l**) Structure of the device. (**m**) Energy band diagram of the device [126]. Copyright 2023, *Nature Photonics*.

4. HgS-CQD-Based Photodetectors

HgS CQDs are a kind of quantum dots with strong constraint; under ambient conditions, the carriers remain stable in the lowest conduction band state [137].

In 2014, Jeong et al. conducted a comprehensive investigation on HgS CQDs [138]. They demonstrated, for the first time, CQDs with stable electron occupancy in the lowest quantum state under ambient conditions, as well as the emergence of intraband luminescence. Figure 9 characterizes the properties of HgS CQDs.

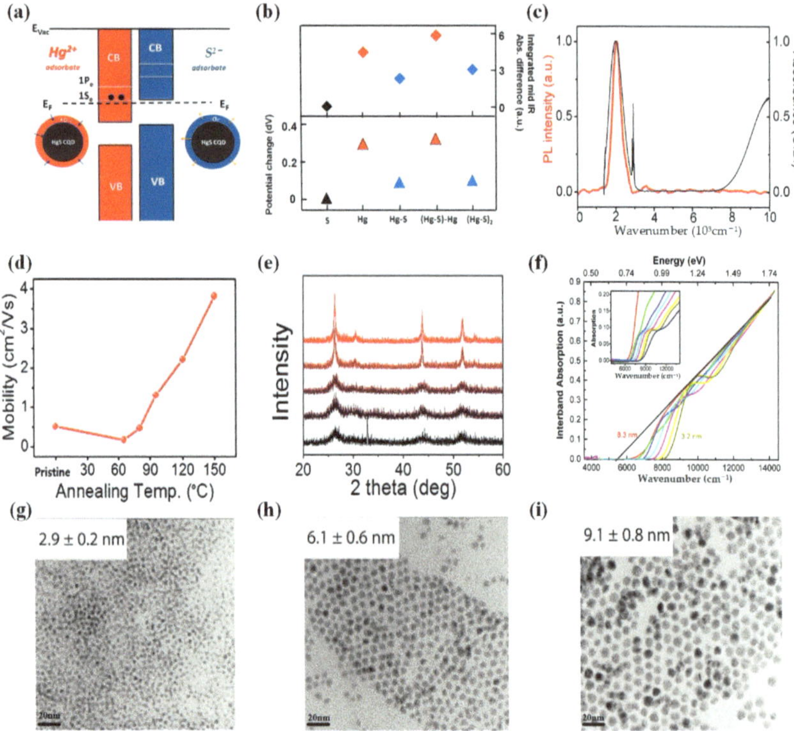

Figure 9. HgS CQD characterization. (**a**) Energy diagrams for Hg^{2+}-doped and S^{2-}-treated HgS CQDs are depicted schematically. (**b**) Relative to the initial measurement of the ethanedithiol cross-linked HgS CQD film, this presentation includes the resting potential difference and the integrated mid-infrared absorption of a HgS colloidal quantum dot (CQD) film. These measurements were taken after alternate exposure to Hg^{2+} and S^{2-} ions, with the resting potential being assessed 40 s after immersion in the electrolyte. (**c**) After photoexcitation at 808 nm, the intraband photoluminescence emission (depicted in red) arises from the $1S_e$–$1P_e$ transition of ambient n-type HgS nanocrystals. The photoluminescence emission spectrum (in red) is overlaid with the absorption spectrum (in black) [138]. Copyright 2014, *J Phys Chem Lett*. (**d**) As a function of annealing temperatures, the linear mobility of HgS CQDs thin-film transistor devices is plotted. (**e**) The XRD spectra of HgS CQDs at the different temperatures are presented [139]. Copyright 2017, *RSC Advances*. (**f**) Interband absorption

post-sulfide treatment of CQDs with different size, normalized at high energy, is presented, as measured in solution. The solid black line represents an extrapolation of the absorption at higher energies, with the x-intercept serving as an estimate for the bulk gap, measured at 0.67 eV. (**g–i**) TEM images of HgS CQDs with different sizes spanning from 2.9 to 9.1 nm [140]. Copyright 2016, *The Journal of Physical Chemistry C*.

Based on the examination of the electron occupancy of the semiconductor conduction band's lowest electronic state ($1S_e$), Kim et al. further investigated the electron mobility of HgS CQDs using field-effect transistors [139] (Figure 10a,b). The optimum carrier mobility was obtained by optimizing annealing temperatures and efficient ligand exchange. Finally, the measured mobility reached an impressive value of 1.29 cm^2/Vs.

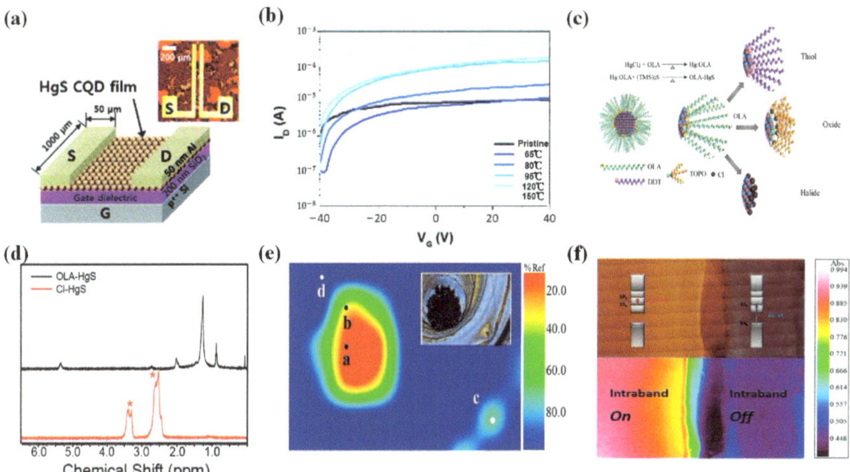

Figure 10. HgS CQD modifications. (**a**) The TFT device structure of HgS CQDs is schematically illustrated, along with an optical image. (**b**) Transfer characteristics of HgS CQDs TFTs, with annealing temperatures ranging from 65 to 150 °C, are displayed [139]. Copyright 2017, *RSC Advances*. (**c**) Versatility in ligands for oleylamine-passivated HgS CQDs is demonstrated, transitioning from amine to thiol, amine to oxide, and amine to halide atomic ligands. (**d**) The bottom spectrum (in red) represents the 1H NMR spectrum of ammonium-chloride-passivated HgS CQDs, indicating the absence of residual oleylamine ligands except for solvent peaks (marked with a star). For reference, the top spectrum corresponds to the NMR spectrum of oleylamine-passivated HgS CQDs. (**e**) The FTIR image of the identical solid HgS CQDs is presented, with an optical image provided in the inset. This figure is obtained by FTIR microscope. The point a and b are two different absorption peaks of the film. (**f**) The electron doping density of two distinct ligand-passivated HgS CQD solid films is monitored using an FT-IR microscope. The top part of the image displays the optical view, whereas the bottom part presents the FTIR image. In the bottom left-hand area (red–yellow), a heavily doped HgS CQD film exhibits mid-IR intraband absorption at 2275 cm^{-1}. Conversely, the bottom right-hand area (blue–purple) represents a sulfide-treated HgS CQD film showing no mid-IR intraband absorption. The film geometry resembles a p-n junction. The image size is 8356 × 700 μm^2 [141]. Copyright 2016, *Journal of Physical Chemistry C*.

To further enhance the performance of HgS CQDs, researchers have proposed an innovative approach for the growth of CQDs. To address the issues of particle aggregation and poor shape control (size distribution) for HgS CQDs, in 2016, Shen et al. utilized a dual-phase method to synthesize HgS CQDs at room temperature, resulting in HgS CQDs with small size dispersion and well-defined optical features [140]. They also employed the same synthetic method to create HgS/CdS core/shell structures. The findings revealed that encapsulating HgS within a CdS shell effectively eliminated the natural n-doping of the HgS core, leading to interband photoluminescence at 1.5 μm with a quantum yield of approximately 5%. Additionally, the core/shell structure significantly enhanced the thermal stability of the HgS core.

The strong binding strength between thiol and HgS CQDs makes it difficult for HgS CQDs to exchange with other ligands species. To solve this problem, Yoon et al. employed the oleamine passivation method to synthesize HgS CQDs, offering a novel approach for producing thiol-free nanocrystals with electrons occupying the lowest quantum state of the conduction band [141] (Figure 10c–f). They achieved heavy doping of HgS CQDs without thiol ligands for the first time, effectively simplifying the process of ligand exchange. Non-thiol ligand-passivated HgS CQDs exhibit strong steady-state band-to-band transitions under ambient conditions. This work provided a direction for the erasable storage, infrared optoelectronics, and infrared free-space optical communication of CQDs.

At present, there is not much research on HgS CQD infrared detectors. The stability HgS CQDs needs to be further improved.

5. Conclusions

Mercury chalcogenide quantum dots, owing to their tunable bandgap, high absorption coefficients, and cost-effective solution processability, are considered promising materials for IR photodetectors. This article provides a comprehensive overview of the research progress in mercury sulfide quantum dot photodetectors, covering aspects such as quantum dot synthesis, material enhancements, device structures, and operating principles. Comprehensive analysis of the detector performance based on these materials reveals several key insights. Compared with other materials, detectors based on HgTe CQDs demonstrate a broader wavelength coverage and relatively superior performance. However, their stability is comparatively lower. HgSe CQDs can leverage intraband transitions for photoelectric detection. In contrast to the inter-band transition in HgTe CQDs, the intraband infrared luminescence attenuation exhibits little or greatly reduced Auger relaxation. This phenomenon arises from the sparse state density in the conduction band. Consequently, the utilization of intraband transitions proves advantageous for enhancing the quantum efficiency of infrared detectors. Nevertheless, achieving precise control over the size and electronic doping state of HgSe CQDs presents a significant challenge. Consequently, the yield of HgSe CQD detectors is not high and their performance lags behind that of HgTe CQD detectors. HgS CQDs are currently hampered by issues such as particle aggregation, limited shape control, and instability in preparation, resulting in a relatively sparse body of related studies. Nonetheless, the ongoing development of new technologies, including core–shell structures, holds the promise of addressing these challenges and enhancing the application potential of HgS CQDs in the field of detectors.

Over the past decade, significant breakthroughs have been achieved in the field of colloidal mercury sulfide quantum dots, ranging from single-point detectors to focal plane arrays. However, the colloidal mercury sulfide quantum dot system still faces several challenges:

1. Material Stability: Mercury sulfide colloidal quantum dots may undergo degradation or deterioration when exposed to prolonged use or high-temperature environments, limiting their widespread applications. Surface modification of quantum dots with organic or inorganic ligands is essential to enhance their stability and improve their dispersion in solutions. This is crucial for the preparation of long-term stable and reliable infrared detectors.
2. Limited Photodetection Range: Mercury sulfide quantum dots are primarily used in the near-infrared, short-wave infrared, and mid-infrared spectral range, with fewer applications in the long-wave infrared range. Expanding their photodetection range is necessary to achieve broader spectral detection capabilities.
3. Operating Temperature: Theoretically, the quantum-mechanical nature of CQDs implies higher operating temperatures for IR photodetectors, since thermal carrier

generation would be significantly reduced compared with the quantum well due to the energy quantization in all three dimensions. Still, cooling is necessary in most CQD photodetectors. Further theoretical and experimental research is needed.

6. Outlook

Mercury chalcogenide CQDs for infrared detectors hold immense promise for future development. Firstly, the simplified fabrication method based on solution manufacturing significantly reduces production costs. Secondly, infrared detectors based on mercury chalcogenide CQDs, with their tunable bandgap and extensive absorption spectra, offer a broader spectral range, making them potentially applicable in various fields. Prospective applications may span biomedical imaging, communication technology, solar cells, and more. Additionally, the flexibility of mercury chalcogenide CQDs in terms of compatibility, allows for application on various substrates, particularly silicon and flexible substrates. Through optimizing synthesis methods, improving material stability and designing novel device structures, achieving higher EQE and faster response times is possible. In summary, mercury-chalcogenide-CQD-based infrared detectors, with their unique properties, demonstrate extensive prospects for applications such as infrared imaging technology, solar cells, biomedical imaging, gas sensing, and environmental monitoring.

Author Contributions: Q.H., H.M., X.X., X.T. and Z.W., investigation; H.M. and X.Z., writing—original draft preparation; X.Z. and M.C., writing—review and editing; M.C., supervision; M.C., project administration; Q.H. and M.C., funding acquisition. All authors have read and agreed to the published version of the manuscript.

Funding: This work is supported by National Natural Science Foundation of China (No. 62105022, No. U22A2081), the Beijing Nova Program (No. Z211100002121069), and the Young Elite Scientists Sponsorship Program by CAST (No. YESS20210142).

Institutional Review Board Statement: Not applicable.

Informed Consent Statement: Not applicable.

Data Availability Statement: Data are contained within the article.

Conflicts of Interest: The authors declare no conflict of interest.

References

1. Rogalski, A. Toward third generation HgCdTe infrared detectors. *J. Alloys Compd.* **2004**, *371*, 53–57. [CrossRef]
2. Rogalski, A.; Antoszewski, J.; Faraone, L. Third-generation infrared photodetector arrays. *J. Appl. Phys.* **2009**, *105*, 091101. [CrossRef]
3. Schmitt, J.; Flemming, H.-C. FTIR-spectroscopy in microbial and material analysis. *Int. Biodeterior. Biodegrad.* **1998**, *41*, 1–11. [CrossRef]
4. Claire, G.; Federico, C.; Deborah, L.S.; Alfred, Y.C. Recent progress in quantum cascade lasers and applications. *Rep. Prog. Phys.* **2001**, *64*, 1533.
5. Lohse, S.E.; Murphy, C.J. Applications of Colloidal Inorganic Nanoparticles: From Medicine to Energy. *J. Am. Chem. Soc.* **2012**, *134*, 15607–15620. [CrossRef] [PubMed]
6. Yadav, P.V.K.; Ajitha, B.; Kumar Reddy, Y.A.; Sreedhar, A. Recent advances in development of nanostructured photodetectors from ultraviolet to infrared region: A review. *Chemosphere* **2021**, *279*, 130473. [CrossRef]
7. Li, Z.; He, Z.; Xi, C.; Zhang, F.; Huang, L.; Yu, Y.; Tan, H.H.; Jagadish, C.; Fu, L. Review on III–V Semiconductor Nanowire Array Infrared Photodetectors. *Adv. Mater. Technol.* **2023**, *8*, 2202126. [CrossRef]
8. Xia, Y.; Geng, C.; Bi, X.; Li, M.; Zhu, Y.; Yao, Z.; Wan, X.; Li, G.; Chen, Y. Biomimetic Flexible High-Sensitivity Near-Infrared II Organic Photodetector for Photon Detection and Imaging. *Adv. Opt. Mater.* **2023**, 2301518. [CrossRef]
9. Li, C.; Wang, H.; Wang, F.; Li, T.; Xu, M.; Wang, H.; Wang, Z.; Zhan, X.; Hu, W.; Shen, L. Ultrafast and broadband photodetectors based on a perovskite/organic bulk heterojunction for large-dynamic-range imaging. *Light Sci. Appl.* **2020**, *9*, 31. [CrossRef]
10. Zhang, C.; Luo, Y.; Maier, S.A.; Li, X. Recent Progress and Future Opportunities for Hot Carrier Photodetectors: From Ultraviolet to Infrared Bands. *Laser Photonics Rev.* **2022**, *16*, 2100714. [CrossRef]
11. Lu, Y.; Warner, J.H. Synthesis and Applications of Wide Bandgap 2D Layered Semiconductors Reaching the Green and Blue Wavelengths. *ACS Appl. Electron. Mater.* **2020**, *2*, 1777–1814. [CrossRef]
12. Hu, W.D.; Chen, X.S.; Ye, Z.H.; Lu, W. A hybrid surface passivation on HgCdTe long wave infrared detector with in-situ CdTe deposition and high-density hydrogen plasma modification. *Appl. Phys. Lett.* **2011**, *99*, 091101. [CrossRef]

13. Tan, H.; Fan, C.; Ma, L.; Zhang, X.; Fan, P.; Yang, Y.; Hu, W.; Zhou, H.; Zhuang, X.; Zhu, X.; et al. Single-Crystalline InGaAs Nanowires for Room-Temperature High-Performance Near-Infrared Photodetectors. *Nano-Micro Lett.* **2016**, *8*, 29–35. [CrossRef] [PubMed]

14. Shen, L.; Qian, H.; Yang, Y.; Ma, Y.; Deng, J.; Zhang, Y. Photoresponse Improvement of InGaAs Nanowire Near-Infrared Photodetectors with Self-Assembled Monolayers. *J. Phys. Chem. C* **2023**, *127*, 11328–11337. [CrossRef]

15. Han, I.S.; Kim, J.S.; Shin, J.C.; Kim, J.O.; Noh, S.K.; Lee, S.J.; Krishna, S. Photoluminescence study of InAs/InGaAs sub-monolayer quantum dot infrared photodetectors with various numbers of multiple stack layers. *J. Lumin.* **2019**, *207*, 512–519. [CrossRef]

16. Zhang, H.; Wang, W.; Yip, S.; Li, D.; Li, F.; Lan, C.; Wang, F.; Liu, C.; Ho, J.C. Enhanced performance of near-infrared photodetectors based on InGaAs nanowires enabled by a two-step growth method. *J. Mater. Chem. C* **2020**, *8*, 17025–17033. [CrossRef]

17. Rogalski, A.; Martyniuk, P.; Kopytko, M. Type-II superlattice photodetectors versus HgCdTe photodiodes. *Prog. Quantum Electron.* **2019**, *68*, 100228. [CrossRef]

18. Wang, X.; Wang, M.; Liao, Y.; Zhang, H.; Zhang, B.; Wen, T.; Yi, J.; Qiao, L. Molecular-beam epitaxy-grown HgCdTe infrared detector: Material physics, structure design, and device fabrication. *Sci. China Phys. Mech. Astron.* **2023**, *66*, 237302. [CrossRef]

19. Batty, K.; Steele, I.; Copperwheat, C. Laboratory and On-sky Testing of an InGaAs Detector for Infrared Imaging. *Publ. Astron. Soc. Pac.* **2022**, *134*, 65001. [CrossRef]

20. Chen, J.; Chen, J.; Li, X.; He, J.; Yang, L.; Wang, J.; Yu, F.; Zhao, Z.; Shen, C.; Guo, H.; et al. High-performance HgCdTe avalanche photodetector enabled with suppression of band-to-band tunneling effect in mid-wavelength infrared. *npj Quantum Mater.* **2021**, *6*, 103. [CrossRef]

21. Rothman, J. Physics and Limitations of HgCdTe APDs: A Review. *J. Electron. Mater.* **2018**, *47*, 5657–5665. [CrossRef]

22. Madejczyk, P.; Manyk, T.; Rutkowski, J. Research on Electro-Optical Characteristics of Infrared Detectors with HgCdTe Operating at Room Temperature. *Sensors* **2023**, *23*, 1088. [CrossRef]

23. Ciura, Ł.; Kopytko, M.; Martyniuk, P. Low-frequency noise limitations of InAsSb-, and HgCdTe-based infrared detectors. *Sens. Actuators A Phys.* **2020**, *305*, 111908. [CrossRef]

24. Cao, F.; Liu, L.; Li, L. Short-wave infrared photodetector. *Mater. Today* **2023**, *62*, 327–349. [CrossRef]

25. Choi, K.K. Detection wavelength of quantum-well infrared photodetectors. *J. Appl. Phys.* **1993**, *73*, 5230–5236. [CrossRef]

26. Billaha, M.A.; Das, M.K. Performance analysis of AlGaAs/GaAs/InGaAs-based asymmetric long-wavelength QWIP. *Appl. Phys. A* **2019**, *125*, 457. [CrossRef]

27. Durmaz, H.; Sookchoo, P.; Cui, X.; Jacobson, R.B.; Savage, D.E.; Lagally, M.G.; Paiella, R. SiGe Nanomembrane Quantum-Well Infrared Photodetectors. *ACS Photonics* **2016**, *3*, 1978–1985. [CrossRef]

28. Liu, L.; Chen, Y.; Huang, Z.; Du, W.; Zhan, P.; Wang, Z. Highly efficient metallic optical incouplers for quantum well infrared photodetectors. *Sci. Rep.* **2016**, *6*, 30414. [CrossRef]

29. Mukherjee, R.; Konar, S. Electromagnetically induced transparency based quantum well infrared photodetectors. *J. Lumin.* **2022**, *251*, 119176. [CrossRef]

30. Wu, H.; Jiang, Y.; Li, J.; Ma, X.; Song, J.; Yu, H.; Fu, D.; Xu, Y.; Zhu, H.; Song, G. Performance analysis of an N-structure type-II superlattice photodetector for long wavelength infrared applications. *J. Alloys Compd.* **2016**, *684*, 663–668. [CrossRef]

31. Ting, D.Z.; Soibel, A.; Gunapala, S.D. Hole effective masses and subband splitting in type-II superlattice infrared detectors. *Appl. Phys. Lett.* **2016**, *108*, 183504. [CrossRef]

32. Rogalski, A.; Martyniuk, P.; Kopytko, M. InAs/GaSb type-II superlattice infrared detectors: Future prospect. *Appl. Phys. Rev.* **2017**, *4*, 031304. [CrossRef]

33. Ting, D.Z.; Rafol, S.B.; Khoshakhlagh, A.; Soibel, A.; Keo, S.A.; Fisher, A.M.; Pepper, B.J.; Hill, C.J.; Gunapala, S.D. InAs/InAsSb Type-II Strained-Layer Superlattice Infrared Photodetectors. *Micromachines* **2020**, *11*, 958. [CrossRef]

34. Ribet-Mohamed, I.; Nghiem, J.; Caes, M.; Guénin, M.; Höglund, L.; Costard, E.; Rodriguez, J.B.; Christol, P. Temporal stability and correctability of a MWIR T2SL focal plane array. *Infrared Phys. Technol.* **2019**, *96*, 145–150. [CrossRef]

35. Gerislioglu, B.; Ahmadivand, A.; Adam, J. Infrared plasmonic photodetectors: The emergence of high photon yield toroidal metadevices. *Mater. Today Chem.* **2019**, *14*, 100206. [CrossRef]

36. Ye, X.; Du, Y.; Wang, M.; Liu, B.; Liu, J.; Jafri, S.H.M.; Liu, W.; Papadakis, R.; Zheng, X.; Li, H. Advances in the Field of Two-Dimensional Crystal-Based Photodetectors. *Nanomaterials* **2023**, *13*, 1379. [CrossRef]

37. Amani, M.; Tan, C.; Zhang, G.; Zhao, C.; Bullock, J.; Song, X.; Kim, H.; Shrestha, V.R.; Gao, Y.; Crozier, K.B.; et al. Solution-Synthesized High-Mobility Tellurium Nanoflakes for Short-Wave Infrared Photodetectors. *ACS Nano* **2018**, *12*, 7253–7263. [CrossRef]

38. Peng, M.; Xie, R.; Wang, Z.; Wang, P.; Wang, F.; Ge, H.; Wang, Y.; Zhong, F.; Wu, P.; Ye, J.; et al. Blackbody-sensitive room-temperature infrared photodetectors based on low-dimensional tellurium grown by chemical vapor deposition. *Sci. Adv.* **2021**, *7*, eabf7358. [CrossRef]

39. Yan, Y.; Xia, K.; Gan, W.; Yang, K.; Li, G.; Tang, X.; Li, L.; Zhang, C.; Fei, G.T.; Li, H. A tellurium short-wave infrared photodetector with fast response and high specific detectivity. *Nanoscale* **2022**, *14*, 13187–13191. [CrossRef] [PubMed]

40. Wang, S.; Ashokan, A.; Balendhran, S.; Yan, W.; Johnson, B.C.; Peruzzo, A.; Crozier, K.B.; Mulvaney, P.; Bullock, J. Room Temperature Bias-Selectable, Dual-Band Infrared Detectors Based on Lead Sulfide Colloidal Quantum Dots and Black Phosphorus. *ACS Nano* **2023**, *17*, 11771–11782. [CrossRef]

41. Xu, K.; Xiao, X.; Zhou, W.; Jiang, X.; Wei, Q.; Chen, H.; Deng, Z.; Huang, J.; Chen, B.; Ning, Z. Inverted Si:PbS Colloidal Quantum Dot Heterojunction-Based Infrared Photodetector. *ACS Appl. Mater. Interfaces* **2020**, *12*, 15414–15421. [CrossRef] [PubMed]
42. Nakotte, T.; Munyan, S.G.; Murphy, J.W.; Hawks, S.A.; Kang, S.; Han, J.; Hiszpanski, A.M. Colloidal quantum dot based infrared detectors: Extending to the mid-infrared and moving from the lab to the field. *J. Mater. Chem. C* **2022**, *10*, 790–804. [CrossRef]
43. Imran, M.; Choi, D.; Parmar, D.H.; Rehl, B.; Zhang, Y.; Atan, O.; Kim, G.; Xia, P.; Pina, J.M.; Li, M.; et al. Halide-Driven Synthetic Control of InSb Colloidal Quantum Dots Enables Short-Wave Infrared Photodetectors. *Adv. Mater.* **2023**, *35*, 2306147.
44. Ouyang, J.; Graddage, N.; Lu, J.; Zhong, Y.; Chu, T.-Y.; Zhang, Y.; Wu, X.; Kodra, O.; Li, Z.; Tao, Y.; et al. Ag2Te Colloidal Quantum Dots for Near-Infrared-II Photodetectors. *ACS Appl. Nano Mater.* **2021**, *4*, 13587–13601. [CrossRef]
45. Chen, W.; Tang, H.; Chen, Y.; Heger, J.E.; Li, N.; Kreuzer, L.P.; Xie, Y.; Li, D.; Anthony, C.; Pikramenou, Z.; et al. Spray-deposited PbS colloidal quantum dot solid for near-infrared photodetectors. *Nano Energy* **2020**, *78*, 105254. [CrossRef]
46. Zhu, T.; Yang, Y.; Zheng, L.; Liu, L.; Becker, M.L.; Gong, X. Solution-Processed Flexible Broadband Photodetectors with Solution-Processed Transparent Polymeric Electrode. *Adv. Funct. Mater.* **2020**, *30*, 1909487. [CrossRef]
47. Xu, Q.; Meng, L.; Sinha, K.; Chowdhury, F.I.; Hu, J.; Wang, X. Ultrafast Colloidal Quantum Dot Infrared Photodiode. *ACS Photonics* **2020**, *7*, 1297–1303. [CrossRef]
48. Huang, J.; Guo, D.; Chen, W.; Deng, Z.; Bai, Y.; Wu, T.; Chen, Y.; Liu, H.; Wu, J.; Chen, B. Sub-monolayer quantum dot quantum cascade mid-infrared photodetector. *Appl. Phys. Lett.* **2017**, *111*, 251104. [CrossRef]
49. Zhang, Y.; Tong, X.; Channa, A.I.; Wang, R.; Zhou, N.; Li, X.; Zhao, H.; Huang, Y.; Wang, Z.M. Effective surface passivation of environment-friendly colloidal quantum dots for highly efficient near-infrared photodetectors. *J. Mater. Chem. C* **2022**, *10*, 7018–7023. [CrossRef]
50. Gong, W.; Wang, P.; Deng, W.; Zhang, X.; An, B.; Li, J.; Sun, Z.; Dai, D.; Liu, Z.; Li, J.; et al. Limiting Factors of Detectivity in Near-Infrared Colloidal Quantum Dot Photodetectors. *ACS Appl. Mater. Interfaces* **2022**, *14*, 25812–25823. [CrossRef]
51. Ming, S.-K.; Taylor, R.A.; McNaughter, P.D.; Lewis, D.J.; O'Brien, P. Tunable structural and optical properties of AgxCuyInS2 colloidal quantum dots. *New J. Chem.* **2022**, *46*, 18899–18910. [CrossRef]
52. Geiregat, P.; Van Thourhout, D.; Hens, Z. A bright future for colloidal quantum dot lasers. *NPG Asia Mater.* **2019**, *11*, 41. [CrossRef]
53. Wang, X.; Koleilat, G.I.; Tang, J.; Liu, H.; Kramer, I.J.; Debnath, R.; Brzozowski, L.; Barkhouse, D.A.R.; Levina, L.; Hoogland, S.; et al. Tandem colloidal quantum dot solar cells employing a graded recombination layer. *Nat. Photonics* **2011**, *5*, 480–484. [CrossRef]
54. Goossens, V.M.; Sukharevska, N.V.; Dirin, D.N.; Kovalenko, M.V.; Loi, M.A. Scalable fabrication of efficient p-n junction lead sulfide quantum dot solar cells. *Cell Rep. Phys. Sci.* **2021**, *2*, 100655. [CrossRef]
55. Speirs, M.J.; Dirin, D.N.; Abdu-Aguye, M.; Balazs, D.M.; Kovalenko, M.V.; Loi, M.A. Temperature dependent behaviour of lead sulfide quantum dot solar cells and films. *Energy Environ. Sci.* **2016**, *9*, 2916–2924. [CrossRef]
56. Bao, J.; Bawendi, M.G. A colloidal quantum dot spectrometer. *Nature* **2015**, *523*, 67–70. [CrossRef] [PubMed]
57. Hasan, M.M.; Moyen, E.; Saha, J.K.; Islam, M.M.; Ali, A.; Jang, J. Solution processed high performance perovskite quantum dots/ZnO phototransistors. *Nano Res.* **2022**, *15*, 3660–3666. [CrossRef]
58. Wang, Y.; Yin, L.; Huang, S.; Xiao, R.; Zhang, Y.; Li, D.; Pi, X.; Yang, D. Silicon-Nanomembrane-Based Broadband Synaptic Phototransistors for Neuromorphic Vision. *Nano Lett.* **2023**, *23*, 8460–8467. [CrossRef]
59. Li, W.; Zhang, P.; Li, H.; Sheng, X.; Li, Y.; Xu, X.; Yi, M.; Shi, N.; Shi, W.; Xie, L.; et al. A tricolour photodetecting memory device based on lead sulfide colloidal quantum dots floating gate. *Org. Electron.* **2019**, *75*, 105111. [CrossRef]
60. Kim, T.J.; Lee, S.-h.; Lee, E.; Seo, C.; Kim, J.; Joo, J. Far-Red Interlayer Excitons of Perovskite/Quantum-Dot Heterostructures. *Adv. Sci.* **2023**, *10*, 2207653. [CrossRef]
61. Park, Y.-S.; Roh, J.; Diroll, B.T.; Schaller, R.D.; Klimov, V.I. Colloidal quantum dot lasers. *Nat. Rev. Mater.* **2021**, *6*, 382–401. [CrossRef]
62. Yun, L.; Zhang, H.; Cui, H.; Ding, C.; Ding, Y.; Yu, K.; Wei, W. Lead Sulfide Quantum Dots Mode Locked, Wavelength-Tunable Soliton Fiber Laser. *IEEE Photonics Technol. Lett.* **2021**, *33*, 119–122. [CrossRef]
63. Yun, L.; Ding, C.; Ding, Y.; Han, D.; Zhang, J.; Cui, H.; Wang, Z.; Yu, K. High-Power Mode-Locked Fiber Laser Using Lead Sulfide Quantum Dots Saturable Absorber. *J. Light. Technol.* **2022**, *40*, 7901–7906. [CrossRef]
64. Shen, X.; Peterson, J.C.; Guyot-Sionnest, P. Mid-infrared HgTe Colloidal Quantum Dot LEDs. *ACS Nano* **2022**, *16*, 7301–7308. [CrossRef] [PubMed]
65. Yin, S.; Ho, C.H.Y.; Ding, S.; Fu, X.; Zhu, L.; Gullett, J.; Dong, C.; So, F. Enhanced Surface Passivation of Lead Sulfide Quantum Dots for Short-Wavelength Photodetectors. *Chem. Mater.* **2022**, *34*, 5433–5442. [CrossRef]
66. Dong, C.; Liu, S.; Barange, N.; Lee, J.; Pardue, T.; Yi, X.; Yin, S.; So, F. Long-Wavelength Lead Sulfide Quantum Dots Sensing up to 2600 nm for Short-Wavelength Infrared Photodetectors. *ACS Appl. Mater. Interfaces* **2019**, *11*, 44451–44457. [CrossRef]
67. Liu, M.; Yazdani, N.; Yarema, M.; Jansen, M.; Wood, V.; Sargent, E.H. Colloidal quantum dot electronics. *Nat. Electron.* **2021**, *4*, 548–558. [CrossRef]
68. Guo, R.; Zhang, M.; Ding, J.; Liu, A.; Huang, F.; Sheng, M. Advances in colloidal quantum dot-based photodetectors. *J. Mater. Chem. C* **2022**, *10*, 7404–7422. [CrossRef]
69. Qin, T.; Mu, G.; Zhao, P.; Tan, Y.; Liu, Y.; Zhang, S.; Luo, Y.; Hao, Q.; Chen, M.; Tang, X. Mercury telluride colloidal quantum-dot focal plane array with planar p-n junctions enabled by in situ electric field–activated doping. *Sci. Adv.* **2023**, *9*, eadg7827. [CrossRef]

70. Gréboval, C.; Chu, A.; Goubet, N.; Livache, C.; Ithurria, S.; Lhuillier, E. Mercury Chalcogenide Quantum Dots: Material Perspective for Device Integration. *Chem. Rev.* **2021**, *121*, 3627–3700. [CrossRef]

71. Tian, Y.; Luo, H.; Chen, M.; Li, C.; Kershaw, S.V.; Zhang, R.; Rogach, A.L. Mercury chalcogenide colloidal quantum dots for infrared photodetection: From synthesis to device applications. *Nanoscale* **2023**, *15*, 6476–6504. [CrossRef] [PubMed]

72. Shen, L.; Pun, E.Y.; Ho, J.C. Recent developments in III–V semiconducting nanowires for high-performance photodetectors. *Mater. Chem. Front.* **2017**, *1*, 630–645. [CrossRef]

73. Sergeeva, K.A.; Fan, K.; Sergeev, A.A.; Hu, S.; Liu, H.; Chan, C.C.; Kershaw, S.V.; Wong, K.S.; Rogach, A.L. Ultrafast Charge Carrier Dynamics and Transport Characteristics in HgTe Quantum Dots. *J. Phys. Chem. C* **2022**, *126*, 19229–19239. [CrossRef]

74. Zhang, H.; Peterson, J.C.; Guyot-Sionnest, P. Intraband Transition of HgTe Nanocrystals for Long-Wave Infrared Detection at 12 µm. *ACS Nano* **2023**, *17*, 7530–7538. [CrossRef] [PubMed]

75. Harrison, M.T.; Kershaw, S.V.; Burt, M.G.; Rogach, A.; Eychmüller, A.; Weller, H. Investigation of factors affecting the photoluminescence of colloidally-prepared HgTe nanocrystals. *J. Mater. Chem.* **1999**, *9*, 2721–2722. [CrossRef]

76. Harrison, M.T.; Kershaw, S.V.; Burt, M.G.; Rogach, A.L.; Kornowski, A.; Eychmüller, A.; Weller, H. Colloidal nanocrystals for telecommunications. Complete coverage of the low-loss fiber windows by mercury telluride quantum dot. *Pure Appl. Chem.* **2000**, *72*, 295–307. [CrossRef]

77. Kershaw, S.V.; Burt, M.; Harrison, M.; Rogach, A.; Weller, H.; Eychmüller, A. Colloidal CdTe/HgTe quantum dots with high photoluminescence quantum efficiency at room temperature. *Appl. Phys. Lett.* **1999**, *75*, 1694–1696. [CrossRef]

78. Keuleyan, S.; Lhuillier, E.; Guyot-Sionnest, P. Synthesis of colloidal HgTe quantum dots for narrow mid-IR emission and detection. *J. Am. Chem. Soc.* **2011**, *133*, 16422–16424. [CrossRef]

79. Lhuillier, E.; Keuleyan, S.; Guyot-Sionnest, P. Optical properties of HgTe colloidal quantum dots. *Nanotechnology* **2012**, *23*, 175705. [CrossRef]

80. Keuleyan, S.; Kohler, J.; Guyot-Sionnest, P. Photoluminescence of Mid-Infrared HgTe Colloidal Quantum Dots. *J. Phys. Chem. C* **2014**, *118*, 2749–2753. [CrossRef]

81. Guyot-Sionnest, P. Electrical Transport in Colloidal Quantum Dot Films. *J. Phys. Chem. Lett.* **2012**, *3*, 1169–1175. [CrossRef] [PubMed]

82. Eyink, K.; Lhuillier, E.; Keuleyan, S.; Guyot-Sionnest, P.; Szmulowicz, F.; Huffaker, D. Transport properties of mid-infrared colloidal quantum dot films. In *Quantum Dots and Nanostructures: Synthesis, Characterization, and Modeling IX*; SPIE: Bellingham, WA, USA, 2012.

83. Keuleyan, S.; Lhuillier, E.; Brajuskovic, V.; Guyot-Sionnest, P. Mid-infrared HgTe colloidal quantum dot photodetectors. *Nat. Photonics* **2011**, *5*, 489–493. [CrossRef]

84. Liu, H.; Keuleyan, S.; Guyot-Sionnest, P. n- and p-Type HgTe Quantum Dot Films. *J. Phys. Chem. C* **2011**, *116*, 1344–1349. [CrossRef]

85. Ba, K.; Wang, J. Advances in solution-processed quantum dots based hybrid structures for infrared photodetector. *Mater. Today* **2022**, *58*, 119–134. [CrossRef]

86. Lhuillier, E.; Keuleyan, S.; Zolotavin, P.; Guyot-Sionnest, P. Mid-infrared HgTe/As$_2$S$_3$ field effect transistors and photodetectors. *Adv. Mater.* **2013**, *25*, 137–141. [CrossRef]

87. Keuleyan, S.E.; Guyot-Sionnest, P.; Delerue, C.; Allan, G. Mercury Telluride Colloidal Quantum Dots: Electronic Structure, Size-Dependent Spectra, and Photocurrent Detection up to 12 µm. *ACS Nano* **2014**, *8*, 8676–8682. [CrossRef] [PubMed]

88. Shen, G.; Chen, M.; Guyot-Sionnest, P. Synthesis of Nonaggregating HgTe Colloidal Quantum Dots and the Emergence of Air-Stable n-Doping. *J. Phys. Chem. Lett.* **2017**, *8*, 2224–2228. [CrossRef]

89. Chen, M.; Lan, X.; Tang, X.; Wang, Y.; Hudson, M.H.; Talapin, D.V.; Guyot-Sionnest, P. High Carrier Mobility in HgTe Quantum Dot Solids Improves Mid-IR Photodetectors. *ACS Photonics* **2019**, *6*, 2358–2365. [CrossRef]

90. Lan, X.; Chen, M.; Hudson, M.H.; Kamysbayev, V.; Wang, Y.; Guyot-Sionnest, P.; Talapin, D.V. Quantum dot solids showing state-resolved band-like transport. *Nat. Mater.* **2020**, *19*, 323–329. [CrossRef]

91. Xue, X.; Chen, M.; Luo, Y.; Qin, T.; Tang, X.; Hao, Q. High-operating-temperature mid-infrared photodetectors via quantum dot gradient homojunction. *Light Sci. Appl.* **2023**, *12*, 2. [CrossRef]

92. Liu, Z.; Wang, P.; Dong, R.; Gong, W.; Li, J.; Dai, D.; Yan, H.; Zhang, Y. Mid-Infrared HgTe Colloidal Quantum Dots In-Situ Passivated by Iodide. *Coatings* **2022**, *12*, 1033. [CrossRef]

93. Yang, J.; Hu, H.; Lv, Y.; Yuan, M.; Wang, B.; He, Z.; Chen, S.; Wang, Y.; Hu, Z.; Yu, M.; et al. Ligand-Engineered HgTe Colloidal Quantum Dot Solids for Infrared Photodetectors. *Nano Lett.* **2022**, *22*, 3465–3472. [CrossRef] [PubMed]

94. Chen, M.; Yu, H.; Kershaw, S.V.; Xu, H.; Gupta, S.; Hetsch, F.; Rogach, A.L.; Zhao, N. Fast, Air-Stable Infrared Photodetectors based on Spray-Deposited Aqueous HgTe Quantum Dots. *Adv. Funct. Mater.* **2014**, *24*, 53–59. [CrossRef]

95. Chen, M.; Lu, H.; Abdelazim, N.M.; Zhu, Y.; Wang, Z.; Ren, W.; Kershaw, S.V.; Rogach, A.L.; Zhao, N. Mercury Telluride Quantum Dot Based Phototransistor Enabling High-Sensitivity Room-Temperature Photodetection at 2000 nm. *ACS Nano* **2017**, *11*, 5614–5622. [CrossRef] [PubMed]

96. Dong, Y.; Chen, M.; Yiu, W.K.; Zhu, Q.; Zhou, G.; Kershaw, S.V.; Ke, N.; Wong, C.P.; Rogach, A.L.; Zhao, N. Solution Processed Hybrid Polymer: HgTe Quantum Dot Phototransistor with High Sensitivity and Fast Infrared Response up to 2400 nm at Room Temperature. *Adv. Sci.* **2020**, *7*, 2000068. [CrossRef] [PubMed]

97. Guyot-Sionnest, P.; Roberts, J.A. Background limited mid-infrared photodetection with photovoltaic HgTe colloidal quantum dots. *Appl. Phys. Lett.* **2015**, *107*, 253104. [CrossRef]

98. Ackerman, M.M.; Tang, X.; Guyot-Sionnest, P. Fast and Sensitive Colloidal Quantum Dot Mid-Wave Infrared Photodetectors. *ACS Nano* **2018**, *12*, 7264–7271. [CrossRef] [PubMed]

99. Yang, J.; Lv, Y.; He, Z.; Wang, B.; Chen, S.; Xiao, F.; Hu, H.; Yu, M.; Liu, H.; Lan, X.; et al. Bi_2S_3 Electron Transport Layer Incorporation for High-Performance Heterostructure HgTe Colloidal Quantum Dot Infrared Photodetectors. *ACS Photonics* **2023**, *10*, 2226–2233. [CrossRef]

100. Marshall, A.R.; Young, M.R.; Nozik, A.J.; Beard, M.C.; Luther, J.M. Exploration of Metal Chloride Uptake for Improved Performance Characteristics of PbSe Quantum Dot Solar Cells. *J. Phys. Chem. Lett.* **2015**, *6*, 2892–2899. [CrossRef]

101. Kirkwood, N.; Monchen, J.O.V.; Crisp, R.W.; Grimaldi, G.; Bergstein, H.A.C.; du Fosse, I.; van der Stam, W.; Infante, I.; Houtepen, A.J. Finding and Fixing Traps in II-VI and III-V Colloidal Quantum Dots: The Importance of Z-Type Ligand Passivation. *J. Am. Chem. Soc.* **2018**, *140*, 15712–15723. [CrossRef]

102. Ackerman, M.M.; Chen, M.; Guyot-Sionnest, P. HgTe colloidal quantum dot photodiodes for extended short-wave infrared detection. *Appl. Phys. Lett.* **2020**, *116*, 083502. [CrossRef]

103. Tang, X.; Ackerman, M.M.; Chen, M.; Guyot-Sionnest, P. Dual-band infrared imaging using stacked colloidal quantum dot photodiodes. *Nat. Photonics* **2019**, *13*, 277–282. [CrossRef]

104. Zhao, P.; Qin, T.; Mu, G.; Zhang, S.; Luo, Y.; Chen, M.; Tang, X. Band-engineered dual-band visible and short-wave infrared photodetector with metal chalcogenide colloidal quantum dots. *J. Mater. Chem. C* **2023**, *11*, 2842–2850. [CrossRef]

105. Tang, X.; Ackerman, M.M.; Guyot-Sionnest, P. Thermal Imaging with Plasmon Resonance Enhanced HgTe Colloidal Quantum Dot Photovoltaic Devices. *ACS Nano* **2018**, *12*, 7362–7370. [CrossRef] [PubMed]

106. Luo, Y.; Zhang, S.; Tang, X.; Chen, M. Resonant cavity-enhanced colloidal quantum-dot dual-band infrared photodetectors. *J. Mater. Chem. C* **2022**, *10*, 8218–8225. [CrossRef]

107. Tang, X.; Ackerman, M.M.; Shen, G.; Guyot-Sionnest, P. Towards Infrared Electronic Eyes: Flexible Colloidal Quantum Dot Photovoltaic Detectors Enhanced by Resonant Cavity. *Small* **2019**, *15*, e1804920. [CrossRef]

108. Tang, X.; Ackerman, M.M.; Guyot-Sionnest, P. Acquisition of Hyperspectral Data with Colloidal Quantum Dots. *Laser Photonics Rev.* **2019**, *13*, 1900165. [CrossRef]

109. Ciani, A.J.; Pimpinella, R.E.; Grein, C.H.; Guyot-Sionnest, P. Colloidal quantum dots for low-cost MWIR imaging. In *Infrared Technology and Applications XLII*; SPIE: Bellingham, WA, USA, 2016.

110. Gréboval, C.; Darson, D.; Parahyba, V.; Alchaar, R.; Abadie, C.; Noguier, V.; Ferré, S.; Izquierdo, E.; Khalili, A.; Prado, Y.; et al. Photoconductive focal plane array based on HgTe quantum dots for fast and cost-effective short-wave infrared imaging. *Nanoscale* **2022**, *14*, 9359–9368. [CrossRef]

111. Alchaar, R.; Khalili, A.; Ledos, N.; Dang, T.H.; Lebreton, M.; Cavallo, M.; Bossavit, E.; Zhang, H.; Prado, Y.; Lafosse, X.; et al. Focal plane array based on HgTe nanocrystals with photovoltaic operation in the short-wave infrared. *Appl. Phys. Lett.* **2023**, *123*, 051108. [CrossRef]

112. Zhang, S.; Bi, C.; Qin, T.; Liu, Y.; Cao, J.; Song, J.; Huo, Y.; Chen, M.; Hao, Q.; Tang, X. Wafer-Scale Fabrication of CMOS-Compatible Trapping-Mode Infrared Imagers with Colloidal Quantum Dots. *ACS Photonics* **2023**, *10*, 673–682. [CrossRef]

113. Izquierdo, E.; Dufour, M.; Chu, A.; Livache, C.; Martinez, B.; Amelot, D.; Patriarche, G.; Lequeux, N.; Lhuillier, E.; Ithurria, S. Coupled HgSe Colloidal Quantum Wells through a Tunable Barrier: A Strategy To Uncouple Optical and Transport Band Gap. *Chem. Mater.* **2018**, *30*, 4065–4072. [CrossRef]

114. Sato, S.A.; Lucchini, M.; Volkov, M.; Schlaepfer, F.; Gallmann, L.; Keller, U.; Rubio, A.; Sato, S.A.; Lucchini, M.; Volkov, M.; et al. Role of intraband transitions in photocarrier generation. *Phys. Rev. B* **2018**, *98*, 035202. [CrossRef]

115. Livache, C.; Goubet, N.; Gréboval, C.; Martinez, B.; Ramade, J.; Qu, J.; Triboulin, A.; Cruguel, H.; Baptiste, B.; Klotz, S.; et al. Effect of Pressure on Interband and Intraband Transition of Mercury Chalcogenide Quantum Dots. *J. Phys. Chem. C* **2019**, *123*, 13122–13130. [CrossRef]

116. Grigel, V.; Sagar, L.K.; De Nolf, K.; Zhao, Q.; Vantomme, A.; De Roo, J.; Infante, I.; Hens, Z. The Surface Chemistry of Colloidal HgSe Nanocrystals, toward Stoichiometric Quantum Dots by Design. *Chem. Mater.* **2018**, *30*, 7637–7647. [CrossRef]

117. Wang, S.; Zhang, X.; Wang, Y.; Guo, T.; Cao, S. Influence of Ink Properties on the Morphology of Long-Wave Infrared HgSe Quantum Dot Films. *Nanomaterials* **2022**, *12*, 2180. [CrossRef] [PubMed]

118. Deng, Z.; Jeong, K.S.; Guyot-Sionnest, P. Colloidal Quantum Dots Intraband Photodetectors. *ACS Nano* **2014**, *8*, 11707–11714. [CrossRef]

119. Hines, M.A.; Guyot-Sionnest, P. Synthesis and Characterization of Strongly Luminescing ZnS-Capped CdSe Nanocrystals. *J. Phys. Chem.* **1996**, *100*, 468–471. [CrossRef]

120. Cao; Banin, U. Growth and Properties of Semiconductor Core/Shell Nanocrystals with InAs Cores. *J. Am. Chem. Soc.* **2000**, *122*, 9692–9702. [CrossRef]

121. Cumberland, S.L.; Hanif, K.M.; Javier, A.; Khitrov, G.A.; Strouse, G.F.; Woessner, S.M.; Yun, C.S. Inorganic Clusters as Single-Source Precursors for Preparation of CdSe, ZnSe, and CdSe/ZnS Nanomaterials. *Chem. Mater.* **2002**, *14*, 1576–1584. [CrossRef]

122. Reiss, P.; Bleuse, J.; Pron, A. Highly Luminescent CdSe/ZnSe Core/Shell Nanocrystals of Low Size Dispersion. *Nano Lett.* **2002**, *2*, 781–784. [CrossRef]

123. Deng, Z.; Guyot-Sionnest, P. Intraband Luminescence from HgSe/CdS Core/Shell Quantum Dots. *ACS Nano* **2016**, *10*, 2121–2127. [CrossRef] [PubMed]

124. Shen, G.; Guyot-Sionnest, P. HgTe/CdTe and HgSe/CdX (X = S, Se, and Te) Core/Shell Mid-Infrared Quantum Dots. *Chem. Mater.* **2018**, *31*, 286–293. [CrossRef]

125. Kamath, A.; Melnychuk, C.; Guyot-Sionnest, P. Toward Bright Mid-Infrared Emitters: Thick-Shell n-Type HgSe/CdS Nanocrystals. *J. Am. Chem. Soc.* **2021**, *143*, 19567–19575. [CrossRef] [PubMed]

126. Shen, X.; Kamath, A.; Guyot-Sionnest, P. Mid-infrared cascade intraband electroluminescence with HgSe–CdSe core–shell colloidal quantum dots. *Nat. Photonics* **2023**, *10*, 1–5. [CrossRef]

127. Martinez, B.; Livache, C.; Notemgnou Mouafo, L.D.; Goubet, N.; Keuleyan, S.; Cruguel, H.; Ithurria, S.; Aubin, H.; Ouerghi, A.; Doudin, B.; et al. HgSe Self-Doped Nanocrystals as a Platform to Investigate the Effects of Vanishing Confinement. *ACS Appl. Mater. Interfaces* **2017**, *9*, 36173–36180. [CrossRef]

128. Martinez, B.; Livache, C.; Meriggio, E.; Xu, X.Z.; Cruguel, H.; Lacaze, E.; Proust, A.; Ithurria, S.; Silly, M.G.; Cabailh, G.; et al. Polyoxometalate as Control Agent for the Doping in HgSe Self-Doped Nanocrystals. *J. Phys. Chem. C* **2018**, *122*, 26680–26685. [CrossRef]

129. Chen, M.; Shen, G.; Guyot-Sionnest, P. State-Resolved Mobility of 1 cm^2/(Vs) with HgSe Quantum Dot Films. *J. Phys. Chem. Lett.* **2020**, *11*, 2303–2307. [CrossRef]

130. Chen, M.; Hao, Q.; Luo, Y.; Tang, X. Mid-Infrared Intraband Photodetector via High Carrier Mobility HgSe Colloidal Quantum Dots. *ACS Nano* **2022**, *16*, 11027–11035. [CrossRef]

131. Chen, M.; Shen, G.; Guyot-Sionnest, P. Size Distribution Effects on Mobility and Intraband Gap of HgSe Quantum Dots. *J. Phys. Chem. C* **2020**, *124*, 16216–16221. [CrossRef]

132. Tang, X.; Wu, G.f.; Lai, K.W.C. Plasmon resonance enhanced colloidal HgSe quantum dot filterless narrowband photodetectors for mid-wave infrared. *J. Mater. Chem. C* **2017**, *5*, 362–369. [CrossRef]

133. Sokolova, D.; Dyomkin, D.V.; Katsaba, A.V.; Bocharova, S.I.; Razumov, V.F. Effect of different ligand types on sensitivity of infrared photodetectors based on colloidal HgSe quantum dots. *Infrared Phys. Technol.* **2022**, *123*, 104188. [CrossRef]

134. Lhuillier, E.; Guyot-Sionnest, P. Recent Progresses in Mid Infrared Nanocrystal Optoelectronics. *IEEE J. Sel. Top. Quantum Electron.* **2017**, *23*, 1–8. [CrossRef]

135. Livache, C.; Martinez, B.; Goubet, N.; Greboval, C.; Qu, J.; Chu, A.; Royer, S.; Ithurria, S.; Silly, M.G.; Dubertret, B.; et al. A colloidal quantum dot infrared photodetector and its use for intraband detection. *Nat. Commun.* **2019**, *10*, 2125. [CrossRef] [PubMed]

136. Khalili, A.; Weis, M.; Mizrahi, S.G.; Chu, A.; Dang, T.H.; Abadie, C.; Gréboval, C.; Dabard, C.; Prado, Y.; Xu, X.Z.; et al. Guided-Mode Resonator Coupled with Nanocrystal Intraband Absorption. *ACS Photonics* **2022**, *9*, 985–993. [CrossRef]

137. Jagtap, A.; Livache, C.; Martinez, B.; Qu, J.; Chu, A.; Gréboval, C.; Goubet, N.; Lhuillier, E. Emergence of intraband transitions in colloidal nanocrystals [Invited]. *Opt. Mater. Express* **2018**, *8*, 1174–1183. [CrossRef]

138. Jeong, K.S.; Deng, Z.; Keuleyan, S.; Liu, H.; Guyot-Sionnest, P. Air-Stable n-Doped Colloidal HgS Quantum Dots. *J. Phys. Chem. Lett.* **2014**, *5*, 1139–1143. [CrossRef]

139. Kim, J.; Yoon, B.; Kim, J.; Choi, Y.; Kwon, Y.-W.; Park, S.K.; Jeong, K.S. High electron mobility of β-HgS colloidal quantum dots with doubly occupied quantum states. *RSC Adv.* **2017**, *7*, 38166–38170. [CrossRef]

140. Shen, G.; Guyot-Sionnest, P. HgS and HgS/CdS Colloidal Quantum Dots with Infrared Intraband Transitions and Emergence of a Surface Plasmon. *J. Phys. Chem. C* **2016**, *120*, 11744–11753. [CrossRef]

141. Yoon, B.; Jeong, J.; Jeong, K.S. Higher Quantum State Transitions in Colloidal Quantum Dot with Heavy Electron Doping. *J. Phys. Chem. C* **2016**, *120*, 22062–22068. [CrossRef]

Disclaimer/Publisher's Note: The statements, opinions and data contained in all publications are solely those of the individual author(s) and contributor(s) and not of MDPI and/or the editor(s). MDPI and/or the editor(s) disclaim responsibility for any injury to people or property resulting from any ideas, methods, instructions or products referred to in the content.

Review

Micro Spectrometers Based on Materials Nanoarchitectonics

Yanyan Qiu [1], Xingting Zhou [1], Xin Tang [1,2], Qun Hao [1,2,*] and Menglu Chen [1,2,*]

[1]   School of Optics and Photonics, Beijing Institute of Technology, Beijing 100081, China
[2]   Yangtze Delta Region Academy, Beijing Institute of Technology, Jiaxing 314019, China
*   Correspondence: qhao@bit.edu.cn (Q.H.); menglu@bit.edu.cn (M.C.)

Abstract: Spectral analysis is an important tool that is widely used in scientific research and industry. Although the performance of benchtop spectrometers is very high, miniaturization and portability are more important indicators in some applications, such as on-site detection and real-time monitoring. Since the 1990s, micro spectrometers have emerged and developed. Meanwhile, with the development of nanotechnology, nanomaterials have been applied in the design of various micro spectrometers in recent years, further reducing the size of the spectrometers. In this paper, we review the research progress of micro spectrometers based on nanomaterials. We also discuss the main limitations and perspectives on micro spectrometers.

Keywords: micro spectrometers; nanomaterials; nanoarchitectonics; spectral analysis

1. Introduction

Elements and their compounds known on Earth show specific spectra; these spectra are thus regarded as the fingerprints that identify the elements themselves. Due to this, the composition, structures, and physical states of substances can be determined by spectral analysis. Spectral analysis has a number of advantages, such as being fast, accurate and non-destructive, leading it to play a remarkable role in substance identification, physicochemical property analysis and content detection. To resolve such spectra, the incident light must be analyzed by optical spectrometers, which decompose composite light mixed with different wavelengths into spectral curves; then, the information regarding wavelengths and intensity is obtained. There are various types of optical spectrometers; in addition to those used in the visible band, there are also ultraviolet and infrared types. Today, spectrometers have a wide range of applications in biosensing [1], industrial inspection [2], mineral exploration [3], environmental monitoring [4], chemical analysis [5], deep space exploration, military technology, etc.

Traditional high-performance benchtop spectrometers, which can provide ultrahigh resolution and a wide spectral range, generally consist of three parts: namely, discrete dispersion elements, mechanical movable parts and detector arrays [6–8]. However, because of this complex structure, spectrometers usually possess many problems such as large volume, high energy consumption and high cost. For some emerging applications, the need for spectrometer miniaturization and integration is much greater than the requirement for high performance. As a result, the development of micro spectrometers has naturally become an important research direction of spectral instruments. With the development of micro nanotechnologies, computer technologies and other fields, micro spectrometers have emerged in recent years [9,10]. Micro spectrometers not only have a low cost, low power consumption and convenience in field detection, but they can also be redeveloped, so the application fields are greatly expanded.

Since the early 1990s, researchers have presented micro spectrometers via various designs and working principles. In the early stage of micro spectrometer development, the main focus lay in system integration. In 1990, D. Goldman et al. first studied and prepared the integrated waveguide miniature spectrometer based on the integrated splitter device

Citation: Qiu, Y.; Zhou, X.; Tang, X.; Hao, Q.; Chen, M. Micro Spectrometers Based on Materials Nanoarchitectonics. *Materials* **2023**, *16*, 2253. https://doi.org/10.3390/ma16062253

Academic Editor: David N. McIlroy

Received: 17 February 2023
Revised: 7 March 2023
Accepted: 8 March 2023
Published: 10 March 2023

Copyright: © 2023 by the authors. Licensee MDPI, Basel, Switzerland. This article is an open access article distributed under the terms and conditions of the Creative Commons Attribution (CC BY) license (https://creativecommons.org/licenses/by/4.0/).

combining planar waveguide and grating [11]. Afterwards, researchers began paying attention to shrinking the system components and system structure. In 1992, Dr. Mike, an American scientist and founder of the Marine company, successfully developed the world's first micro-optical fiber spectrometer, namely the micro-optical fiber spectrometer S1000. This micro spectrometer was compact: one-tenth the size of a conventional spectrometer. As a result, production costs were greatly reduced, and great progress was made in the portability of the spectrometer. Since then, micro spectrometers have been developed [12–16] and gradually used in industry. However, in some applications, the spectrometer needs to be further shrunk. Yet, reducing the size of the optical and detection elements leads to a significant decrease in the spectral resolution, sensitivity and dynamic detection range of the spectrometer. Thus, traditional scale methods are facing technical challenges.

The development of nanomaterials [17–22] provides researchers with a new method of thinking to shrink the spectrometer. Nanomaterials, such as zero-dimensional nanomaterials (quantum dots), one-dimensional nanomaterials (nanowires or nanorods) and two-dimensional nanomaterials (graphene, black phosphorus, etc.) have low cost, high carrier mobility, adjustable bandgaps and a precisely controlled materials growth process. In particular, nanomaterials exhibit many novel optical and electrical properties due to the quantum confinement effect; therefore, they show attractive prospects in electrical and optical devices and have attracted wide attention and research all over the world. Thus, it is of great significance to use this new nanomaterial for spectral filters or the photodetector units of spectrometers. In recent years, researchers have designed a variety of spectrometers by using nanomaterials, some of which do not even require optical components, allowing the spectrometers to be further miniaturized.

In this article, we review the recent relevant results on micro spectrometers based on nanomaterials and nanoarchitectonics. Depending on the dimension of nanomaterials used for the spectrometer's filter or detector units, the micro spectrometers can be divided into three categories: zero-dimensional nanomaterials-based micro spectrometers, one-dimensional nanomaterials-based micro spectrometers and two-dimensional nanomaterials-based micro spectrometers. The progress in micro spectrometers based on materials nanoarchitectonics is summarized in Figure 1.

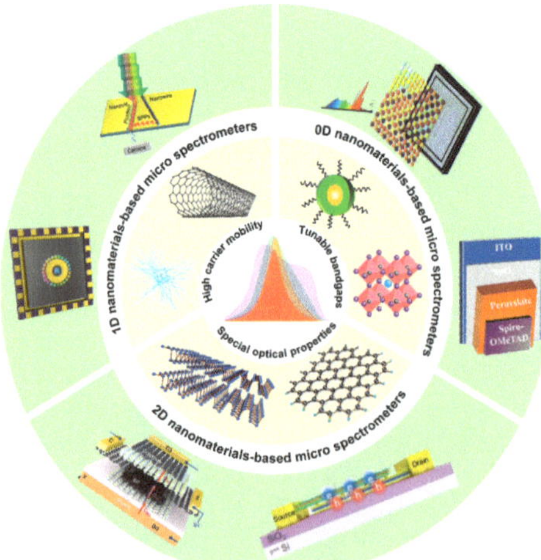

Figure 1. Progress in micro spectrometers based on nanomaterials. The image in the center presents the spectrum. The light-gold annulus around it shows the representative nanomaterials. The green annulus at the outermost area demonstrates some representative structures of these micro spectrometers based on nanomaterials.

2. Types of Micro Spectrometers

Micro spectrometers constructed by nanomaterials have different structures and operating modes from traditional spectrometers, as well as a series of advantages, such as structure, cost, integration, stability and so on. There have been many reports on micro spectrometers based on nanomaterials at home and abroad. In this review, we divide the type of micro spectrometers based on nanomaterials.

2.1. Zero-Dimensional Nanomaterials-Based Micro Spectrometers

The typical zero-dimensional (0D) nanomaterial representative is quantum dots (QDs). QDs have the characteristics of broadband absorption and narrowband emission, and the absorption spectra can be precisely tuned by varying their composition and size. Thus, QDs have fine spectral responses over a wide wavelength range. For example, by tuning the particle size, the optical response of cadmium chalcogenide QDs [23–27] could cover the visible wavelength, while lead [28–31] or mercury chalcogenide QDs [32–41] could cover infrared regions to the terahertz region. Due to this property, QDs are excellent for identifying colors or the spectra of matter.

In 1997, Jimenez et al. proposed a sensitive QD spectrometer which was created in the InAs/GaAs/Al$_x$Ga$_{1-x}$As material system. This device consisted of two main parts. An inhomogeneous QD plane was used to obtain the spectra information, while a resonant-tunneling device was used to carry out the spectra readout. It could integrate into an array because its size was small, similar to that of a common semiconductor photodetector [42]. In 2015, Zhang et al. designed an embedded micro spectrometer based on a 64-pixel QD photodetector array, showing a good spectral response in the 500–1000 nm band, especially 700–900 nm. This detector was a GaAs n-i-n photodetector with a double-barrier AlAs, which implanted an InAs QDs into its thin quantum well. The spectrometer had high sensitivity and could detect weak light down to 10^{-14} W, which was more suitable for the spectral detection of biological micro-regions [43]. Similarly, in 2018, Wang et al. also made use of InAs QDs and a In$_{0.15}$Ga$_{0.85}$As quantum well to design an AlAs double-barrier structure. The sensitivity of the photodetector based on this structure was significantly increased. They combined this 64-pixel QD photodetector with a readout circuit to create an android-based micro spectrometer. The QD micro spectrometer consisted of two main structures, namely the optical platform and the optoelectronic system. It had been experimentally proven that the proposed spectrometer was more sensitive than a commercial charge-coupled device (CCD) spectrometer. Moreover, this spectrometer had promising applications in the detection of micro-regions in biological samples [44].

In addition, colloidal QDs (CQDs) can be synthesized by solution treatment and be processed, molded and integrated with a liquid state, which will greatly miniaturize spectrometers at a low cost. Hence, CQDs are an attractive nanomaterial for micro spectrometers.

In 2015, Bao et al. demonstrated a micro QD spectrometer, which used CQDs as a spectral filter array for the first time and integrated it with a CCD. They chose 195 spectra-tunable cadmium sulfide (CdS) and cadmium selenide (CdSe) CQDs materials, which were obtained by adjusting the size or composition of CQDs and had different wavelength-selective transmittance in the selected spectral range. The CQD filters and the integrated QDs spectrometer are shown in Figure 2a. The spectral measurement of this micro spectrometer was based on the principle of wave-division multiplexing, computationally reconstructing the original spectrum by measuring the total transmission intensity of each given CQD filter. A spectral resolution of ~3.2 nm was achieved in the spectra range of 390–690 nm [45]. This work successfully demonstrates that CQDs are ideal broadband filter materials for use in micro spectrometers.

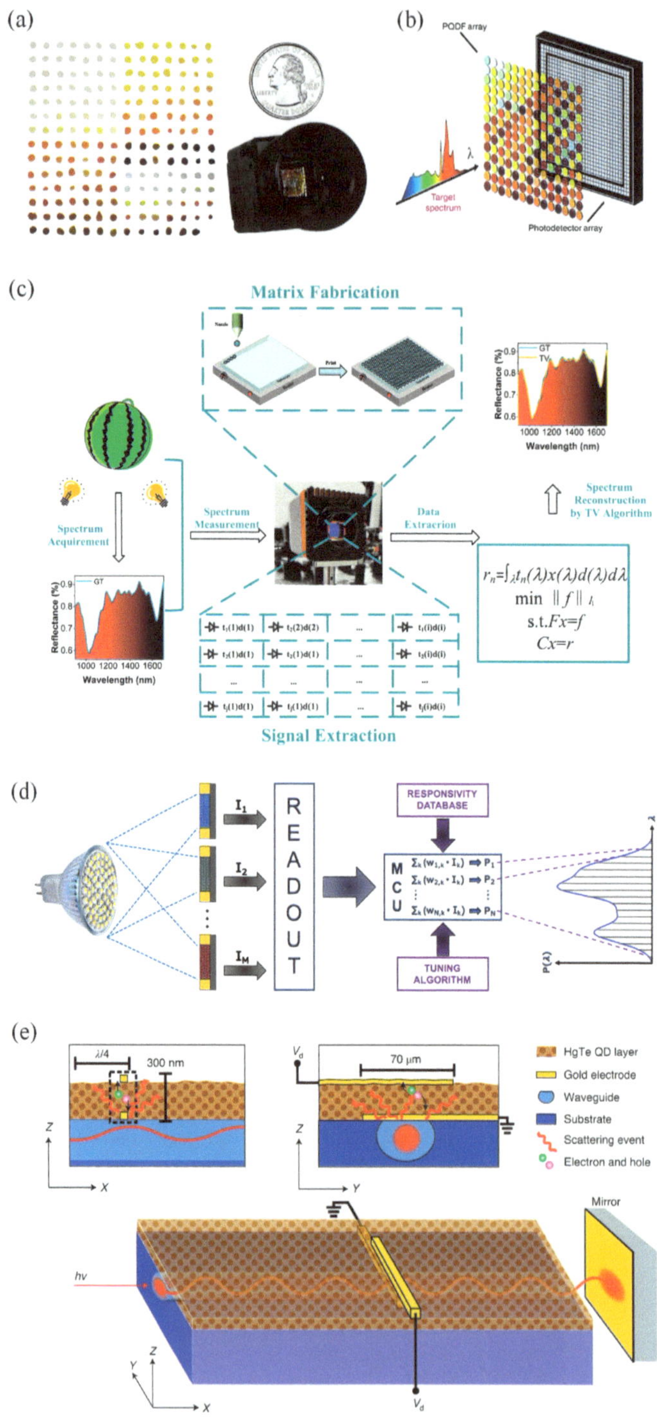

Figure 2. Architecture and schematic diagram of 0D nanomaterials-based micro spectrometers.
(**a**) CQD filters and an integrated quantum dot spectrometer. On the left are CQD filters. A total of

195 CQD materials in the form of filters. On the right is a QD micro spectrometer made from the 195 CQD filters and a CCD array detector, which is in the form of a digital camera with electronics and circuits, comparable in size to a US quarter [45]. Copyright 2015, *Nature*. (**b**) The architecture of the proposed PQDF-based hyperspectrometer [46]. Copyright 2020, *Light Science & Applications*. (**c**) Schematic diagram of the developed NIR QD spectrometer, showing the whole process including the acquirement of a real sample's spectrum, construction of a NIR QD spectrometer, measurement of the spectrum, extraction of the signal and reconstruction of the spectrum by compressive sensingbased total-variation algorithm [47]. Copyright 2021, *Advanced Optical Materials*. (**d**) Schematic diagram of the proposed algorithm-based spectrometer [48]. Copyright 2021, *Photonics and Nanostructures–Fundamentals and Applications*. (**e**) Schematic of the developed waveguide spectrometer. Waveguide spectrometer with a monolithically integrated photoconductor and the respective cross sections. The subwavelength photodetector, created on top of a buried and leaky optical waveguide, contains one bottom gold electrode operating as a scattering center, a photoactive HgTe CQD layer and a top gold electrode [49]. Copyright 2022, *Nature Photonics*.

In 2019, Tang et al. took advantage of the excellent optical properties of HgTe CQDs to design a CQD hyperspectral sensor, which could acquire hyperspectral image cubes in the short-wave infrared range. The device consisted of HgTe CQD photovoltaic sensors and distributed Bragg reflector filter arrays. The spectral measurement of this hyperspectral sensor showed a spectral resolution of up to 180 [50]. In 2020, Zhu et al. reported a broadband perovskite QD (PQD) spectrometer with a spectra bandwidth of 750 nm (250–1000 nm) and a spectral resolution of ~1.6 nm, and both parameters were superior to that of human visualization. In this study, 361 non-emissive in situ fabricated PQD-embedded films (PQDFs) were selected to form a filter array with which a silicon-based photodetector array was integrated. Further, a total-variation optimization algorithm based on compressive sensing was used to reconstruct the spectral information. The architecture of the developed PQDF-based hyperspectrometer is shown in Figure 2b [46]. To address the problem of a lack of high-resolution NIR (near-infrared) QD filters, in 2021, Li et al. proposed a NIR QD spectrometer. They combined 195 PbS and PbSe CQD filters with a NIR CMOS (complementary metal oxide semiconductor) detector and utilized the compressive sensing-based total-variation algorithm, achieving a spectra resolution of 6 nm in the NIR wavelength range between 900 and 1700 nm [47]. The schematic diagram of the proposed NIR QD spectrometer is shown in Figure 2c.

In 2021, Venettacci et al. fabricated an algorithm-based spectrometer covering the whole visible-NIR range by utilizing narrowband PbS CQDs. They drop-coated the PbS CQD dispersions onto a planar glass substrate as a filter and then placed it over a photoconductive detector whose active layer material was also PbS CQDs, thus forming a spectra-selective detector. They prepared seven batches of PbS CQD dispersions and combined the filters with photodetectors, eventually obtaining 28 devices with different responsivity spectra. The schematic diagram of the proposed spectrometer is shown in Figure 2d. For the spectral reconstruction and the peak wavelength identification spectral reconstruction, they achieved a maximum resolution of 40 nm and a mean error of 5 nm [48]. In June 2022, a single-dot spectrometer based on the 5% LiCl-doped perovskite film was proposed by Guo et al. For the structure of this single-dot spectrometer, the top electrode was a silver probe, the active layer was the perovskite film (~500 nm thickness), and the materials located above and below it were the Spiro-OMeTAD layer and the SnO2 layer, respectively. A 5 nm spectral resolution was achieved in the working area footprint of $440 \times 440\ \mu m^2$. Moreover, this spectrometer showed a good linear response in the irradiance range of 0.42 $\mu W\ cm^{-2}$ to 120 $mW\ cm^{-2}$. They showed an on-chip integrated spectral imaging system and obtained spectral information by spectral reconstruction [51].

The interferometric platforms of Fourier transform spectrometers have been shrunk in recent years, but their signal detection still relies on external imaging sensors [52,53]. To further address the challenge of the sensor miniaturization of waveguide spectrometers, in October 2022, Grotevent et al. integrated a piece of subwavelength HgTe CQD photodetector into a LiNbO$_3$ waveguide to demonstrate an ultra-compact (below 100 μm × 100 μm × 100 μm) miniature Fourier transform waveguide spectrometer (Figure 2e). This short-wave infrared spectrometer, operating at room temperature, had a spectral response at a maximum wavelength of 2 μm and a moderate spectral resolution of 50 cm^{-1} [49]. The design of this ultra-compact spectrometer offers a new idea for the miniaturization of the Fourier transform spectrometer, which is expected to be applied to hyperspectral cameras, consumer electronics, etc., in the future.

Overall, the QDs spectrometer represents a great breakthrough in the miniaturization of spectrometers. QDs are highly tunable, tiny and light-sensitive semiconductor crystals. In theory, the use of QDs can miniaturize the spectrometer without affecting its spectral resolution, spectral range and efficiency. However, micro spectrometers based on more environmentally friendly CQD materials, such as ZnSe and InP, should be developed. In the future, such tiny QDs spectrometers may be used in personalized medicine, microfluidic chip laboratories, fluid detection [54] and other fields.

2.2. One-Dimensional Nanomaterials-Based Micro Spectrometers

One-dimensional (1D) nanomaterials, such as nanowires and nanotubes, have shown attractive prospects in electrical and optical devices [55–58] in recent years. Since the nanowire and nanotube diameter reduces in size, singularities in the electronic density of states develop at special energies, called van Hove singularities. The electronic density of states is large and bears closer resemblance to molecules and atoms, but sharply differs from crystalline solids. Thus, within the limits of small diameters, novel quantum phenomena associated with this characteristic 1D density of state features are observed. Moreover, with the development of nanotechnology, the growth technology for nanowires and nanotubes has been advanced to a level in which the required composition, structure and heterojunction can be easily synthesized. As a result, the difficulty of micro spectrometer integration can be greatly reduced. Using 1D nanomaterials can even combine the spectral splitting unit and detection unit, which will greatly reduce the volume of the spectrometer. In recent years, there has been much research on 1D nanomaterials-based spectrometers.

In 2011, Cavalier et al. integrated a SiN single-mode ribbed waveguide interferometer with an array of 24 superconducting epi-NbN nanowire single-photon detectors, exhibiting an infrared SWIFTS (Stationary Wave Integrated Fourier Transform Spectrometer) device. The nanowire arrays they designed had a width of 40 nm, a thickness of 4 nm and a space of 120 nm. After a colored light of around a 1.55 μm wavelength was introduced to the optical waveguide, the nanowire arrays would produce a counter-propagative stationary interferogram. Using the SNSPD (Superconducting Nanowire Single-Photon Detectors) operating in single-photon counting mode, the modulation of the source bandwidth was detected. With 4.2 K optical power modulation, a spectral resolution of 170 nm and a spectral width of 2 μm were achieved. The operating principles of the SWIFTS-SNSPD device are shown in Figure 3a [59]. In 2017, French et al. demonstrated a hyperspectral imaging spectrometer making use of 1.7 μm thin-layered gallium phosphide (GaP) nanowires as a multiply scattering medium in combination with a compressive sensing approach. This scattering medium was only a few microns thick and was characterized by uniformity and high dispersion. In the illuminated area, there were 100 separate transmission channels in the nanowire mats, which was sufficient for spectral information encoding. When light passed through the nanowire mats' scattering medium, a series of speckled patterns appeared, which were then imaged onto the camera [60]. In June 2019, Uulu et al. first presented a Fourier transform plasmon resonance (FTPR) nanospectrometer. In this work, a metal nanowire or ridge was deposited beside the nanoslit, tilting five degrees to the nanoslit principal axis (Figure 3b). In this way, the SPPs excited by the Pt nanowire at the interface between the metal and the dielectric would combine with the incident light whose polarization was p-polarized, thus generating the interference fringes of SPP waves. After the transmitted SPPs decoupled from the metal surface, they would be collected by an objective in the far field and finally detected by a CCD sensor. By applying fast Fourier transform (FFT) to the fringe pattern of the SPPs, the spectral information of the incoming light was acquired. This nanospectrometer, based on the interference of the plasmon wave in the subwavelength metal nanoslit, can be used to sense the refractive variation and measure plasmon–exciton hybridization in a small region [61].

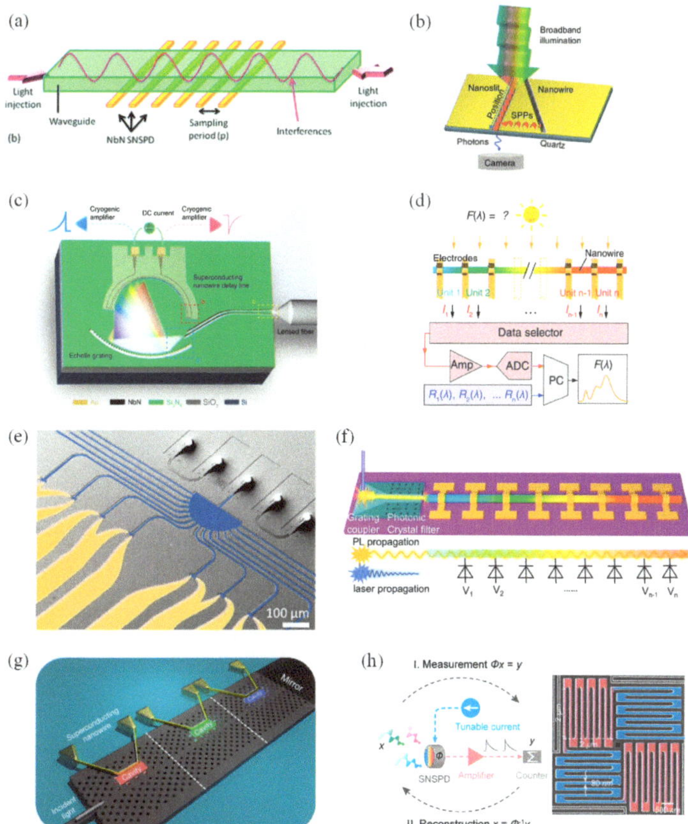

Figure 3. Architecture and schematic diagram of 1D nanomaterials-based micro spectrometers. (**a**) Operating principles of the SWIFTS−SNSPD device. An on−chip−located SNSPD detects a part of the light localized above, within the waveguide [59]. Copyright 2011, *AIP Advances*. (**b**) Schematic drawing of the FTPR nanospectrometer composed of subwavelength slit-nanowire plasmonic interferometer fabricated by focused ion beam (FIB) etching and metal nanowire (or ridge) deposition [61]. Copyright 2019, *Applied Physics Letters*. (**c**) The schematic sketch of the broadband spectrometer. The on-chip focusing echelle grating is used as a wavelength-discriminating component, while the superconducting nanowire acts simultaneously as a single-photon detector and a slow microwave delay line to continuously map the dispersed photons [62]. Copyright 2019, *Nature communications*. (**d**) Operational schematic of the proposed nanowire spectrometer [63]. Copyright 2019, *Science*. (**e**) A top view of the designed spectrometer in a false-color scanning electron microscope (SEM) image [64]. Copyright 2020, *Nano letter*. (**f**) Schematic of the monolithic spectrometer [65]. Copyright 2020, *Advanced Optical Materials*. (**g**) Architecture of the single-photon spectrometer [66]. Copyright 2020, *Physical Review Applied*. (**h**) Working principles diagram of the superconducting nanowire single-detector spectrometer. On the right is an SEM of the SNSPD, which meanders in perpendicular directions to reduce photon polarization sensitivity (painted with two different colors) [67]. Copyright 2021, *Nano letter*.

In September 2019, Cheng et al. reported their efforts in the first broadband on-chip single-photon spectrometer. They used a continuous NbN nanowire with a length of 7 mm to function as a multi-channel spectral-resolving single-photon detector, which was integrated with an on-chip dispersive echelle grating (Figure 3c). The spectrometer covered visible and infrared bands from 600 nm to 2000 nm, providing more than 200 equivalent wavelength detection channels with further scalability [62]. In the same month, Yang et al. used special CdS_xSe_{1-x} nanowires with a graded bandgap to develop a micro spectrometer with a size of only tens of microns. In their work, they used a process similar to that used to make computer chips to fabricate an array of photodetectors on the nanowires. Since each detector had varying responses to different colors of light, the team reconstructed the spectral information that needed to be measured from the equations of response functions by solving the inverse problem. The operational schematic of the nanowire spectrometer is shown in Figure 3d. The photosensitivity of the photodetector units reached 1.4×10^4 AW^{-1}, with a response speed of ~1.5 ms and a recovery time of ~3.5 ms. With further development, this spectrometer may be used for disease surveillance and food safety testing [63].

Instead of varying composition in a precise manner with the axial position along the nanowire, as in Yang's work [63], in December 2019, Meng et al. achieved wavelength selectivity by controlling the nanowire radius. They reported a micro spectrometer based on structurally colored nanowires for the first time. This micro spectrometer consisted of an array of vertical silicon nanowire (Si NW) (doped p+/i(n−)/n+) photodetectors formed above an array of planar photodetectors (doped n+/i(n−)/p+), covering the entire visible spectrum from 400–800 nm. The NWs with wavelength selectivity could be regarded as acting as a filter for the planar photodetector. They utilized the recursive least squares algorithm (for broadband spectra) and lasso regression (for narrowband spectra) to reconstruct the incident spectra [68].

Arrayed waveguide gratings compatible with waveguide-integrated single-photon detectors have limitations in bandwidth, thus only providing a limited number of spectral channels. Therefore, to enable high-resolution broadband optical operation, in March 2020, Hartmann et al. showed a fully integrated on-chip spectrometer based on tailored disorder for broadband light scattering. The device, fabricated on a silicon nitride platform (Si_3N_4), contained fiber-to-chip couplers, the nanophotonic structure and single-photon detectors. The light was launched by an input waveguide into a semicircular scattering region with a radius of 100 μm and randomly distributed air pores with a radius of 125 nm. The light diffused and was coupled into 16 output waveguides; eventually, it was detected by 16 SNSPDs, respectively (Figure 3e). Spectral-to-spatial mapping by the transmission matrix at the system using the SNSPDs allowed the reconstruction of a given signal. The resolution of this spectrometer reached 4 nm at NIR wavelengths and was improved to 30 pm at shorter wavelengths. Moreover, such a spectrometer had high sensitivity and was able to reconstruct the detection signal even at a very low input power of −111.5 dBm [64].

In April 2020, Zheng and co-workers, presented a monolithic spectrometer with a dispersive photodiode array integrated into a single composition-graded CdS_xSe_{1-x} nanowire (Figure 3f). The device operated in a waveguide mode. The nanowire, enhancing light–matter interaction, was the CdS of a 2.42 eV (optical) bandgap on one end and the CdSe of a 1.74 eV bandgap on the other, and it functioned as a wavelength selective component. A 5 nm spectral resolution and 10^{13} Jones room-temperature detectivity were exhibited. This presentation has made new progress in high-resolution and sensitive detection in on-chip spectroscopy [65]. In July 2020, Yun et al. demonstrated a design for a single-photon spectrometer using SNSPDs in cascaded photonic crystal (PC) structures (Figure 3g). The structure of this spectrometer contained a wavelength division multiplexing filter with the ability to resolve wavelengths, which was composed of a bus waveguide coupled to a planar array of PC nanocavities with different resonant wavelengths. Since the superconducting nanowires were located next to a single cavity, there was an evanescent coupling between the two, which could allow the nanowires to absorb the light captured in the cavity, and thus, detect the presence of a single photon. Through their analysis, absorption efficiencies reached ~80% in the structure with four cascaded cavities or two cascaded cavities, and the spectral resolution achieved was about 1 nm [66].

Designing a spectrometer that does not require wavelength-multiplexed optics can effectively reduce complexity and physical footprint. In 2021, Wu's research group from Nanjing University reported an optics-free single-detector spectrometer using a superconducting nanowire. They used the nonlinear spectral modulation properties of an SNSPD and computational spectral algorithms to realize spectral detection. This micro spectrometer had a wide spectral response capability covering 660 nm–1900 nm and realized a spectral resolution of sub-10 nm at the telecom. The operating principles diagram of the spectrometer is demonstrated in Figure 3h. They also combined a LiDAR with this spectrometer to demonstrate the multi-spectral LiDAR capabilities [67].

An SNSPD is suitable for detecting the spectrum of weak light. In 2023, Zheng et al. reported their efforts on a photon-counting reconstructive spectrometer that was formed on a silicon-on-insulator (SOI) substrate with a silicon layer thickness of 340 nm. The team fabricated the SNSPDs on metasurfaces with different structural parameters; then, the spectral response of the SNSPDs was modulated according to the corresponding metasurfaces. Combined with the compressive sensing algorithm, the spectrometer achieved spectral reconstruction of monochromatic light in the wavelength range of 1500–1600 nm with a spectral resolution of 2 nm. At the same time, the measurement time in this band was greatly reduced, and the detection efficiency was 1.4–3.2% [69].

One-dimensional nanomaterials, especially nanowires, have been gradually used in micro spectrometers because of their excellent light absorption ability and support for simple line array construction. As can be seen from the reports summarized above, some nanowire spectrometers no longer require optical components, which greatly reduces the size of the spectrometer. In the future, such miniature spectrometers are expected to be used in applications such as electronic consumer and wearable electronic devices. However, for use in practical production, more research work is still needed to control the size and growth of high-quality 1D nanomaterials and to address the large-area integrated positioning problem of 1D nanomaterials.

2.3. Two-Dimensional Nanomaterials-Based Micro Spectrometers

Since the success of the mechanical stripping of graphene in 2004, the exploration and research of 2D materials have developed rapidly. Two-dimensional nanomaterials have been at the forefront of research due to their thickness at the atomic scale and their unique optoelectronic properties. In recent years, 2D nanomaterials, such as hexagonal boron nitride (hBN), transition group metal sulfides, and black phosphorus, (BP) have been discovered and developed rapidly, greatly expanding the scope of applications of 2D materials. They show many novel applications and bright prospects in the field of photoelectric detection [70–74]. The optical detection of 2D materials has advantages due to their strong light–matter interactions, sharp atomic interfaces and electrically tunable optical responses. For example, BP [71,75–77] is a 2D semiconductor material with a controllable band gap and high carrier mobility. The band gap (0.3–1.5 eV) of thin-layer BP can be adjusted by the thickness. Monolayer BP has a wide photoresponse range (1–5 μm) and can strongly interact with electromagnetic waves. Therefore, 2D materials are demonstrated as an intriguing material for photodetection and have great application value in spectrometers.

In 2016, Wang et al. designed a graphene-based spectrometer operating in the mid-infrared wavelength range. The proposed spectrometer used Fabry–Pérot cavities as filters. The transmission wavelength of the spectrometer at room temperature varied with the voltage applied to this cavity. Meanwhile, graphene acted as a surface waveguide, which could spread the SPPs excitated by periodic grating structures into the Fabry–Pérot cavities (Figure 4a). This structure had a micron-sized footprint and was compatible with CMOS processing [78]. In 2018, Liu et al. showed a micro spectrometer with a plasmon-tunable graphene photodetector (Figure 4b). This photodetector, based on a graphene/monolayer MoS_2 vertical heterostructure, had a photoresponsivity of up to 10^7 AW^{-1} at room temperature. While the device had merged mini-filters and detectors needed in a spectrometer, it could still be arranged as a two-dimensional array with different bias voltages to detect spectral information in the spectrum range of 6–16 µm. In this way, more spectral information would be collected, which was convenient to realize spectral reconstruction [79].

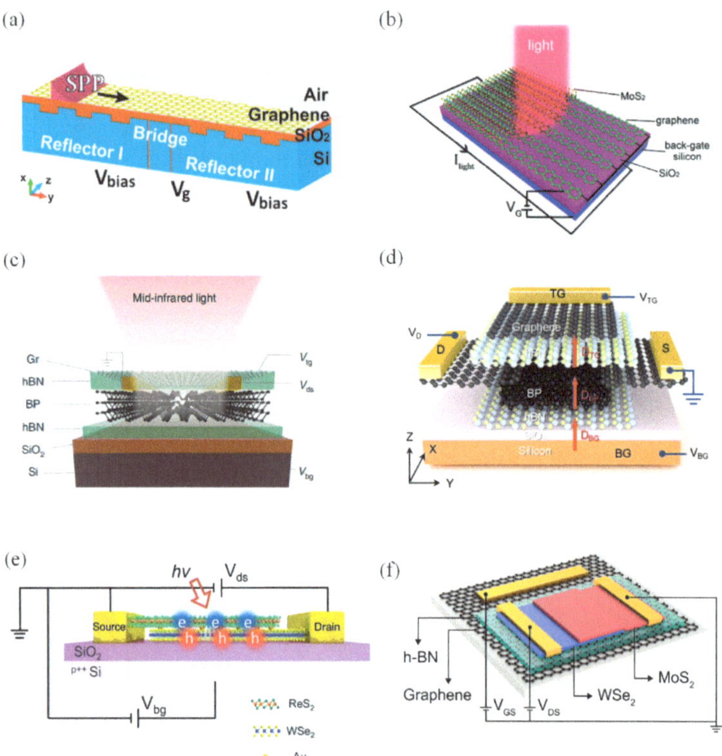

Figure 4. Architecture and schematic diagram of 2D nanomaterials-based micro spectrometers. (**a**) Schematic diagram of the graphene-based Fabry–Pérot spectrometer. Surface plasmon polaritons (SPPs) in graphene propagate across two Bragg reflectors and a bridge between them that are modulated by different voltages. The arrow at the top of the graphene layer indicates the direction of SPPs propagation [78]. Copyright 2016, *Scientific Reports*. (**b**) Schematic of the graphene ribbons and MoS_2 vertical heterostructure photodetector [79]. Copyright 2018, *Nanoscale*. (**c**) Schematic of the BP spectrometer. Gr, graphene [80]. Copyright 2021, *Nature Photonics*. (**d**) Schematic diagram of a typical dual-gate BP transistor. The applied top and bottom gate voltages (V_{TG} and V_{BG}, respectively) can control the carrier density and the electric displacement field in the sample [81]. Copyright 2022, *Applied Physics Letters*. (**e**) Schematic view of the 2D-vdWH spectrometer. V_{ds} and V_{bg} are bias voltage and back gate voltage, respectively. '*hv*' and the red arrow indicate the incident light [82]. Copyright 2022, *Nature communications*. (**f**) Schematic of the MoS_2/WSe_2 heterojunction spectrometer [83]. Copyright 2022, *Science*.

Taking advantage of the tunable energy band of BP in the mid-infrared band, in 2021, Yuan et al. produced a BP spectrometer in the 2–9 µm spectral range. This spectrometer was based on a single tunable BP photodetector (Figure 4c); thus, its active area footprint was very compact at only 9×16 µm^2. The implementation concept of the spectrometer was based on the strong photoresponse and Stark effect of the BP photodetectors. There was an hBN/BP/hBN heterostructure in this device's structure, which was capped by a monolayer of graphene. The source-drain electrodes consisted of chromium/gold, while the silicon substrate and the mid-infrared transparent graphene operated as the back and top gates, respectively. This compact single-detector spectrometer had medium spectral resolution and was more suitable for sensing applications of spectral information [80]. In June 2022, Zheng et al. used a dual-gate BP phototransistor to create a BP monolithic spectrometer. This single-detector spectrometer contained a 10 nm BP channel, and the top and bottom were encapsulated by two hBN sheets (roughly 10–20 nm in size). Meanwhile, few-layer graphene sheets (5 nm) functioned as the source, drain and top gate electrode of the phototransistor (Figure 4d). This single-detector spectrometer operated in the broadband mid-infrared range with the ultra-fine spectral and temporal resolution, which were ~2 cm^{-1} and 2 ms, respectively [81].

Van der Waals junctions (vdW) are artificial materials formed by stacking two-dimensional materials in a specific order, which have excellent optoelectronic properties. This heterojunction provides highly tunable functions that can solve the problem of an inadequate energy band structure modulation of a single 2D material, therefore allowing the realization of high-resolution broadband spectral sensing. In July 2022, Deng et al. designed a near-infrared micro spectrometer using a 2D van der Waals heterostructure (2D-vdWH). They chose ReS_2 and WSe_2 materials and introduced Au atoms into this 2D heterostructure to form a ReS_2/Au/WSe_2 sandwich structure. The heavy metal Au atoms enhanced the excited state transition dipole moments between the two-dimensional heterogeneous layer; therefore, the interlayer coupling was significantly enhanced. This structure allowed the spectrometer to achieve an electrically tunable infrared photoresponse in the wavelength range from 1.15 to 1.47 μm. Combined with the regression algorithm, this spectrometer, with an active area of only 6 μm, could accomplish spectral reconstruction and spectral imaging. The schematic drawing of the 2D-vdWH spectrometer is demonstrated in Figure 4e. However, this spectrometer has limited resolution and operates at cryogenic conditions [82]. Later, in October 2022, Yoon et al. demonstrated a micro spectrometer with a single vdW junction that could operate under ambient conditions (Figure 4f). In their work, a MoS_2/WSe_2 heterojunction was chosen, and to insulate and passivate it, they encapsulated it with hBN layers on the top and bottom. This high-performance spectrometer could distinguish the peak wavelength of monochromatic light and resolve the broadband spectrum, with an accuracy and resolution of 0.36 nm and 3 nm, respectively. The proposed single-junction spectrometer, operating in the wavelength range of ~405–845 nm, was compact; it had an active area of ~22 mm by 8 mm and was able to provide scalability. Its concept could be extended to other junctions and had the advantage of compatibility with CMOS and photonic integrated circuits. It was expected to have a broad application prospect in smartphones, satellite devices, etc. [83].

Two-dimensional nanomaterials possess excellent optical detection characteristics. As seen in the above demonstrations, several types of 2D nanomaterials have been successfully used in micro spectrometers. Moreover, most of these spectrometers have a micron footprint. However, due to the ultra-thin thickness of 2D nanomaterials, their light absorption performance may be poor, so further work is needed to optimize device design and manufacturing techniques. Moreover, there is an urgent need for a micro-spectrometer via 2D nanomaterials with a good light response, wide band and high detection sensitivity. Therefore, micro spectrometers via 2D nanomaterials will face many challenges in the future.

In conclusion, over the past three decades, spectrometers have been moving toward miniaturization. Especially, researchers have designed a variety of spectrometer structures by utilizing nanomaterials and combined them with computational reconstruction algorithms to develop many novel micro spectrometers. The performance and structures of micro spectrometers based on materials nanoarchitectonics discussed above are shown in Table 1.

Table 1. Progress in micro spectrometers based on nanomaterials.

Year	Nanomaterials Used	Spectral Range	Spectral Resolution	Footprint	Ref.
1997	InAs/GaAs QDs	-	-	-	[42]
2015	InAs QDs	500–1000 nm	-	-	[43]
2018	InAs QDs	500–930 nm	-	-	[44]
2015	CdS/CdSe CQDs	390–690 nm	~3.2 nm	8.5×6.8 mm^2	[45]
2019	HgTe CQDs	1.53–2.08 μm	~180	-	[50]
2020	Perovskite CQDs	250–1000 nm	~1.6 nm	7×7 cm^2	[46]
2021	PbS/PbSe CQDs	900–1700 nm	~6 nm	$55 \times 55 \times 82$ mm^3	[47]
2021	PbS CQDs	400–1600 nm	~40 nm	-	[48]
2022	Perovskite	350–750 nm	~5 nm	440×440 μm^2	[51]
2022	HgTe CQDs	SWIR (Max. 2 μm)	~50 cm^{-1}	$100 \times 100 \times 100$ μm^3	[49]
2011	epi-NbN nanowires	VIS-Mid-IR	~170 nm	6×8 mm^2	[59]
2017	GaP nanowires	~610–670 nm	-	Thickness: Micron-scale	[60]
2019	Pt nanowires	~600–800 nm	-	Length: 200 μm	[61]
2019	NbN nanowires	600–2000 nm	~7 nm	12×20 mm^2	[62]
2019	Single CdS$_x$Se$_{1-x}$ nanowire	500–630 nm	~10 nm	75×0.5 μm^2	[63]
2019	Multiple Silicon nanowires	450–800 nm	~6 nm	2×2 mm^2	[68]
2020	Superconducting nanowire	800–1550 nm	~30 pm /4 nm	Micron-scale	[64]
2020	CdS$_x$Se$_{1-x}$ nanowire	~560–620 nm	~5 nm	Length: 50 μm	[65]
2020	Superconducting nanowire	-	~1 nm	Height: 220 nm	[66]
2021	Superconducting nanowire	660–1900 nm	~6 nm	6×6 μm^2	[67]
2023	Superconducting nanowire	1350–1629 nm	~2 nm	-	[69]
2016	Graphene	Mid-IR	-	Micron-scale	[78]
2018	Graphene ribbons/MoS$_2$	6–16 μm	-	-	[79]
2021	Black phosphorus	2–9 μm	~90 nm	9×16 μm^2	[80]
2022	Black phosphorus	Mid-IR	2 cm^{-1}	-	[81]
2022	ReS$_2$, WSe$_2$	1.15–1.47 μm	~20 nm	6×4 μm^2	[82]
2022	MoS$_2$, WSe$_2$	405–845 nm	~3 nm	22×8 μm^2	[83]

3. Conclusions

Using nanomaterials in miniature spectrometers has greatly reduced the cost and is successfully moving spectroscopic technology out of the laboratory and into daily applications—for example, for disease detection and food detection. Nevertheless, there are still some problems to be considered:

1. The detection wavelengths of nanomaterials-based micro spectrometers are mainly concentrated in the visible and near-infrared spectral ranges. Although some of the reported micro spectrometers reach the infrared range, their spectral resolution is relatively low and far below the human visual resolution. In addition, the operating band of most micro spectrometers also needs to be broadened.
2. As the detector and the photon collection area are further reduced, the sensitivity of each instrument design is an increasingly important factor. Therefore, the sensitivity of the detector will need to be higher.
3. Many of the micro spectrometers mentioned above operate at cryogenic conditions, and thus, have limited operations.

According to the above, there is still room to improve the nanomaterials-based micro spectrometer. A wide band, high resolution and high sensitivity are important for micro spectrometers, and for their future development, focus must lie in the following aspects:

1. Broaden the spectral detection range of nanomaterials-based micro spectrometers. Micro spectrometers that respond to multiple wavelength bands may offer great flexibility and enable the development of various spectroscopic applications. In addition, key parameters such as spectral resolution, responsiveness and efficiency also need to be considered.
2. Improve the detection sensitivity of micro spectrometers. For applications requiring low light-level detection, the sensitivity of the spectrometer is an increasingly important factor. For example, improving the signal-to-noise ratio is an effective way to achieve this.
3. Focus on developing micro spectrometers operating at ambient conditions. The operation of the spectrometer needs to be carried out in certain conditions; if these conditions are not met, the instrument will not be used effectively. Operating in ambient conditions can increase the operational flexibility of the spectrometer.
4. Some of the micro spectrometers based on nanomaterials use toxic nanomaterials, such as cadmium and mercury. Although the performance of such materials is high, instruments based on such materials cannot be applied in biological applications and cannot be mass-produced. Therefore, the use of environmentally friendly nanomaterials for micro spectrometers is an important trend.

To summarize, micro spectrometers are developing towards a broadband range, high resolution, high signal-to-noise ratio, high integration, low cost, rapid detection and other directions. Researchers have geared their persistent exploration and efforts towards new principles, new technology and new materials. Nanomaterials have potential applications in spectrometers due to their excellent optical properties. In the future, there may be a strong trend for nanomaterials to replace traditional materials in the field of micro spectrometers. By utilizing nanomaterials, optical component-free micro spectrometers based on photodetectors will become an important development direction.

Author Contributions: Conceptualization, M.C. and Q.H.; Investigation, Y.Q. and X.Z.; Writing—original draft preparation, Y.Q.; Writing—review and editing, M.C. and X.T.; Project administration, M.C. and Q.H.; Funding acquisition, M.C. and Q.H.; All authors have read and agreed to the published version of the manuscript.

Funding: This research was funded by the National Key R&D Program of China and National Natural Science Foundation of China (2021YFA0717600 and NSFC No. 62105022, NSFC No. U22A2081). M.C. is also sponsored by the Beijing Nova Program (No. Z211100002121069) and Young Elite Scientists Sponsorship Program by CAST (No. YESS20210142).

Institutional Review Board Statement: Not applicable.

Informed Consent Statement: Not applicable.

Data Availability Statement: Not applicable.

Conflicts of Interest: The authors declare no conflict of interest.

References

1. Wang, Y.; Liu, X.; Chen, P.; Tran, N.T.; Zhang, J.; Chia, W.S.; Boujday, S.; Liedberg, B. Smartphone Spectrometer for Colorimetric Biosensing. *Analyst* **2016**, *141*, 3233–3238. [CrossRef] [PubMed]
2. Hashimoto, A.; Kameoka, T. Applications of Infrared Spectroscopy to Biochemical, Food, and Agricultural Processes. *Appl. Spectrosc. Rev.* **2008**, *43*, 416–451. [CrossRef]
3. Vane, G.; Goetz, A.F.H. Terrestrial Imaging Spectroscopy. *Remote Sens. Environ.* **1988**, *24*, 1–29. [CrossRef]
4. Pons, S.; Aymerich, I.F.; Torrecilla, E.; Piera, J. Monolithic spectrometer for environmental monitoring applications. In Proceedings of the OCEANS 2007—Europe, IEEE, Aberdeen, UK, 18–21 June 2007.
5. Bacon, C.P.; Mattley, Y.; DeFrece, R. Miniature Spectroscopic Instrumentation: Applications to Biology and Chemistry. *Rev. Sci. Instrum.* **2004**, *75*, 1–16. [CrossRef]

6. Cadusch, J.J.; Meng, J.; Craig, B.J.; Shrestha, V.R.; Crozier, K.B. Visible to Long-Wave Infrared Chip-Scale Spectrometers Based on Photodetectors with Tailored Responsivities and Multispectral Filters. *Nanophotonics* **2020**, *9*, 3197–3208. [CrossRef]
7. Yang, Z.; Albrow-Owen, T.; Cai, W.; Hasan, T. Miniaturization of Optical Spectrometers. *Science* **2021**, *371*, eabe0722. [CrossRef]
8. Li, A.; Yao, C.; Xia, J.; Wang, H.; Cheng, Q.; Penty, R.; Fainman, Y.; Pan, S. Advances in cost-effective integrated spectrometers. *Light Sci. Appl.* **2022**, *11*, 174. [CrossRef]
9. Deng, W.; You, C.; Zhang, Y. Spectral Discrimination Sensors Based on Nanomaterials and Nanostructures: A Review. *IEEE Sens. J.* **2021**, *21*, 4044–4060. [CrossRef]
10. Liu, Q.; Xuan, Z.; Wang, Z.; Zhao, X.; Yin, Z.; Li, C.; Chen, G.; Wang, S.; Lu, W. Low-cost micro-spectrometer based on a nano-imprint and spectral-feature reconstruction algorithm. *Opt. Lett.* **2022**, *47*, 2923–2926. [CrossRef]
11. Goldman, D.S.; White, P.L.; Anheier, N.C. Miniaturized Spectrometer Employing Planar Waveguides and Grating Couplers for Chemical Analysis. *Appl. Opt.* **1990**, *29*, 4583–4589. [CrossRef]
12. Manzardo, O.; Herzig, H.P.; Marxer, C.R.; de Rooij, N.F. Miniaturized time-scanning Fourier transform spectrometer based on silicon technology. *Opt. Lett.* **1999**, *24*, 1705–1707. [CrossRef] [PubMed]
13. Wolffenbuttel, R.F. State-of-the-art in integrated optical microspectrometers. *IEEE Trans. Instrum. Meas.* **2004**, *53*, 197–202. [CrossRef]
14. Chang, C.C.; Lee, H.N. On the estimation of target spectrum for filter-array based spectrometers. *Opt. Express* **2008**, *16*, 1056–1061. [CrossRef] [PubMed]
15. Savage, N. Spectrometers. *Nat. Photonics* **2009**, *3*, 601–602. [CrossRef]
16. Yesilkoy, F.; Arvelo, E.R.; Jahani, Y.; Liu, M.; Tittl, A.; Cevher, V.; Kivshar, Y.; Altug, H. Ultrasensitive hyperspectral imaging and biodetection enabled by dielectric metasurfaces. *Nat. Photonics* **2019**, *13*, 390–396. [CrossRef]
17. Patolsky, F.; Lieber, C.M. Nanowire Nanosensors. *Mater. Today* **2005**, *8*, 20–28. [CrossRef]
18. Lu, W.; Lieber, C.M. Semiconductor Nanowires. *J. Phys. D Appl. Phys.* **2006**, *39*, R387. [CrossRef]
19. Yan, R.; Gargas, D.; Yang, P. Nanowire Photonics. *Nat. Photonics* **2009**, *3*, 569–576. [CrossRef]
20. Adams, F.C.; Barbante, C. Nanoscience, Nanotechnology and Spectrometry. *Spectrochim. Acta Part B Spectrosc.* **2013**, *86*, 3–13. [CrossRef]
21. Bhimanapati, G.R.; Lin, Z.; Meunier, V.; Jung, Y.; Cha, J.; Das, S.; Xiao, D.; Son, Y.; Strano, M.S.; Cooper, V.R.; et al. Recent Advances in Two-Dimensional Materials beyond Graphene. *ACS Nano* **2015**, *9*, 11509–11539. [CrossRef]
22. Efros, A.L.; Brus, L.E. Nanocrystal Quantum Dots: From Discovery to Modern Development. *ACS Nano* **2021**, *15*, 6192–6210. [CrossRef] [PubMed]
23. Oertel, D.C.; Bawendi, M.G.; Arango, A.C.; Bulović, V. Photodetectors based on treated CdSe quantum-dot films. *Appl. Phys. Lett.* **2005**, *87*, 213505. [CrossRef]
24. Shin, S.W.; Lee, K.H.; Park, J.S.; Kang, S.J. Highly transparent, visible-light photodetector based on oxide semiconductors and quantum dots. *ACS Appl. Mater. Inter.* **2015**, *7*, 19666–19671. [CrossRef]
25. Kim, J.; Kwon, S.M.; Kang, Y.K.; Kim, Y.H.; Lee, M.J.; Han, K.; Facchetti, A.; Kim, M.; Park, S.K. A skin-like two-dimensionally pixelized full-color quantum dot photodetector. *Sci. Adv.* **2019**, *5*, eaax8801. [CrossRef] [PubMed]
26. Zhou, W.; Shang, Y.; García de Arquer, F.P.; Xu, K.; Wang, R.; Luo, S.; Xiao, X.; Zhou, X.; Huang, R.; Sargent, E.H.; et al. Solution-processed upconversion photodetectors based on quantum dots. *Nat. Electron.* **2020**, *3*, 251–258. [CrossRef]
27. Zi, Y.; Zhu, J.; Wang, M.; Hu, L.; Hu, Y.; Wageh, S.; Al-Hartomy, O.A.; Al-Ghamdi, A.; Huang, W.; Zhang, H. CdS@ CdSe core/shell quantum dots for highly improved self-powered photodetection performance. *Inorg. Chem.* **2021**, *60*, 18608–18613. [CrossRef]
28. Pietryga, J.M.; Schaller, R.D.; Werder, D.; Stewart, M.H.; Klimov, V.I.; Hollingsworth, J.A. Pushing the band gap envelope: Mid-infrared emitting colloidal PbSe quantum dots. *J. Am. Chem. Soc.* **2004**, *126*, 11752–11753. [CrossRef]
29. Schornbaum, J.; Winter, B.; Schießl, S.P.; Gannott, F.; Katsukis, G.; Guldi, D.M.; Spiecker, E.; Zaumseil, J. Epitaxial growth of PbSe quantum dots on MoS$_2$ nanosheets and their near-infrared photoresponse. *Adv. Funct. Mater.* **2014**, *24*, 5798–5806. [CrossRef]
30. Ramiro, I.; Özdemir, O.; Christodoulou, S.; Gupta, S.; Dalmases, M.; Torre, I.; Konstantatos, G. Mid-and long-wave infrared optoelectronics via intraband transitions in PbS colloidal quantum dots. *Nano Lett.* **2020**, *20*, 1003–1008. [CrossRef]
31. García de Arquer, F.P.; Talapin, D.V.; Klimov, V.I.; Arakawa, Y.; Bayer, M.; Sargent, E.H. Semiconductor quantum dots: Technological progress and future challenges. *Science* **2021**, *373*, eaaz8541. [CrossRef]
32. Keuleyan, S.; Lhuillier, E.; Brajuskovic, V.; Guyot-Sionnest, P. Mid-Infrared HgTe Colloidal Quantum Dot Photodetectors. *Nat. Photonics* **2011**, *5*, 489–493. [CrossRef]
33. Keuleyan, S.; Lhuillier, E.; Guyot-Sionnest, P. Synthesis of Colloidal HgTe Quantum Dots for Narrow Mid-IR Emission and Detection. *J. Am. Chem. Soc.* **2011**, *133*, 16422–16424. [CrossRef]
34. Jeong, K.S.; Deng, Z.; Keuleyan, S.; Liu, H.; Guyot-Sionnest, P. Air-stable n-doped colloidal HgS quantum dots. *J. Phys. Chem. Lett.* **2014**, *5*, 1139–1143. [CrossRef]
35. Deng, Z.; Jeong, K.S.; Guyot-Sionnest, P. Colloidal quantum dots intraband photodetectors. *ACS Nano* **2014**, *8*, 11707–11714. [CrossRef]
36. Ackerman, M.M.; Tang, X.; Guyot-Sionnest, P. Fast and Sensitive Colloidal Quantum Dot Mid-Wave Infrared Photodetectors. *ACS Nano* **2018**, *12*, 7264–7271. [CrossRef]

37. Tang, X.; Ackerman, M.M.; Guyot-Sionnest, P. Thermal Imaging with Plasmon Resonance Enhanced HgTe Colloidal Quantum Dot Photovoltaic Devices. *ACS Nano* **2018**, *12*, 7362–7370. [CrossRef] [PubMed]

38. Tang, X.; Ackerman, M.M.; Chen, M.; Guyot-Sionnest, P. Dual-Band Infrared Imaging Using Stacked Colloidal Quantum Dot Photodiodes. *Nat. Photonics* **2019**, *13*, 277–282. [CrossRef]

39. Ackerman, M.M.; Chen, M.; Guyot-Sionnest, P. HgTe Colloidal Quantum Dot Photodiodes for Extended Short-Wave Infrared Detection. *Appl. Phys. Lett.* **2020**, *116*, 083502. [CrossRef]

40. Tang, X.; Chen, M.; Kamath, A.; Ackerman, M.M.; Guyot-Sionnest, P. Colloidal Quantum-Dots/Graphene/Silicon Dual-Channel Detection of Visible Light and Short-Wave Infrared. *ACS Photonics* **2020**, *7*, 1117–1121. [CrossRef]

41. Chen, M.; Hao, Q.; Luo, Y.; Tang, X. Mid-infrared intraband photodetector via high carrier mobility HgSe colloidal quantum dots. *ACS Nano* **2022**, *16*, 11027–11035. [CrossRef] [PubMed]

42. Jimenez, J.L.; Fonseca, L.R.C.; Brady, D.J.; Leburton, J.P.; Wohlert, D.E.; Cheng, K.Y. The Quantum Dot Spectrometer. *Appl. Phys. Lett.* **1997**, *71*, 3558–3560. [CrossRef]

43. Zhang, S.H.; Wang, M.J.; Jin, X.B.; Wang, W.; Lu, H.D.; Ning, W.; Guo, F.M.; Shen, J.H. Micro-Spectrometer Based on 64-Pixel High-Sensitivity Quantum Dot Detector Array. *Micro Nano Lett.* **2015**, *10*, 573–576. [CrossRef]

44. Wang, M.J.; Cao, R.S.; Liang, K.K.; Qin, H.H. Quantum Dots Photodetector and Micro-Area Spectra Application. *Ferroelectrics* **2018**, *523*, 67–73. [CrossRef]

45. Bao, J.; Bawendi, M.G. A Colloidal Quantum Dot Spectrometer. *Nature* **2015**, *523*, 67–70. [CrossRef]

46. Zhu, X.; Bian, L.; Fu, H.; Wang, L.; Zou, B.; Dai, Q.; Zhang, J.; Zhong, H. Broadband Perovskite Quantum Dot Spectrometer beyond Human Visual Resolution. *Light Sci. Appl.* **2020**, *9*, 73. [CrossRef] [PubMed]

47. Li, H.; Bian, L.; Gu, K.; Fu, H.; Yang, G.; Zhong, H.; Zhang, J. A Near-Infrared Miniature Quantum Dot Spectrometer. *Adv. Opt. Mater.* **2021**, *9*, 2100376. [CrossRef]

48. Venettacci, C.; de Iacovo, A.; Giansante, C.; Colace, L. Algorithm-Based Spectrometer Exploiting Colloidal PbS Quantum Dots. *Photonics Nanostruct.* **2021**, *43*, 100861. [CrossRef]

49. Grotevent, M.J.; Yakunin, S.; Bachmann, D.; Romero, C.; Vázquez de Aldana, J.R.; Madi, M.; Calame, M.; Kovalenko, M.V.; Shorubalko, I. Integrated Photodetectors for Compact Fourier-Transform Waveguide Spectrometers. *Nat. Photonics* **2022**, *17*, 59–64. [CrossRef]

50. Tang, X.; Ackerman, M.M.; Guyot-Sionnest, P. Acquisition of hyperspectral data with colloidal quantum dots. *Laser Photonics Rev.* **2019**, *13*, 1900165. [CrossRef]

51. Guo, L.; Sun, H.; Wang, M.; Wang, M.; Min, L.; Cao, F.; Tian, W.; Li, L. A Single-Dot Perovskite Spectrometer. *Adv. Mater.* **2022**, *34*, 2200221. [CrossRef]

52. Le Coarer, E.; Blaize, S.; Benech, P.; Stefanon, I.; Morand, A.; Lérondel, G.; Leblond, G.; Kern, P.; Fedeli, J.M.; Royer, P. Wavelength-Scale Stationary-Wave Integrated Fourier-Transform Spectrometry. *Nat. Photonics* **2007**, *1*, 473–478. [CrossRef]

53. Pohl, D.; Reig Escalé, M.; Madi, M.; Kaufmann, F.; Brotzer, P.; Sergeyev, A.; Guldimann, B.; Giaccari, P.; Alberti, E.; Meier, U.; et al. An Integrated Broadband Spectrometer on Thin-Film Lithium Niobate. *Nat. Photonics* **2020**, *14*, 24–29. [CrossRef]

54. García-Merino, J.A.; Torres-Torres, D.; Carrillo-Delgado, C.; Trejo-Valdez, M.; Torres-Torres, C. Optofluidic and strain measurements induced by polarization-resolved nanosecond pulses in gold-based nanofluids. *Optik* **2019**, *182*, 443–451. [CrossRef]

55. Tian, B.; Kempa, T.J.; Lieber, C.M. Single nanowire photovoltaics. *Chem. Soc. Rev.* **2009**, *38*, 16–24. [CrossRef] [PubMed]

56. Kim, C.J.; Lee, H.S.; Cho, Y.J.; Yang, J.E.; Lee, R.R.; Lee, J.K.; Jo, M.H. On-Nanowire Band-Graded Si: Ge Photodetectors. *Adv. Mater.* **2011**, *23*, 1025–1029. [CrossRef] [PubMed]

57. Ren, P.; Hu, W.; Zhang, Q.; Zhu, X.; Zhuang, X.; Ma, L.; Fan, X.; Zhou, H.; Liao, L.; Duan, X.; et al. Band-selective infrared photodetectors with complete-composition-range InAs$_x$P$_{1-x}$ alloy nanowires. *Adv. Mater.* **2014**, *26*, 7444–7449. [CrossRef]

58. Hu, X.; Liu, H.; Wang, X.; Zhang, X.; Shan, Z.; Zheng, W.; Li, H.; Wang, X.; Zhu, X.; Jiang, Y.; et al. Wavelength selective photodetectors integrated on a single composition-graded semiconductor nanowire. *Adv. Opt. Mater.* **2018**, *6*, 1800293. [CrossRef]

59. Cavalier, P.; Villégier, J.C.; Feautrier, P.; Constancias, C.; Morand, A. Light Interference Detection On-Chip by Integrated SNSPD Counters. *AIP Adv.* **2011**, *1*, 042120. [CrossRef]

60. French, R.; Gigan, S.; Muskens, O.L. Speckle-Based Hyperspectral Imaging Combining Multiple Scattering and Compressive Sensing in Nanowire Mats. *Opt. Lett.* **2017**, *42*, 1820–1823. [CrossRef]

61. Uulu, D.A.; Ashirov, T.; Polat, N.; Yakar, O.; Balci, S.; Kocabas, C. Fourier Transform Plasmon Resonance Spectrometer Using Nanoslit-Nanowire Pair. *Appl. Phys. Lett.* **2019**, *114*, 251101. [CrossRef]

62. Cheng, R.; Zou, C.L.; Guo, X.; Wang, S.; Han, X.; Tang, H.X. Broadband On-Chip Single-Photon Spectrometer. *Nat. Commun.* **2019**, *10*, 4104. [CrossRef]

63. Yang, Z.; Albrow-Owen, T.; Cui, H.; Alexander-Webber, J.; Gu, F.; Wang, X.; Wu, T.-C.; Zhuge, M.; Williams, C.; Wang, P.; et al. Single-Nanowire Spectrometers. *Science* **2019**, *365*, 1017–1020. [CrossRef] [PubMed]

64. Hartmann, W.; Varytis, P.; Gehring, H.; Walter, N.; Beutel, F.; Busch, K.; Pernice, W. Broadband Spectrometer with Single-Photon Sensitivity Exploiting Tailored Disorder. *Nano Lett.* **2020**, *20*, 2625–2631. [CrossRef] [PubMed]

65. Zheng, B.; Li, L.; Wang, J.; Zhuge, M.; Su, X.; Xu, Y.; Yang, Q.; Shi, Y.; Wang, X. On-Chip Measurement of Photoluminescence with High Sensitivity Monolithic Spectrometer. *Adv. Opt. Mater.* **2020**, *8*, 2000191. [CrossRef]

66. Yun, Y.; Vetter, A.; Stegmueller, R.; Ferrari, S.; Pernice, W.H.P.; Rockstuhl, C.; Lee, C. Superconducting-Nanowire Single-Photon Spectrometer Exploiting Cascaded Photonic Crystal Cavities. *Phys. Rev. Appl.* **2020**, *13*, 014061. [CrossRef]

67. Kong, L.; Zhao, Q.; Wang, H.; Guo, J.; Lu, H.; Hao, H.; Guo, S.; Tu, X.; Zhang, L.; Jia, X.; et al. Single-Detector Spectrometer Using a Superconducting Nanowire. *Nano Lett.* **2021**, *21*, 9625–9632. [CrossRef]
68. Meng, J.; Cadusch, J.J.; Crozier, K.B. Detector-Only Spectrometer Based on Structurally Colored Silicon Nanowires and a Reconstruction Algorithm. *Nano Lett.* **2020**, *20*, 320–328. [CrossRef]
69. Zheng, J.; Xiao, Y.; Hu, M.; Zhao, Y.; Li, H.; You, L.; Feng, X.; Liu, F.; Cui, K.; Huang, Y.; et al. Photon counting reconstructive spectrometer combining metasurfaces and superconducting nanowire single-photon detectors. *Photonics Res.* **2023**, *11*, 234–244. [CrossRef]
70. Ray, A.; Kopelman, R.; Chon, B.; Briggman, K.; Hwang, J. Scattering based hyperspectral imaging of plasmonic nanoplate clusters towards biomedical applications. *J. Biophotonics* **2016**, *9*, 721–729. [CrossRef]
71. Chen, X.; Lu, X.; Deng, B.; Sinai, O.; Shao, Y.; Li, C.; Yuan, S.; Tran, V.; Watanabe, K.; Taniguchi, T.; et al. Widely tunable black phosphorus mid-infrared photodetector. *Nat. Commun.* **2017**, *8*, 1672. [CrossRef]
72. Wang, L.; Huang, L.; Tan, W.C.; Feng, X.; Chen, L.; Huang, X.; Ang, K.W. 2D photovoltaic devices: Progress and prospects. *Small Methods* **2018**, *2*, 1700294. [CrossRef]
73. Varghese, A.; Saha, D.; Thakar, K.; Jindal, V.; Ghosh, S.; Medhekar, N.V.; Ghosh, S.; Lodha, S. Near-direct bandgap WSe$_2$/ReS$_2$ type-II pn heterojunction for enhanced ultrafast photodetection and high-performance photovoltaics. *Nano Lett.* **2020**, *20*, 1707–1717. [CrossRef] [PubMed]
74. Shrestha, V.R.; Craig, B.; Meng, J.; Bullock, J.; Javey, A.; Crozier, K.B. Mid-to long-wave infrared computational spectroscopy with a graphene metasurface modulator. *Sci. Rep.* **2020**, *10*, 5377. [CrossRef]
75. Xia, F.; Wang, H.; Jia, Y. Rediscovering Black Phosphorus as an Anisotropic Layered Material for Optoelectronics and Electronics. *Nat. Commun.* **2014**, *5*, 4458. [CrossRef] [PubMed]
76. Wang, X.; Lan, S. Optical Properties of Black Phosphorus. *Adv. Opt. Photonics* **2016**, *8*, 618–655. [CrossRef]
77. Xia, F.; Wang, H.; Hwang, J.C.M.; Neto, A.H.C.; Yang, L. Black phosphorus and its isoelectronic materials. *Nat. Rev. Phys.* **2019**, *1*, 306–317. [CrossRef]
78. Wang, X.; Chen, C.; Pan, L.; Wang, J. A graphene-based Fabry-Pérot spectrometer in mid-infrared region. *Sci. Rep.* **2016**, *6*, 32616. [CrossRef]
79. Liu, Y.; Gong, T.; Zheng, Y.; Wang, X.; Xu, J.; Ai, Q.; Guo, J.; Huang, W.; Zhou, S.; Liu, Z.; et al. Ultra-Sensitive and Plasmon-Tunable Graphene Photodetectors for Micro-Spectrometry. *Nanoscale* **2018**, *10*, 20013–20019. [CrossRef] [PubMed]
80. Yuan, S.; Naveh, D.; Watanabe, K.; Taniguchi, T.; Xia, F. A Wavelength-Scale Black Phosphorus Spectrometer. *Nat. Photonics* **2021**, *15*, 601–607. [CrossRef]
81. Zheng, B.; Wang, J.; Huang, T.; Su, X.; Shi, Y.; Wang, X. Single-detector black phosphorus monolithic spectrometer with high spectral and temporal resolution. *Appl. Phys. Lett.* **2022**, *120*, 251102. [CrossRef]
82. Deng, W.; Zheng, Z.; Li, J.; Zhou, R.; Chen, X.; Zhang, D.; Lu, Y.; Wang, C.; You, C.; Li, S.; et al. Electrically Tunable Two-Dimensional Heterojunctions for Miniaturized near-Infrared Spectrometers. *Nat. Commun.* **2022**, *13*, 4627. [CrossRef] [PubMed]
83. Yoon, H.H.; Fernandez, H.A.; Nigmatulin, F.; Cai, W.; Yang, Z.; Cui, H.; Ahmed, F.; Cui, X.; Uddin, G.; Minot, E.D.; et al. Miniaturized Spectrometers with a Tunable van Der Waals Junction. *Science* **2022**, *378*, 296–299. [CrossRef] [PubMed]

Disclaimer/Publisher's Note: The statements, opinions and data contained in all publications are solely those of the individual author(s) and contributor(s) and not of MDPI and/or the editor(s). MDPI and/or the editor(s) disclaim responsibility for any injury to people or property resulting from any ideas, methods, instructions or products referred to in the content.

Article

Spatial Shifts of Reflected Light Beam on Hexagonal Boron Nitride/Alpha-Molybdenum Trioxide Structure

Song Bai [1], Yubo Li [1], Xiaoyin Cui [1], Shufang Fu [1,*], Sheng Zhou [2], Xuanzhang Wang [1] and Qiang Zhang [1,*]

[1] Key Laboratory for Photonic and Electronic Bandgap Materials, Ministry of Education, School of Physics and Electronic Engineering, Harbin Normal University, Harbin 150025, China; hsdbs2000@126.com (S.B.); yuboli1998@163.com (Y.L.); 13104688811@163.com (X.C.); xzwang696@126.com (X.W.)

[2] Department of Basic Courses, Guangzhou Maritime University, Guangzhou 510725, China; zhousheng_wl@126.com

* Correspondence: shufangfu75@163.com (S.F.); hsdzq80@126.com (Q.Z.)

Abstract: This investigation focuses on the Goos–Hänchen (GH) and Imbert–Fedorov (IF) shifts on the surface of the uniaxial hyperbolic material hexagonal boron nitride (hBN) based on the biaxial hyperbolic material alpha-molybdenum (α-MoO$_3$) trioxide structure, where the anisotropic axis of hBN is rotated by an angle with respect to the incident plane. The surface with the highest degree of anisotropy among the two crystals is selected in order to analyze and calculate the GH- and IF-shifts of the system, and obtain the complex beam-shift spectra. The addition of α-MoO$_3$ substrate significantly amplified the GH shift on the system's surface, as compared to silica substrate. With the p-polarization light incident, the GH shift can reach $381.76\lambda_0$ at about 759.82 cm^{-1}, with the s-polarization light incident, the GH shift can reach $288.84\lambda_0$ at about 906.88 cm^{-1}, and with the c-polarization light incident, the IF shift can reach $3.76\lambda_0$ at about 751.94 cm^{-1}. The adjustment of the IF shift, both positive and negative, as well as its asymmetric nature, can be achieved by manipulating the left and right circular polarization light and torsion angle. The aforementioned intriguing phenomena offer novel insights for the advancement of sensor technology and optical encoder design.

Keywords: Goos–Hänchen shift; Imbert–Fedorov shift; α-MoO$_3$

Citation: Bai, S.; Li, Y.; Cui, X.; Fu, S.; Zhou, S.; Wang, X.; Zhang, Q. Spatial Shifts of Reflected Light Beam on Hexagonal Boron Nitride/Alpha-Molybdenum Trioxide Structure. *Materials* **2024**, *17*, 1625.

https://doi.org/10.3390/ma17071625

Academic Editor: Joo Yull Rhee

Received: 30 November 2023
Revised: 26 December 2023
Accepted: 27 February 2024
Published: 2 April 2024

Copyright: © 2024 by the authors. Licensee MDPI, Basel, Switzerland. This article is an open access article distributed under the terms and conditions of the Creative Commons Attribution (CC BY) license (https://creativecommons.org/licenses/by/4.0/).

1. Introduction

The study of the kinematic effect and physical mechanism of hyperbolic materials has gained increasing popularity in recent years [1–4]. The momentum space exhibits hyperbolic behavior on the equifrequency plane. The signs of the principal components of the dielectric tensor are significant [5], expressed as diag (ε_x, ε_y, ε_z). When $\varepsilon_x \neq \varepsilon_y = \varepsilon_z$, the material is uniaxial hyperbolic material, where the characteristic feature of this material is the presence of hexagonal boron nitride [6–8]. The system possesses two distinct Reststrahlen frequency bands (RB), or RB-I corresponding to the longitudinal principal-value of permittivity and RB-II related to the transverse principal values. When $\varepsilon_x \neq \varepsilon_y \neq \varepsilon_z$, the material is biaxial hyperbolic material, where the characteristic feature of this material is the presence of alpha-molybdenum (α-MoO$_3$). This crystal exhibits van der Waals bonding and also possesses an asymmetric crystalline structure, with three dielectric tensor principal values corresponding to three Reststrahlen frequency bands [9,10]. In contrast to hBN, the three Reststrahlen frequency bands of alpha-molybdenum trioxide exhibit overlapping regions that are theoretically more anisotropic. Due to their distinctive optical properties, these materials have found extensive applications in various fields such as surface phonon polaritons [11–17], nanoimaging technology [18–20], sensors [21], and waveguides [22,23].

The existing studies indicate that the beam reflected and transmitted from the surface of the medium will exhibit a certain degree of spatial shift. The GH shift occurs in the incident plane, while the IF shift is perpendicular to the incident plane [24]. The GH shift, initially proposed by Newton, refers to a specific displacement between the incident and reflected points of a beam in the incident plane during total reflection. This phenomenon was subsequently experimentally confirmed by Goos and Hänchen in 1947 [25]. The shift primarily arises from the dispersion properties of the reflection coefficient and refraction coefficient. The phenomenon of the shift occurs when the incident beam is fully reflected at the interface, resulting in both a transverse shift within the incident plane and a longitudinal shift perpendicular to the incident plane. The shift primarily arises from the spin-orbit interaction [26]. The two types of shift mentioned above have found wide applications in various fields, including biosensors [27], temperature sensors [28], and high-sensitivity solution concentration measurement [29]. Consequently, they have garnered significant attention from scholars. Deng et al. [30] conducted a study on the beam shift occurring at the surface of a graphene-ITO/TiO_2/ITO sandwich structure. The GH shift of a Gaussian beam incident on a symmetric structure containing two polar dielectric layers separated by a spacer layer was theoretically calculated by Gupta et al. [31]. The study conducted by Li et al. [32] examined the occurrence of the GH shift and IF shift on the surface of an hBN system that was coated with black phosphorus.

The van der Waals structures offer a platform for the design of quantum materials through the manipulation of weakly bonded atomic layers, employing techniques such as twisting and stacking [33,34]. The utilization of hyperbolic materials is widespread due to their unique structural characteristics, which effectively enhance their in-plane anisotropy through torsional forces. The topological polaron and photon magic angle of the twist bilayer α-MoO_3 molecular layer were experimentally determined by Hu et al. [35]. The tunable phonon polaron in the twisted molybdenum trioxide system was experimentally determined by Chen et al. [36].

In this paper, we have chosen an α-MoO_3-based hBN thin film system and successfully achieved enhanced anisotropy within the hBN plane through torsion. The hBN is a natural uniaxial hyperbolic material, while α-MoO_3 is a natural biaxial hyperbolic material. The Reststrahlen frequency band of these two materials partially overlap. Consequently, the two materials are combined to form a heterostructure. The optical properties at the system's surface differ significantly from those of a single medium. These interface optical properties can be observed through beam shift on the system's surface. The shift of the hBN surface beam has been investigated using both the transmission matrix method and the center beam method. The α-MoO_3 substrate demonstrates enhanced anisotropy and enables greater beam shift compared to the SiO_2 substrate. Studies on the beam shift using α-MoO_3 as a substrate have been limited. Therefore, beam shifts on the surface of the hBN/α-MoO_3 heterostructure is studied. The GH shift on the surface of the hBN/α-MoO_3 system is enhanced by approximately 15 times compared to graphene/hBN [37], without requiring an increase in bias voltage for the GH shift adjustment.

2. Theoretical Model

The hBN/α-MoO_3 material structure is fabricated, as depicted in Figure 1. The x–z plane has been designated as the incident plane. The x–y plane is selected as the anisotropic plane. The solid blue line represents both the incident light and the reflected light, with the incidence angle set to θ. The hBN film has a thickness of d, and its anisotropic axis is torsioned by an angle φ relative to the direction of the incident plane. The three crystallographic directions of the semi-infinite α-MoO_3 substrate are denoted as x, y, and z, respectively. Herein, the x direction corresponds to the [100] crystallographic direction, the y direction corresponds to the [010] crystallographic direction, and the z direction corresponds to the [001] crystallographic direction. The polarization angle (β) of incident light determines its polarization state. When $\beta = 0°$, the incident light is p-polarization. At $\beta = 90°$, it becomes s-polarization. For $\beta = 45°$, the incident light exhibits circular polarization. Δ_X and Δ_Y denote the GH and IF shifts, respectively. Under this circumstance, the reflected light beam on the material surface will generate a transverse GH shift and longitudinal IF shift in comparison to geometrically reflected light.

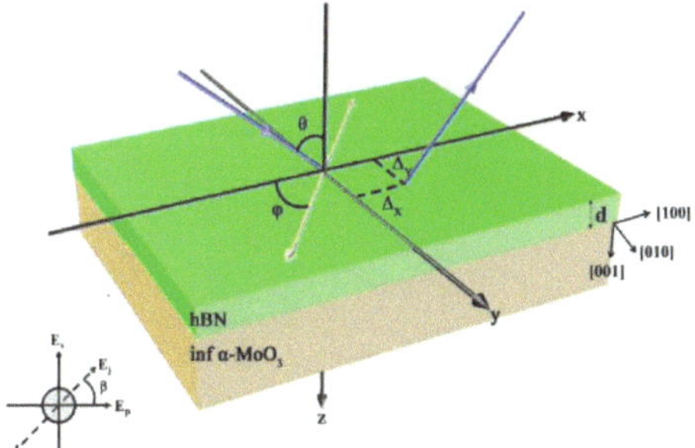

Figure 1. The configuration diagram illustrates the structure of a semi-infinite alpha-molybdenum trioxide substrate with an hBN film coating.

The hBN film is twisted at a specific angle with respect to the incident plane. The dielectric tensor matrix is represented by its equivalent form

$$\varepsilon = \varepsilon_0 \begin{pmatrix} \varepsilon_{xx} & \varepsilon_{xy} & 0 \\ \varepsilon_{xy} & \varepsilon_{yy} & 0 \\ 0 & 0 & \varepsilon_c \end{pmatrix}, \tag{1}$$

where $\varepsilon_{xx} = \varepsilon_l \cos^2 \varphi + \varepsilon_t \sin^2 \varphi$, $\varepsilon_{yy} = \varepsilon_t \cos^2 \varphi + \varepsilon_l \sin^2 \varphi$, and $\varepsilon_{xy} = (\varepsilon_t - \varepsilon_l) \cos \varphi \sin \varphi$.

The determination of $k_\pm$ in hBN and $k_{o,e}$ in α-MoO$_3$ can be achieved by solving Maxwell's equations and wave equations

$$k_\pm^2 = \frac{1}{2a}(-b \pm \sqrt{b^2 - 4ac}), \tag{2}$$

$$k_o = \sqrt{\varepsilon_y f^2 - k_x^2}, \tag{3}$$

$$k_e = \sqrt{(\varepsilon_x \varepsilon_z f^2 - \varepsilon_x k_x^2)/\varepsilon_z}. \tag{4}$$

The detailed solution procedure is in Appendix A.

The electromagnetic boundary condition dictates that the parallel components of both electric and magnetic fields remain continuous across any interface, thereby enabling the determination of the transmission matrix T. The detailed solution procedure can be found in Appendix B. We obtain $E_{x,y}^{I,R} = TF_{x,y} = T_1 T_2^{-1} F_{x,y}$. The representation of each element in the transmission matrix T is denoted as t_{mn}. The expressions of the reflection field components are

$$E_y^R = \frac{t_{11}t_{23} - t_{21}t_{13}}{t_{11}t_{33} - t_{13}t_{31}} E_x^I + \frac{t_{21}t_{33} - t_{23}t_{31}}{t_{11}t_{33} - t_{13}t_{31}} E_y^I = a_{11} E_x^I + a_{22} E_{y'}^I \tag{5}$$

$$E_x^R = \frac{t_{11}t_{43} - t_{41}t_{13}}{t_{11}t_{33} - t_{13}t_{31}} E_x^I + \frac{t_{41}t_{33} - t_{43}t_{31}}{t_{11}t_{33} - t_{13}t_{31}} E_y^I = a_{21} E_x^I + a_{22} E_{y'}^I \tag{6}$$

while the reflective electric field (z-component) can be denoted by $E_z^R = \tan(\beta) E_x^R$.

The amplitude of the electric field for the transverse electric (TE) wave with s-polarization is accurately represented as $E_{y'}^R$, while in the reflected beam, the two components of the field amplitude for the transverse magnetic (TM) wave with p-polarization are denoted by E_x^R and E_z^R. The reflective electric field is represented in matrix form using the O-xyz coordinate system. To accurately address the GH and IF shifts in the reflected beam, it is crucial to apply rotational transformations to both the incident and reflected central waves. According to the geometric correlation, the relationship between the transformation of the incident and reflected electric field can be obtained as $E_x^I = E_p^I \cos(\beta)$, $E_y^I = E_s^I$, $E_z^I = -E_p^I \sin(\beta)$, $E_p^R = -E_x^R \cos(\beta) - E_z^R \sin(\beta)$, $E_s^R = E_y^R$, and $E_z^R = \cot(\beta) E_x^R$. Consequently, the correlation equation between the incident and reflective fields in the beam coordinate systems can be formulated as

$$\begin{pmatrix} E_p^R \\ E_s^R \end{pmatrix} = \begin{pmatrix} -a_{11} & -\frac{a_{12}}{\cos(\beta)} \\ a_{21}\cos(\beta) & a_{22} \end{pmatrix} \begin{pmatrix} E_p^I \\ E_s^I \end{pmatrix}. \tag{7}$$

The subscripts "s" and "p" denote the s-polarization and p-polarization of the incident and reflected central waves, respectively. The previous results were obtained using the central plane wave. However, if we consider the paraxial wave, the incident beam is confined to a narrow range of plane waves around the central wave in momentum space. The incident beam is assumed to be a Gaussian beam, which deviates from the law of reflection. Equation (7) provides that all elements of the reflection coefficient matrix are dependent on θ and φ. To do so, we can expand the elements of the coefficient matrix to the first order of θ_p in Equations (5) and (6) to solely consider θ_p while ignoring φ_s. If this dependence arises from the anisotropy of the system, it can be anticipated that both GH and IF shifts may occur for a linearly polarization incident beam. By applying Taylor expansion to Equation (7), we can derive

$$\begin{pmatrix} E_p^R \\ E_s^R \end{pmatrix} = \begin{pmatrix} -(a_{11} + a'_{11}\beta_p) & -\frac{a_{12}+a'_{12}\beta_p}{\cos(\beta)} \\ (a_{21} + a'_{21}\beta_p)\cos(\beta) & a_{22} + a'_{22}\beta_p \end{pmatrix} \begin{pmatrix} E_p^I \\ E_s^I \end{pmatrix}. \tag{8}$$

Hence, the explanation of the GH shift provided by

$$\Delta_{\mathrm{GH}} = \frac{1}{2\pi k_0 Q}\mathrm{Im}\left\langle E^{\mathrm{R}} \left| \frac{\partial}{\partial \theta_p} E^{\mathrm{R}} \right. \right\rangle, \tag{9}$$

with $E^{\mathrm{R}} = E_p^R + E_s^R$ and $Q = \left| E^{\mathrm{R}} \right|^2$.

The formula for the IF shift is derived as follows. The polarization direction of the incident or reflected beam can be directly altered by making a slight adjustment to the orientation of the incident plane. The matrix components in Equation (7), where only φ_s is considered and θ_p is omitted, are also affected. For smaller rotation φ_s, the reflected electric field can be written as

$$\begin{pmatrix} E_p^R \\ E_s^R \end{pmatrix} = \begin{pmatrix} s_{11} & s_{12} \\ s_{21} & s_{22} \end{pmatrix} \begin{pmatrix} E_p^I \\ E_s^I \end{pmatrix}. \tag{10}$$

Elements in this coefficient matrix are obtained to be

$$s_{11} = -a_{11} + \varphi_s\left\{ [a_{12}/\cos(\theta) + a_{21}\cos(\theta)]/\tan(\theta) - a'_{11} \right\}, \tag{11}$$

$$s_{12} = -a_{12}/\cos(\theta) + \varphi_s\left[(-a_{11} + a_{22})/\tan(\theta) - a'_{12}/\cos(\beta) \right], \tag{12}$$

$$s_{21} = a_{21}\cos(\theta) + \varphi_s\left[(a_{11} - a_{22})/\tan(\theta) + a'_{21}\cos(\theta) \right], \tag{13}$$

$$s_{22} = a_{22} + \varphi_s\left\{ a'_{22} + [a_{12}/\cos(\theta) + a_{21}\cos(\theta)]/\tan(\theta) \right\}. \tag{14}$$

The IF shift is computed numerically using Equation (10) and its derivative, which is defined as

$$\Delta_{\mathrm{IF}} = -\frac{1}{2\pi k_0 Q \sin(\theta)}\mathrm{Im}\left\langle \mathbf{E}^{\mathrm{R}} \left| \frac{\partial}{\partial \varphi_s} \mathbf{E}^{\mathrm{R}} \right. \right\rangle. \tag{15}$$

If the surface layer and substrate are replaced by an isotropic material, the matrix elements of Equation (6) will become $a_{12} = a_{21} = 0$, $a_{11} = -f_p$, and $a_{22} = f_s$ (where f_p and f_s

are the reflective coefficients of the p and s waves). For the high-symmetry configurations ($\varphi = 0, 90^o$), the same results can be obtained $a_{12} = a_{21} = 0$, $a_{11} = -f_p$, and $a_{22} = f_s$. Likewise, the expression of the GH and IF shifts upon the isotropic media can be obtained with the same method. Consequently, Equations (10) and (15) can be simplified to a well-known result, which was cited in the relevant reference [24].

3. Results and Discussions

The Lorentz model of hBN and α-MoO$_3$ is provided to enhance the analysis of beam shift on the system's surface, which is respectively expressed as

$$\varepsilon_i = \varepsilon_{\infty,i}\left[1 + \left(\omega_{LO,i}^2 - \omega_{TO,i}^2\right)/\left(\omega_{TO,i}^2 - \omega^2 - i\omega\Gamma\right)\right], \tag{16}$$

and

$$\varepsilon_j = \varepsilon_{\infty,j}\left[1 + \left(\omega_{LO,j}^2 - \omega_{TO,j}^2\right)/\left(\omega_{TO,j}^2 - \omega^2 - i\omega\Gamma_j\right)\right]. \tag{17}$$

The parameters are presented in Tables 1 and 2.

Table 1. The parameters of hBN [38].

	t	l
$\varepsilon_{\infty,i}$	4.52	4.95
$\omega_{TO,i}$ (cm^{-1})	1610	825
$\omega_{LO,i}$ (cm^{-1})	1360	760
Γ (cm^{-1})	2	

Table 2. The parameters of α-MoO$_3$ [39,40].

	x	y	z
$\varepsilon_{\infty,j}$	4.0	5.2	2.4
$\omega_{TO,j}$ (cm^{-1})	972	851	1004
$\omega_{TO,j}$ (cm^{-1})	820	545	958
Γ_j (cm^{-1})	4	4	2

The Reststrahlen frequency band (RB) is defined as the frequency interval in which the product of the transverse dielectric real part and the longitudinal dielectric real part is less than 0. The material exhibits hyperbolic behavior within the residual frequency range and ellipsoidal behavior outside this range. When the real part of the dielectric in only one direction within the residual frequency band is negative, it manifests as a type I hyperbolic material. Conversely, when the real part of the dielectric in both directions within the residual frequency band is negative, it corresponds to a type II hyperbolic material. The relationship between the dielectric real part of hBN and α-MoO$_3$ with frequency is numerically simulated based on the given parameters. The relationship between the real dielectric part of hBN and frequency change is illustrated in Figure 2a. Here, ε_t and ε_l represent the transverse and longitudinal dielectric constants, respectively. It can be observed that the hBN crystal exhibits two Reststrahlen frequency bands within the mid-infrared region. The range of ω is 760 cm^{-1} < ω < 825 cm^{-1} (green region), indicating the presence of a type I hyperbolic material. Similarly, the range of ω is 1361 cm^{-1} < ω < 1610 cm^{-1} (pink region), suggesting the existence of a type II hyperbolic material. The relationship between the real part of α-MoO$_3$'s dielectric and frequency is illustrated in Figure 2b, while ε_x, ε_y, and ε_z represent the dielectric constants of α-MoO$_3$ in three orthogonal directions. The alpha-MoO$_3$ exhibits five residual frequency bands in comparison to hBN. These include a Type I hyperbolic material with frequencies ranging from 545 cm^{-1} to 820 cm^{-1} (green region), a Type II hyperbolic material with frequencies ranging from 820 cm^{-1} to 851 cm^{-1} (blue region), another Type I hyperbolic material with frequencies ranging from 851 cm^{-1} to 958 cm^{-1} (yellow region), a Type II hyperbolic material with frequencies ranging from 958 cm^{-1} to 972 cm^{-1} (pink region), and finally, a Type I hyperbolic material with frequencies ranging from 972 cm^{-1} to 1004 cm^{-1} (purple region). The region where the real part of the dielectric function approaches zero is referred to as the epsilon-near-zero (ENZ) region, while the region where the dielectric function tends toward infinity is known as the epsilon-near-pole (ENP) region. This will provide crucial theoretical support for our subsequent analysis of the beam shift.

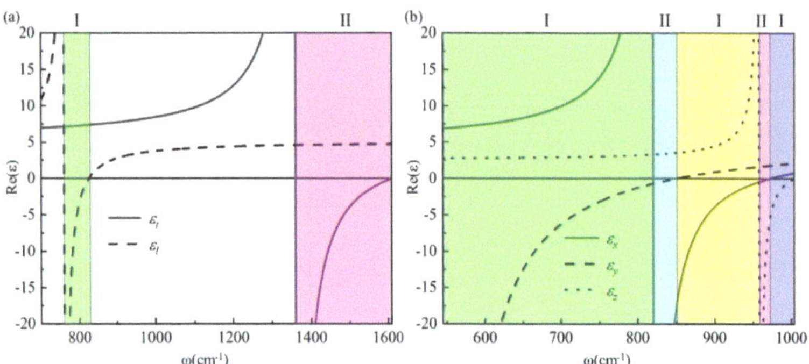

Figure 2. Illustrates the relationship between the real part of the dielectric constant and frequency for: (**a**) hBN crystal; (**b**) α-MoO$_3$ crystal.

The physical mechanism of the beam shift is investigated by employing Comsol Multiphysics to simulate the beam shift of α-MoO$_3$ covered by hBN near the ENP region, as depicted in Figure 3a. The displacement of the incident light and reflected light centers can be observed, accompanied by a significant enhancement and localization of the electric field intensity. The hybridization of hBN and α-MoO$_3$ phonon polaron gives rise to the formation of a surface polaron with enhanced professionalism. The |E| profile of the hBN/α-MoO$_3$ surface incident and reflected beams is illustrated in Figure 3b. The solid green line depicts the electric field of the incident light. The solid blue line depicts the electric field of the reflected light in a precise and technical manner. The presence of a disparity in height can be observed at the summit of |E| when x = 48 μm. The highly localized electric field strength of the hBN/α-MoO$_3$ surface enables the construction and integration of relatively compact devices.

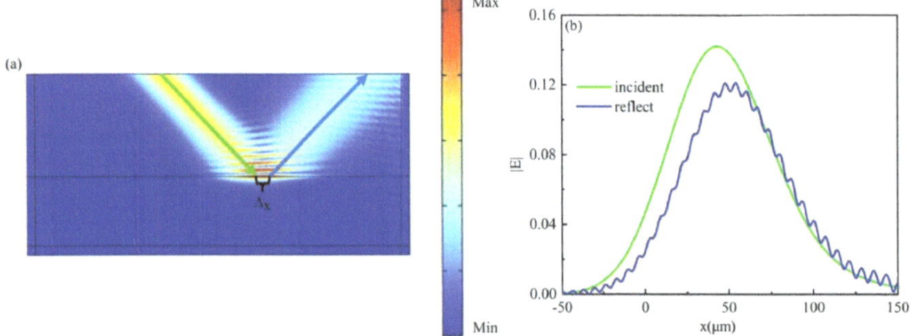

Figure 3. (**a**) The simulation of the GH shift on the hBN/MoO3 surface. (**b**) The distribution of electric field for incident and reflected light on the surface of the ENP region.

The special case is chosen when the torsion angle $\varphi = 0°$, under which circumstance the wave vectors in hBN can be expressed as $k_+ = \sqrt{\varepsilon_t f^2 - k_x^2}$ and $k_- = \sqrt{\varepsilon_t f^2 - \varepsilon_t k_x^2/\varepsilon_l}$. The surface GH shift of hBN/SiO$_2$ and hBN/α-MoO$_3$ systems is investigated in the current study. The solid lines of varying colors in Figure 4 depict different thicknesses of hBN. The relationship between the GH shift on the hBN/SiO$_2$ surface and the frequency change is simulated in Figure 4a. In this case, the GH shift on the system's surface reaches up to $1.8\lambda_0$. The frequency-dependent variation of the GH shift on the hBN/α-MoO$_3$ surface under the p-wave incident is illustrated in Figure 4b. The surface GH shift of the system exhibits rapid transitions from negative to positive within the frequency range of 759.5 cm$^{-1} < \omega < 760$ cm^{-1}. The GH shift can achieve a maximum value of $381.76\lambda_0$ at the frequency $\omega = 759.82$ cm^{-1}. At this juncture, in conjunction with Figure 2a, it can be observed that the value of $\mathrm{Re}(\varepsilon_l) \approx \infty$ at this frequency is situated within the ENP region of hBN. The combination of Figure 2a,b reveals an overlap in the residual frequency bands between hBN and α-MoO$_3$ within this particular region. Compared with Figure 4a,b, the GH shift on the surface of the hBN/α-MoO$_3$ system is enhanced by about 212 times. The surface of the hBN/SiO$_2$ system under s-wave irradiation is depicted in Figure 4c, illustrating the observed GH shift. The GH shift value generated on the system's surface is negligible. However, due to the pronounced anisotropy of the α-MoO$_3$ substrate, the surface of the system also exhibits a significant GH shift under s-wave irradiation, as illustrated in Figure 4d. Under s-wave irradiation, the GH shift on the surface of the hBN/α-MoO$_3$ system exhibits a transition from negative to positive within the frequency range of 906.5 cm$^{-1} < \omega < 907.25$ cm^{-1}, with a peak value of $288.84\lambda_0$ observed at $\omega = 906.88$ cm^{-1}. This frequency falls within the Reststrahlen frequency band of α-MoO$_3$.

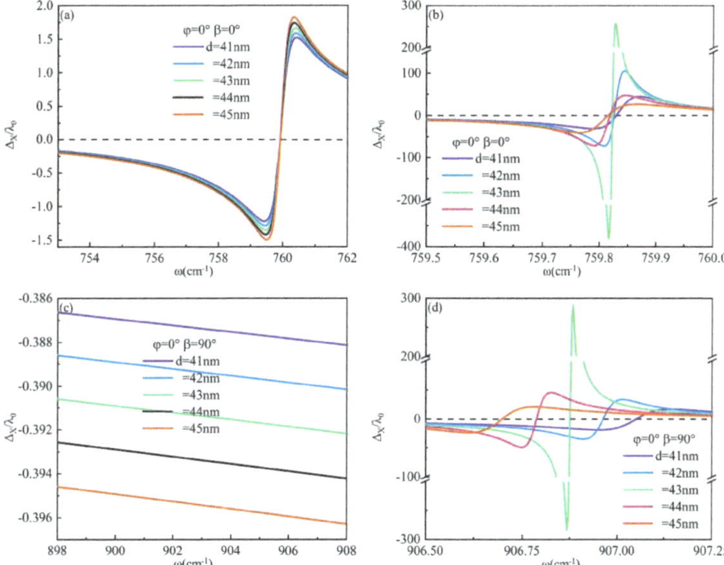

Figure 4. The relationship between the GH shift on the system surface and frequency when $\varphi = 0°$: (**a**) p-polarization light incident on hBN/SiO$_2$; (**b**) p-polarization light incident on hBN/α-MoO$_3$; (**c**) s-polarization light incident on hBN/SiO$_2$; (**d**) s-polarization light incident on hBN/α-MoO$_3$.

In the analysis of the beam shift, it is crucial to consider not only the distinct residual frequency bands of the material but also place significant emphasis on the reflectivity parameter. The reflection coefficient of p-polarization light is as follows: $R_p = (\varepsilon f \cos\beta - k_z)/(\varepsilon f \cos\beta - k_z)$. The reflection coefficient of s-polarization light is being studied: $R_s = (f \cos\beta - k_z)/(f \cos\beta + k_z)$. The critical angle is defined as $\sin^2\beta < \varepsilon$, where ε is a positive value between 0 and 1. The

Brewster angle can be defined as the angle of incidence at which the reflectance $R \approx 0$ corresponds to a specific frequency. The reflectance of the p-wave and s-wave exhibits frequency-dependent variations for an hBN thickness of 43 nm, as illustrated in Figure 4. Analysis of Figure 5a,b reveals that the reflectance $R \approx 0$ coincides with the occurrence of the GH shift peak at frequencies $\omega = 759.82$ cm^{-1} and $\omega = 906.88$ cm^{-1}, respectively, for incident p-polarization and s-polarization light, both aligning with the Brewster angle.

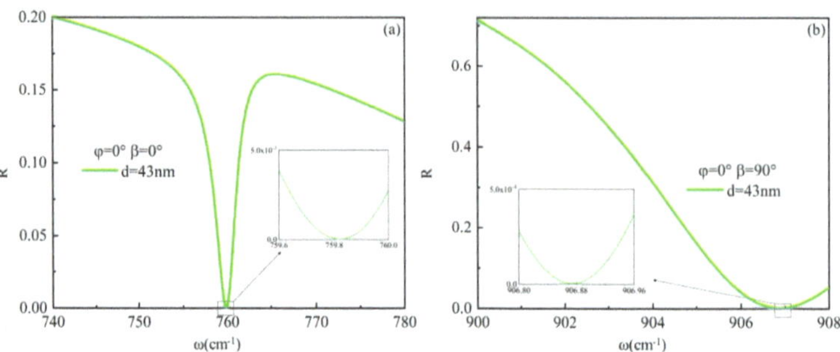

Figure 5. When $\varphi = 0°$ and $d = 43$ nm, the reflectance of the system surface varies with frequency under: (**a**) p-polarization light incidence; (**b**) s-polarization light incidence.

The previous discussion focused on the GH shift of p-polarization and s-polarization light, followed by a numerical simulation of the IF shift of the system's surface. The IF shift of the system surface is primarily attributed to the interaction of spin orbital angular momentum, which arises from the interaction between elliptically polarization beams comprising p-polarization light and s-polarization light. Consequently, our focus lies in discussing the IF shift under circularly polarization light incidence. When the polarization angle $\beta = 45°$, it corresponds to right-handed circularly polarization light. Conversely, when $\beta = -45°$, it corresponds to left-handed circularly polarization light. The frequency variation of the IF shift for incident right-handed circular-polarization light and left-handed circular-polarization light on hBN surfaces with varying thicknesses is investigated under two specific torsion angle conditions, $\varphi = 0°$ and $\varphi = 90°$, as illustrated in Figure 6. The dielectric tensor of hBN is represented as diag(ε_l, ε_t, ε_t) and diag(ε_t, ε_l, ε_t) at two torsion angles, respectively. Similarly, the dielectric tensor of α-MoO$_3$ is expressed as diag(ε_x, ε_y, ε_z), both of which are denoted by diagonal matrices. Under identical conditions, it can be observed that the positive and negative symmetry of the IF shift induced by right-handed circularly polarization light and left-handed circularly polarization light is attributed to the material's dielectric tensor in the form of a diagonal matrix.

We conducted numerical simulations to investigate the reflectance of the system under circularly polarization light incidence, as depicted in Figure 7. Our findings indicate that both right-handed and left-handed circularly polarization light have no impact on the reflectance of the system under identical conditions. Moreover, due to a small incident angle θ, when the torsion angle φ is $0°$ and the hBN thickness is 100 nm at a frequency $\omega = 852$ cm^{-1}, the reflectance R approaches approximately 0.5. In this scenario, the induced IF shift can reach up to $1.55\lambda_0$. Conversely, when $\varphi = 90°$ with an hBN thickness of 500 nm and frequency $\omega = 852$ cm^{-1}, we observe a similar reflectance $R \approx 0.5$ leading to an IF shift of approximately $2.53\lambda_0$. Notably, the high reflectivity of the material eliminates the need for weak measurement technology during observation, highlighting its practical application value.

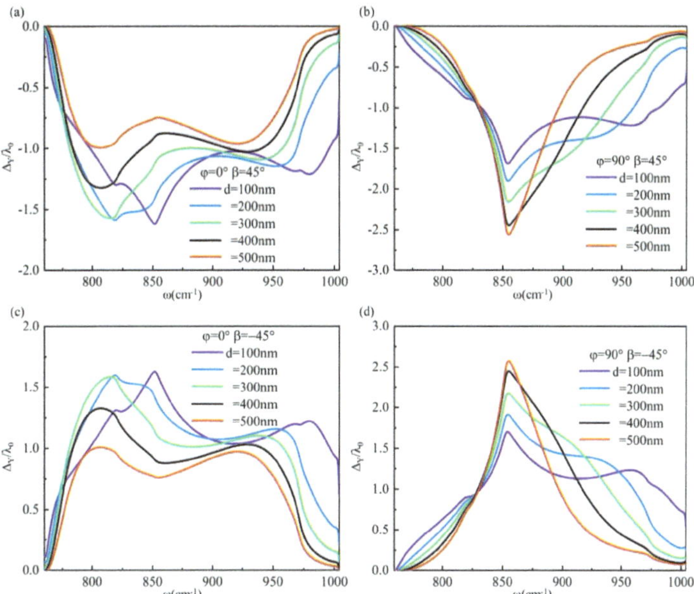

Figure 6. The frequency-dependent variation of the shift of an hBN surface with different thickness is observed at incidence angles $\theta = 10°$: (**a**) $\varphi = 0°$, $\beta = 45°$; (**b**) $\varphi = 90°$, $\beta = 45°$; (**c**) $\varphi = 0°$, $\beta = -45°$; and (**d**) $\varphi = 90°$, $\beta = -45°$.

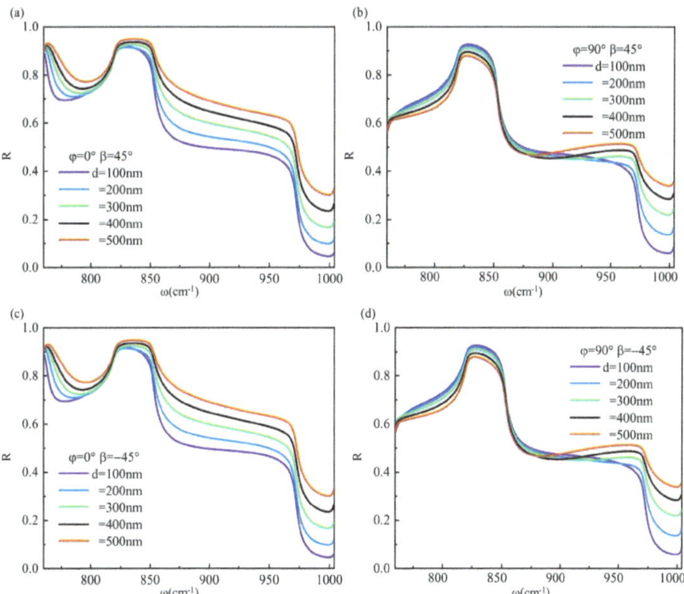

Figure 7. When the hBN thickness of the system varies, the surface reflectance R exhibits frequency-dependent changes under an incidence angle $\theta = 10°$: (**a**) $\varphi = 0°$, $\beta = 45°$; (**b**) $\varphi = 90°$, $\beta = 45°$; (**c**) $\varphi = 0°$, $\beta = -45°$; (**d**) $\varphi = 90°$, $\beta = -45°$.

As mentioned above, the positive and negative symmetry of the IF shift generated by right-handed and left-handed circular-polarization light under identical conditions is attributed to the diagonal form of the dielectric tensor matrix of the material. The expression form of the dielectric matrix can be altered through twisting. Subsequently, hBN with thicknesses of 100 nm and 500 nm, respectively, were selected. By varying the torsion angle, we explored the positive and negative symmetry of the IF shift as depicted in Figure 8. It is observed that altering the torsion angle of hBN's anisotropic axis resulted in a nondiagonal dielectric tensor matrix for hBN, and therefore eliminated the positive and negative symmetry in the IF shift on system surfaces under identical conditions. This unique optical phenomenon provides novel insights for designing and fabricating optical encoders. The surface anisotropy of the system is enhanced due to the specific twisting angle of hBN's main optical axis. The IF shift is observed in the reststrahlen frequency band of α-MoO$_3$, as depicted in Figure 8b,d. The IF shift with frequency and twisting angle is depicted in Figure 8b when right-handed circular polarizing light is incident. Narrow peaks are observed within the Reststrahlen frequency band of α-MoO$_3$, specifically between frequencies 700 cm^{-1} < ω < 750 cm^{-1} and 850 cm^{-1} < ω < 900 cm^{-1}. The IF shift with frequency and twisting angle is depicted in Figure 8d when left-handed circular polarizing light is incident on the surface of the system. Narrow peaks were observed 800 cm^{-1} < ω < 850 cm^{-1}. The frequency range in question is situated at the interface between the Reststrahlen frequency band of hBN and α-MoO$_3$, specifically within the ENZ region of α-MoO$_3$. It should be noted that α-MoO$_3$ exhibits Type II hyperbolic material behavior with a notable degree of anisotropy.

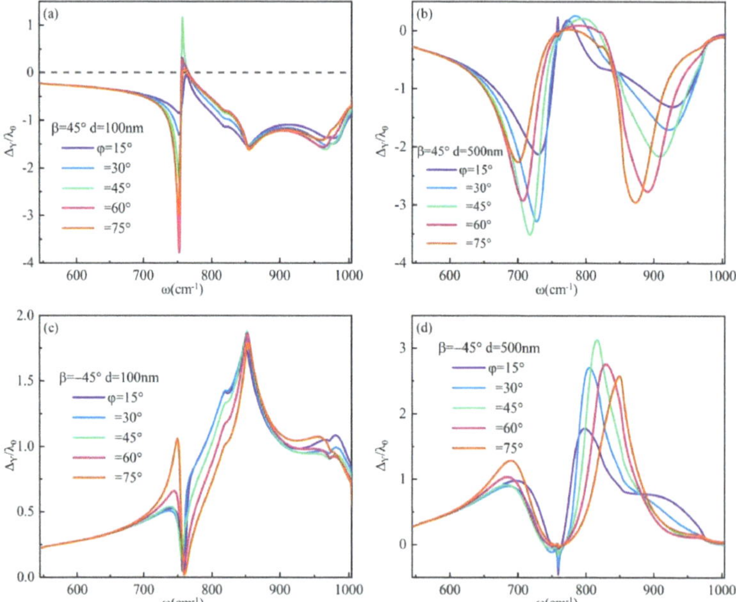

Figure 8. The IF shift of the system surface exhibits frequency-dependent variations when the torsion angle is altered, with an incident angle $\theta = 10°$: (**a**) $\beta = 45°$, $d = 100$ nm; (**b**) $\beta = 45°$, $d = 500$ nm; (**c**) $\beta = -45°$, $d = 100$ nm; (**d**) $\beta = -45°$, $d = 500$ nm.

Next, we investigate the maximum value of the system's IF shift, as depicted in Figure 9a,b. It is observed that for an hBN thickness of 100 nm, torsion angle $\varphi = 60°$, incident angle $\theta = 10°$, and right-circularly polarization incident light, the system exhibits an IF shift of 3.76λ_0 at frequency $\omega = 751.94$ cm^{-1}, thereby enhancing the in-plane hyperbolism of hBN. Additionally, when the hBN thickness is increased to 500 nm with a torsion angle

$\varphi = 45°$ and left-handed circularly polarization incident light, the IF shift can reach $3.18\lambda_0$ at frequency $\omega = 820$ cm^{-1} within the Reststrahlen frequency band of Type II hyperbolic material α-MoO$_3$. Based on Figure 2a,b analysis reveals that this frequency corresponds to an overlap between the Reststrahlen frequency band of hBN and α-MoO$_3$.

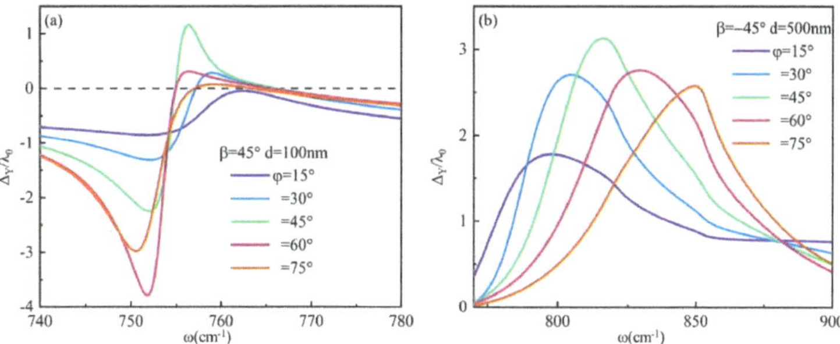

Figure 9. When the torsion angle is adjusted, the frequency-dependent shift of the interface on the system's surface varies. The incident angle θ is set at 10°: (a) $\beta = 45°$, $d = 100$ nm; (b) $\beta = -45°$, $d = 500$ nm.

4. Conclusions

This present study investigates the GH shift and IF shift on the hBN/α-MoO$_3$ by using the transmission matrix method and the central beam method. The formula for beam shift on the structure's surface is derived using a torsion coordinate system, followed by numerical simulation. Under p-polarization light incidence, GH shift reaches its maximum value of $381.76\lambda_0$ at frequency $\omega = 759.82$ cm^{-1}, with a corresponding reflectance R ≈ 0 and an incidence angle equal to the Brewster angle. We also examine the IF shift of right-handed and left-handed circularly polarization light reflected on the hBN/α-MoO$_3$ system's surface. When the dielectric tensor of the system is diagonal, both right-handed and left-handed circularly polarization light exhibits positive and negative symmetry under identical conditions with matching frequencies. However, this symmetry ceases to exist when torsion coordinates alter the dielectric tensor to a nondiagonal matrix. By applying torsion at $d = 100$ nm and $\varphi = 60°$, an IF shift of $3.76\lambda_0$ can be achieved. These findings offer theoretical guidance for advancing novel nano-optical devices and optical encoders.

Author Contributions: Conceptualization, S.B. and X.C.; methodology, S.B.; software, S.B., Y.L. and S.Z.; validation, S.F., X.W. and Q.Z.; formal analysis, Q.Z.; investigation, Q.Z.; resources, Q.Z.; data curation, S.B.; writing—original draft preparation, S.B. and X.C.; writing—review and editing, Q.Z. and S.F.; visualization, Q.Z.; supervision, Q.Z.; project administration, X.W.; funding acquisition, Q.Z. All authors have read and agreed to the published version of the manuscript.

Funding: Project supported by the Natural Science Foundation of Heilongjiang Province (Grant No. LH2020A014), the Harbin Normal University Postgraduate Innovative Research Project (Grant Nos. HSDSSCX2023-16).

Data Availability Statement: The references in this paper have been explained.

Conflicts of Interest: The authors declare no conflict of interest.

Appendix A. Derivation of Wave Vector $k_{\pm}$ and $k_{o,e}$

The induced electric field, as described by Maxwell's equations, can be expressed in the following manner:

$$\mathcal{M}E = \begin{pmatrix} k_z^2 - \varepsilon_{xx}f^2 & -\varepsilon_{xy}f^2 & k_xk_z \\ -\varepsilon_{xy}f^2 & k_x^2 + k - \varepsilon_{yy}f^2 & 0 \\ k_xk_z & 0 & k_x^2 - \varepsilon_c f^2 \end{pmatrix} \begin{pmatrix} E_x \\ E_y \\ E_z \end{pmatrix} = 0, \tag{A1}$$

The value of f in Equation (A1) is $\mu_0 \varepsilon_0 \omega^2$. The satisfaction of Equation (A1) is a necessary to be satisfied $\det(\mathcal{M}) = 0$. The electric field components of the hBN in the x and z directions can be represented by the electric field components in the y direction under this condition, thereby indicating that the equation k_z^2 can be transformed into a quadratic equation with one variable. The wave vector in the z direction of hBN needs to be solved. Its solution k_z is read as

$$k_{\pm}^2 = \frac{1}{2a}\left(-b \pm \sqrt{b^2 - 4ac}\right), \tag{A2}$$

where $a = \varepsilon_t; b = \varepsilon_{xx}\left(k_x^2 - \varepsilon_t f^2\right) + \varepsilon_t\left(k_x^2 - \varepsilon_{yy} f^2\right); c = \varepsilon_{xx}\left(k_x^2 - \varepsilon_t f^2\right)\left(k_x^2 - \varepsilon_{yy} f^2\right) + \varepsilon_{xy}\varepsilon_{yx} f^2\left(k_x^2 - \varepsilon_t f^2\right)$. The three components of the electric field in hBN intersect: $E_x^{\pm} = \left(k_x^2 + k_{\pm}^2 - \varepsilon_{yy} f^2\right)E_y^{\pm}/\varepsilon_{yx} f^2 = \Lambda_{\pm}E_y^{\pm}; E_z^{\pm} = k_{\pm}k_x\Lambda_{\pm}E_y^{\pm}/\left(k_x^2 - \varepsilon_t f^2\right)$.

The α-MoO$_3$ substrate, being a semi-infinite crystal, imposes restrictions on the incident wave vectors, limiting them to only two options: k_o and k_e. The formulas for both are

$$k_o = \sqrt{\varepsilon_y f^2 - k_x^2}, \tag{A3}$$

and

$$k_e = \sqrt{\left(\varepsilon_x \varepsilon_z f^2 - \varepsilon_x k_x^2\right)/\varepsilon_z}. \tag{A4}$$

The wave vectors k_o and k_e correspond to the transverse electric (TE) and transverse magnetic (TM) modes of α-MoO$_3$, respectively. The TE(s-polarization) mode is perpendicular to the incident plane while the TM(p-polarization) mode lies within it.

Appendix B. Determine the Transmission Matrix for z = 0 and z = d

The electric field components in the x and y directions in the air and α-MoO$_3$ base, as well as the electric field component in the y direction in the hBN, need to be proposed as

$$E_{x,y} = E_{x,y}^I e^{ik_z z} + E_{x,y}^R e^{-ik_z z}, \ (z < 0) \tag{A5a}$$

$$E_y = A e^{ik_+ z} + B e^{-ik_+ z} + C e^{ik_- z} + D e^{-ik_- z}, \ (0 < z < d) \tag{A5b}$$

$$E_x = F_x e^{ik_e z}, E_y = F_y e^{ik_o z}. \ (z > d) \tag{A5c}$$

The term "z = 0" denotes the interface between the vacuum and hBN; The term "z = d" denotes the interface between the hBN and bulk α-MoO$_3$. In accordance with the electromagnetic boundary conditions, the correlation transmission matrix of the incident field and the transmitted field can be derived.

From Equation (A5), the magnetic fields in different directions can be obtained from $\mathbf{H} = \nabla \times \mathbf{E}/i\mu_0\omega$. The electromagnetic boundary conditions dictate that the parallel components of both electric and magnetic fields remain uninterrupted across any interface. The boundary value relationship between the vacuum and hBN can be obtained as follows:

$$E_y^I + E_y^R = A + B + C + D, \tag{A6a}$$

$$E_y^I - E_y^R = \frac{k_+}{k_z}(A - B) + \frac{k_-}{k_z}(C - D), \tag{A6b}$$

$$E_x^I + E_x^R = \Lambda_+(A + B) + \Lambda_-(C + D), \tag{A6c}$$

$$E_x^I - E_x^R = \lambda_+(A - B) + \lambda_-(C - D), \tag{A6d}$$

where $\lambda_{\pm} = \Lambda_{\pm}k_{\pm}k_z\varepsilon_z/\left(\varepsilon_z f^2 - k_x^2\right)$. The interface between the hBN layer and the α-MoO$_3$ crystal reveals

$$A e^{ik_+ d} + B e^{-ik_+ d} + C e^{ik_- d} + D e^{-ik_- d} = F_y, \tag{A7a}$$

$$\frac{k_+}{k_o}\left(A e^{ik_+ d} - B e^{-ik_+ d}\right) + \frac{k_-}{k_o}\left(C e^{ik_- d} - D e^{-ik_- d}\right) = F_y, \tag{A7b}$$

$$\Lambda_+\left(A e^{ik_+ d} + B e^{-ik_+ d}\right) + \Lambda_-\left(C e^{ik_- d} + D e^{-ik_- d}\right) = F_x, \tag{A7c}$$

$$\gamma_+\left(Ae^{ik_+d} + Be^{-ik_+d}\right) + \gamma_-\left(Ce^{ik_-d} + De^{-ik_-d}\right) = F_x, \tag{A7d}$$

where $\gamma_\pm = \Lambda_\pm k_\pm \varepsilon_t \left(\varepsilon_z f^2 - k_x^2\right)/k_e \varepsilon_z \left(\varepsilon_t f^2 - k_x^2\right)$. Equations (A6) and (A7) are transformed into matrix forms for the sake of convenience. The electric field's amplitude is ascertained by

$$\begin{pmatrix} E_y^I \\ E_y^R \\ E_x^I \\ E_x^R \end{pmatrix} = \frac{1}{2}\begin{pmatrix} 1+\frac{k_+}{k_z} & 1-\frac{k_+}{k_z} & 1+\frac{k_-}{k_z} & 1-\frac{k_-}{k_z} \\ 1-\frac{k_+}{k_z} & 1+\frac{k_+}{k_z} & 1-\frac{k_-}{k_z} & 1+\frac{k_-}{k_z} \\ \Lambda_+ + \lambda_+ & \Lambda_+ - \lambda_+ & \Lambda_- + \lambda_- & \Lambda_- - \lambda_- \\ \Lambda_+ - \lambda_+ & \Lambda_+ + \lambda_+ & \Lambda_- - \lambda_- & \Lambda_- + \lambda_- \end{pmatrix}\begin{pmatrix} A \\ B \\ C \\ D \end{pmatrix} = T_1\begin{pmatrix} A \\ B \\ C \\ D \end{pmatrix}, \tag{A8}$$

$$\begin{pmatrix} F_y \\ 0 \\ F_x \\ 0 \end{pmatrix} = \frac{1}{2}\begin{pmatrix} 1+\frac{k_+}{k_o} & 1-\frac{k_+}{k_o} & 1+\frac{k_-}{k_o} & 1-\frac{k_-}{k_o} \\ 1-\frac{k_+}{k_o} & 1+\frac{k_+}{k_o} & 1-\frac{k_-}{k_o} & 1+\frac{k_-}{k_o} \\ \Lambda_+ + \gamma_+ & \Lambda_+ - \gamma_+ & \Lambda_- - \gamma_- & \Lambda_- - \gamma_- \\ \Lambda_+ - \gamma_+ & \Lambda_+ + \gamma_+ & \Lambda_- - \gamma_- & \Lambda_- - \gamma_- \end{pmatrix}\begin{pmatrix} e^{ik_+d} \\ e^{-ik_+d} \\ e^{ik_-d} \\ e^{-ik_-d} \end{pmatrix}\begin{pmatrix} A \\ B \\ C \\ D \end{pmatrix} = T_2\begin{pmatrix} A \\ B \\ C \\ D \end{pmatrix}, \tag{A9}$$

with $\lambda_\pm = k_\pm k_z \varepsilon_t \Lambda_\pm /(\varepsilon_t f^2 - k_x^2)$ and $\gamma_\pm = k_\pm \varepsilon_t \Lambda_\pm (\varepsilon_z f^2 - k_x^2)/k_e \varepsilon_z (\varepsilon_z f^2 - k_x^2)$. Eliminating the ABCD yields, we obtain $E_{x,y}^{I,R} = TF_{x,y} = T_1 T_2^{-1} F_{x,y}$.

References

1. Korzeb, K.; Gajc, M.; Pawlak, D.A. Compendium of Natural Hyperbolic Materials. *Opt. Express* **2015**, *23*, 25406. [CrossRef] [PubMed]
2. Palermo, G.; Sreekanth, K.V.; Strangi, G. Hyperbolic Dispersion Metamaterials and Metasurfaces. *EPJ Appl. Metamater.* **2020**, *7*, 11. [CrossRef]
3. Pustovit, V.N.; Zelmon, D.E.; Eyink, K.; Urbas, A.M. Optical Phase Transitions in Bilayer Semiconductor Hyperbolic Metamaterials. *Photonics Nanostructures Fundam. Appl.* **2022**, *51*, 101049. [CrossRef]
4. Díaz-Valencia, B.F.; Moncada-Villa, E.; Gómez, F.R.; Porras-Montenegro, N.; Mejía-Salazar, J.R. Bulk Plasmon Polariton Modes in Hyperbolic Metamaterials for Giant Enhancement of the Transverse Magneto-Optical Kerr Effect. *Molecules* **2022**, *27*, 5312. [CrossRef] [PubMed]
5. Drachev, V.P.; Podolskiy, V.A.; Kildishev, A.V. Hyperbolic Metamaterials: New Physics behind a Classical Problem. *Opt. Express* **2013**, *21*, 15048. [CrossRef] [PubMed]
6. Caldwell, J.D.; Kretinin, A.V.; Chen, Y.; Giannini, V.; Fogler, M.M.; Francescato, Y.; Ellis, C.T.; Tischler, J.G.; Woods, C.R.; Giles, A.J.; et al. Sub-Diffractional Volume-Confined Polaritons in the Natural Hyperbolic Material Hexagonal Boron Nitride. *Nat. Commun.* **2014**, *5*, 5221. [CrossRef] [PubMed]
7. Maestre, C.; Li, Y.; Garnier, V.; Steyer, P.; Roux, S.; Plaud, A.; Loiseau, A.; Barjon, J.; Ren, L.; Robert, C.; et al. From the Synthesis of hBN Crystals to Their Use as Nanosheets in van Der Waals Heterostructures. *2D Mater.* **2022**, *9*, 035008. [CrossRef]
8. Giles, A.J.; Dai, S.; Glembocki, O.J.; Kretinin, A.V.; Sun, Z.; Ellis, C.T.; Tischler, J.G.; Taniguchi, T.; Watanabe, K.; Fogler, M.M.; et al. Imaging of Anomalous Internal Reflections of Hyperbolic Phonon-Polaritons in Hexagonal Boron Nitride. *Nano Lett.* **2016**, *16*, 3858–3865. [CrossRef]
9. Wei, C.; Abedini Dereshgi, S.; Song, X.; Murthy, A.; Dravid, V.P.; Cao, T.; Aydin, K. Polarization Reflector/Color Filter at Visible Frequencies via Anisotropic α-MoO₃. *Adv. Opt. Mater.* **2020**, *8*, 2000088. [CrossRef]
10. Abedini Dereshgi, S.; Folland, T.G.; Murthy, A.A.; Song, X.; Tanriover, I.; Dravid, V.P.; Caldwell, J.D.; Aydin, K. Lithography-Free IR Polarization Converters via Orthogonal in-Plane Phonons in α-MoO₃ Flakes. *Nat. Commun.* **2020**, *11*, 5771. [CrossRef]
11. Bergeron, A.; Gradziel, C.; Leonelli, R.; Francoeur, S. Probing Hyperbolic and Surface Phonon-Polaritons in 2D Materials Using Raman Spectroscopy. *Nat. Commun.* **2023**, *14*, 4098. [CrossRef]
12. Konečná, A.; Li, J.; Edgar, J.H.; García De Abajo, F.J.; Hachtel, J.A. Revealing Nanoscale Confinement Effects on Hyperbolic Phonon Polaritons with an Electron Beam. *Small* **2021**, *17*, 2103404. [CrossRef]
13. Feres, F.H.; Mayer, R.A.; Wehmeier, L.; Maia, F.C.B.; Viana, E.R.; Malachias, A.; Bechtel, H.A.; Klopf, J.M.; Eng, L.M.; Kehr, S.C.; et al. Sub-Diffractional Cavity Modes of Terahertz Hyperbolic Phonon Polaritons in Tin Oxide. *Nat. Commun.* **2021**, *12*, 1995. [CrossRef]
14. Maurya, K.C.; Caligiuri, V.; Pillai, A.I.K.; Garbrecht, M.; Krahne, R.; Saha, B. Radiative Volume Plasmon and Phonon-Polariton Resonances in TiN-Based Plasmonic/Polar-Dielectric Hyperbolic Optical Metamaterials. *Appl. Phys. Lett.* **2023**, *122*, 221704. [CrossRef]
15. Wang, X.-G.; Hao, S.-P.; Fu, S.-F.; Zhang, Q.; Wang, X.-Z. Ghost Surface Polaritons in Naturally Uniaxial Hyperbolic Materials. *J. Opt. Soc. Am. B* **2023**, *40*, 1667. [CrossRef]
16. Zhang, Y.; Wang, X.; Zhang, D.; Fu, S.; Zhou, S.; Wang, X.-Z. Unusual Spin and Angular Momentum of Dyakonov Waves at the Hyperbolic-Material Surface. *Opt. Express* **2020**, *28*, 19205. [CrossRef] [PubMed]

17. Pavlidis, G.; Schwartz, J.J.; Matson, J.; Folland, T.; Liu, S.; Edgar, J.H.; Caldwell, J.D.; Centrone, A. Experimental Confirmation of Long Hyperbolic Polariton Lifetimes in Monoisotopic (10B) Hexagonal Boron Nitride at Room Temperature. *APL Mater.* **2021**, *9*, 091109. [CrossRef] [PubMed]

18. Folland, T.G.; Fali, A.; White, S.T.; Matson, J.R.; Liu, S.; Aghamiri, N.A.; Edgar, J.H.; Haglund, R.F.; Abate, Y.; Caldwell, J.D. Reconfigurable Infrared Hyperbolic Metasurfaces Using Phase Change Materials. *Nat. Commun.* **2018**, *9*, 4371. [CrossRef]

19. Barcelos, I.D.; Canassa, T.A.; Mayer, R.A.; Feres, F.H.; De Oliveira, E.G.; Goncalves, A.-M.B.; Bechtel, H.A.; Freitas, R.O.; Maia, F.C.B.; Alves, D.C.B. Ultrabroadband Nanocavity of Hyperbolic Phonon–Polaritons in 1D-Like α-MoO$_3$. *ACS Photonics* **2021**, *8*, 3017–3026. [CrossRef]

20. Vasilantonakis, N.; Wurtz, G.A.; Podolskiy, V.A.; Zayats, A.V. Refractive Index Sensing with Hyperbolic Metamaterials: Strategies for Biosensing and Nonlinearity Enhancement. *Opt. Express* **2015**, *23*, 14329. [CrossRef] [PubMed]

21. Alsalhin, A.A.; Lail, B.A. A Systematic Investigation of Coupling Between Dark and Bright Polaritons for Infrared Sensing Design. *IEEE Sens. J.* **2021**, *21*, 14899–14905. [CrossRef]

22. Janaszek, B.; Tyszka-Zawadzka, A.; Szczepański, P. Influence of Spatial Dispersion on Propagation Properties of Waveguides Based on Hyperbolic Metamaterial. *Materials* **2021**, *14*, 6885. [CrossRef]

23. Zaitsev, A.D.; Demchenko, P.S.; Kablukova, N.S.; Vozianova, A.V.; Khodzitsky, M.K. Frequency-Selective Surface Based on Negative-Group-Delay Bismuth–Mica Medium. *Photonics* **2023**, *10*, 501. [CrossRef]

24. Bliokh, K.Y.; Aiello, A. Goos–Hänchen and Imbert–Fedorov Beam Shifts: An Overview. *J. Opt.* **2013**, *15*, 014001. [CrossRef]

25. Goos, F.; Hänchen, H. Ein Neuer Und Fundamentaler Versuch Zur Totalreflexion. *Ann. Der Phys.* **1947**, *436*, 333–346. [CrossRef]

26. Takayama, O.; Sukham, J.; Malureanu, R.; Lavrinenko, A.V.; Puentes, G. Photonic Spin Hall Effect in Hyperbolic Metamaterials at Visible Wavelengths. *Opt. Lett.* **2018**, *43*, 4602. [CrossRef] [PubMed]

27. Zakirov, N.; Zhu, S.; Bruyant, A.; Lérondel, G.; Bachelot, R.; Zeng, S. Sensitivity Enhancement of Hybrid Two-Dimensional Nanomaterials-Based Surface Plasmon Resonance Biosensor. *Biosensors* **2022**, *12*, 810. [CrossRef]

28. Wang, X.; Yin, C.; Sun, J.; Li, H.; Wang, Y.; Ran, M.; Cao, Z. High-Sensitivity Temperature Sensor Using the Ultrahigh Order Mode-Enhanced Goos-Hänchen Effect. *Opt. Express* **2013**, *21*, 13380. [CrossRef]

29. Cai, Y.-P.; Wan, R.-G. Bistable Reflection and Beam Shifts with Excitation of Surface Plasmons in a Saturable Absorbing Medium. *Opt. Express* **2022**, *30*, 20725. [CrossRef]

30. Ding, Y.; Deng, D.; Zhou, X.; Zhen, W.; Gao, M.; Zhang, Y. Barcode Encryption Based on Negative and Positive Goos-Hänchen Shifts in a Graphene-ITO/TiO2/ITO Sandwich Structure. *Opt. Express* **2021**, *29*, 41164. [CrossRef]

31. Sinha Biswas, S.; Remesh, G.; Achanta, V.G.; Banerjee, A.; Ghosh, N.; Dutta Gupta, S. Enhanced Beam Shifts Mediated by Bound States in Continuum. *J. Opt.* **2023**, *25*, 095401. [CrossRef]

32. Li, Y.; Song, H.; Zhang, Y.; Zhou, S.; Fu, S.; Zhang, Q.; Wang, X.-Z. Spatial Shifts of Reflected Light Beam on Black Phosphorus/Hexagonal Boron Nitride Structure. *Opt. Laser Technol.* **2023**, *159*, 108968. [CrossRef]

33. Novoselov, K.S.; Mishchenko, A.; Carvalho, A.; Castro Neto, A.H. 2D Materials and van Der Waals Heterostructures. *Science* **2016**, *353*, aac9439. [CrossRef]

34. Ribeiro-Palau, R.; Zhang, C.; Watanabe, K.; Taniguchi, T.; Hone, J.; Dean, C.R. Twistable Electronics with Dynamically Rotatable Heterostructures. *Science* **2018**, *361*, 690–693. [CrossRef]

35. Hu, G.; Ou, Q.; Si, G.; Wu, Y.; Wu, J.; Dai, Z.; Krasnok, A.; Mazor, Y.; Zhang, Q.; Bao, Q.; et al. Topological Polaritons and Photonic Magic Angles in Twisted α-MoO3 Bilayers. *Nature* **2020**, *582*, 209–213. [CrossRef] [PubMed]

36. Chen, M.; Lin, X.; Dinh, T.H.; Zheng, Z.; Shen, J.; Ma, Q.; Chen, H.; Jarillo-Herrero, P.; Dai, S. Configurable Phonon Polaritons in Twisted α-MoO$_3$. *Nat. Mater.* **2020**, *19*, 1307–1311. [CrossRef] [PubMed]

37. Wu, S.; Song, H.; Li, Y.; Fu, S.; Wang, X. Tunable Spin Hall Effect via Hybrid Polaritons around Epsilon-near-Zero on Graphene-hBN Heterostructures. *Results Phys.* **2022**, *35*, 105383. [CrossRef]

38. Dai, S.; Fei, Z.; Ma, Q.; Rodin, A.S.; Wagner, M.; McLeod, A.S.; Liu, M.K.; Gannett, W.; Regan, W.; Watanabe, K.; et al. Tunable Phonon Polaritons in Atomically Thin van Der Waals Crystals of Boron Nitride. *Science* **2014**, *343*, 1125–1129. [CrossRef] [PubMed]

39. Eda, K. Longitudinal-Transverse Splitting Effects in IR Absorption Spectra of MoO$_3$. *J. Solid. State Chem.* **1991**, *95*, 64–73. [CrossRef]

40. Py, M.A.; Schmid, P.E.; Vallin, J.T. Raman Scattering and Structural Properties of MoO$_3$. *Nuov. Cim. B* **1977**, *38*, 271–279. [CrossRef]

Disclaimer/Publisher's Note: The statements, opinions and data contained in all publications are solely those of the individual author(s) and contributor(s) and not of MDPI and/or the editor(s). MDPI and/or the editor(s) disclaim responsibility for any injury to people or property resulting from any ideas, methods, instructions or products referred to in the content.

Article

The Effect of the Solution Flow and Electrical Field on the Homogeneity of Large-Scale Electrodeposited ZnO Nanorods

Yanmin Zhao [1,2,*], Kexue Li [1,*], Ying Hu [1,2], Xiaobing Hou [1], Fengyuan Lin [1], Jilong Tang [1], Xin Tang [3], Xida Xing [4], Xiao Zhao [5], Haibin Zhu [5], Xiaohua Wang [1,2] and Zhipeng Wei [1]

[1] State Key Laboratory of High Power Semiconductor Lasers, Changchun University of Science and Technology, 7089 Wei-Xing Road, Changchun 130022, China; houxiaobingdavid@163.com (X.H.); linfengyuan_0116@163.com (F.L.)
[2] Zhongshan Institute of Changchun University of Science and Technology, Zhongshan 528437, China
[3] School of Optoelectronics, Beijing Institute of Technology, Beijing 100081, China
[4] Shandong North Optical Electronic Co., Ltd., Taian 271000, China
[5] Northeast Industry Group Co., Ltd., Changchun 130212, China
* Correspondence: qzymq@126.com (Y.Z.); likx@cust.edu.cn (K.L.)

Abstract: In this paper, we demonstrate the significant impact of the solution flow and electrical field on the homogeneity of large-scale ZnO nanorod electrodeposition from an aqueous solution containing zinc nitrate and ammonium nitrate, primarily based on the X-ray fluorescence results. The homogeneity can be enhanced by adjusting the counter electrode size and solution flow rate. We have successfully produced relatively uniform nanorod arrays on an $8 \times 10\ cm^2$ i-ZnO-coated fluorine-doped tin oxide (FTO) substrate using a compact counter electrode and a vertical stirring setup. The as-grown nanorods exhibit similar surface morphologies and dominant, intense, almost uniform near-band-edge emissions in different regions of the sample. Additionally, the surface reflectance is significantly reduced after depositing the ZnO nanorods, achieving a moth-eye effect through subwavelength structuring. This effect of the nanorod array structure indicates that it can improve the utilization efficiency of light reception or emission in various optoelectronic devices and products. The large-scale preparation of ZnO nanorods is more practical to apply and has an extremely broad application value. Based on the research results, it is feasible to prepare large-scale ZnO nanorods suitable for antireflective coatings and commercial applications by optimizing the electrodeposition conditions.

Keywords: zinc oxide nanorod arrays; large-scale; electrodeposition; homogeneous nanorod arrays

Citation: Zhao, Y.; Li, K.; Hu, Y.; Hou, X.; Lin, F.; Tang, J.; Tang, X.; Xing, X.; Zhao, X.; Zhu, H.; et al. The Effect of the Solution Flow and Electrical Field on the Homogeneity of Large-Scale Electrodeposited ZnO Nanorods. *Materials* **2024**, *17*, 1241. https://doi.org/10.3390/ma17061241

Academic Editor: Masato Sone

Received: 11 December 2023
Revised: 18 February 2024
Accepted: 24 February 2024
Published: 8 March 2024

Copyright: © 2024 by the authors. Licensee MDPI, Basel, Switzerland. This article is an open access article distributed under the terms and conditions of the Creative Commons Attribution (CC BY) license (https://creativecommons.org/licenses/by/4.0/).

1. Introduction

Antireflective coatings (ARCs) play a crucial role in enhancing the optical absorption or light extraction efficiency of optoelectronic devices, such as solar cells, solar energy heat collectors, LDs, LEDs, PDs, and others. Numerous efforts have been made to explore high energy conversion efficiency, as a significant portion of incident or emitted light is lost due to surface reflection. In crystalline silicon solar cells, for instance, approximately 36% of the incident or emitted light is directly reflected, leading to substantial energy utilization loss [1,2]. To address this issue, it is important to enhance the light trapping or output by minimizing undesirable surface reflection losses through the use of ARCs. Achieving an effective antireflective effect involves satisfying two conditions: firstly, antireflective materials should be transparent, avoiding additional optical absorption, and secondly, the refractive index (n) of the materials should be between that of air and the device materials. Conventional ARCs reduce reflectivity by utilizing destructive interference with the waves reflected from the interface of the ARC with the top of devices and their surroundings. Typically, these ARCs are thin films with a quarter-wavelength thickness and a certain refractive index (RI) to minimize the reflectivity. But for ARC thin films, the antireflection effect works only at a single wavelength and at the normal incidence of light. In contrast to bulk materials, nano-scale structures (e.g., NWs, NRs, nanotips, nanocolumns, nanopores) can increase the incident light path and enhance the light absorption. These structures can be equivalent to a gradient change in the refractive index in the direction of incident light according to the formula of effective medium approximation, and a good light trapping effect can be obtained by adjusting the morphology of the nanostructure [3]. This property attracts the exploration of nanostructures as effective ARCs. In this work, we pay attention to the large-scale ZnO NRs fabricated.

Common materials for ARCs include MgF_2, CaF_2, SiN, ZnS, and ZnO. Among these materials, zinc oxide is a promising candidate for antireflective coatings due to its abundance in the Earth's crust and suitability for various manufacturing processes. ZnO is cost-effective, environmentally friendly, and non-toxic. Moreover, it possesses good transparency, a wide band gap (Eg~3.37 eV at 300 K), an appropriate refractive index (n~1.9), and the ability to form textured coatings via anisotropic growth. Furthermore, ZnO offers several advantages. It is a semiconductor with a large exciton binding energy (60 meV), and high-quality films and nanostructures can be prepared at relatively low temperatures (less than 700 °C). It is more resistant to radiation damage and has a high mechanical strength, with a Young's modulus of 25–150 GPa and a high fracture strain property. Additionally, it exhibits a thermal expansion coefficient from 5×10^{-6} to $8 \times 10^{-6}/°C$ within the temperature range of 25–400 °C. ZnO is a favorable material for nanostructure functionality [4–6]. Among all kinds of nanostructures, quasi-one-dimensional ZnO nanostructures, such as nanorods (NRs) and nanowires (NWs), have garnered significant attention for their potential applications in high-technology fields, including electronics and optoelectronics. These structures find applications in light-emitting diodes [7,8], photodetectors [9,10], photodiodes [11], gas sensors [12], and solar cells [13]. For ARC applications, ZnO nanorod arrays, featuring a high length-to-diameter ratio, a high surface area-to-volume ratio, and a unique morphology, as well as perfect crystalline structures, are expected to significantly improve light absorption and are considered a new generation of ARCs. ZnO nanorod (NR) arrays offer several advantageous properties, including good transparency for the solar spectrum and a subwavelength structure size, as well as an appropriate refractive index of ~2, leading to continuously varying refractive index profiles in the arrays [14,15]. Thus, they can effectively suppress surface reflection and enhance light coupling through phenomena such as zero-order grating characteristics or the moth-eye effect [16,17]. Nanostructures of ZnO with an RI profile between air and bulk ZnO are one of the most promising candidates [18–20] for ARCs, which are physically and technically compatible with the window layer. Consequently, these characteristics suggest that ZnO NRs are emerging as promising materials for solar cell antireflective coatings and solar thermal selective surfaces [21–26], demonstrating remarkable prospects.

In the solar cell region, their effectiveness as antireflective coatings (ARCs) is supported by the related research. Aé, L. demonstrated that the short-circuit current of a solar cell can realize a relative increase of nearly 6% before and after the deposition of ZnO NRs [18]. Furthermore, Bai, A. and Tang, Y. fabricated a ZnO NR ARC on $Cu(In,Ga)Se_2$ thin-film solar cells, successfully reducing the weighted reflectance from 8.6% to 3.5% [27]. In order to realize practical application, deposition at a large scale is necessary and valuable, paving the way for the industrialized production of ZnO NR ARCs. For production, various methods have been employed to prepare ZnO NRs, for example, chemical vapor deposition [28,29], the vapor transport method [30], thermal evaporation [31], and the vapor–liquid–solid technique [32]. Compared to these methods, electrodeposition is simpler and more compatible with low-temperature substrates due to its moderate operating temperature [33–38]. It is also an energy-efficient and cost-effective method for applications in large-scale fabrication in industry thanks to the use of aqueous solutions in an open atmosphere. But for large-scale electrodeposition, achieving homogeneity in the ZnO NRs is crucial yet challenging. Therefore, it is important to analyze and determine the factors affecting the homogeneity of deposition, facilitating the production of a large-area uniform

ZnO NR ARC and aiding future industrialization efforts. Among the influencing factors for homogenous large-scale deposition, the solution flow and the electric field should play significant roles in zinc electrodeposition. The morphology appears to be susceptible to the flow field and also dependent on the flow pattern [39,40]. R.D. Naybour's results have clearly shown that the hydrodynamic condition of the electrolyte is an important factor in controlling the morphology of the electrodeposits [41]. Naybour pointed out that changes in the surface morphology of electrodeposited zinc occur because there is a large amount of microturbulence in the electrolyte during deposition [42]. Milan M. considered that toroidal vortices and local fluctuations existed in the electrolyte, bringing a higher bulk concentration of the reacting species into regions of the electrode surface [42], thereby providing for faster local and spatially periodic growth of the deposit. These local concentration fluctuations could contribute to the local current density change [43], resulting in local growth rate differences.

In our experiments, in order to confirm the influence of the solution flow and electric field, ZnO NRs were deposited onto a cathode electrode from aqueous electrolyte solution; the ZnO deposition should also be affected by the solution flow. Besides random thermal motion, concentration gradient diffusion, and solution flow transportation, ions also are driven by the electric field between the counter and working electrodes in electrodeposition, which also will affect the deposition outcomes. Altering the shape of the counter electrode and the distance between the working and counter electrodes can adjust the electrical field distribution, thereby impacting the transportation behavior of the reacting species. Regarding the electrical field factor, we established the significant influence of the electrical field on the homogeneity of the ZnO NRs by merely changing the counter electrode size at a fixed distance. In situations with a small counter electrode, the electrical field balanced the solution flow factor in our electrochemical setup, resulting in relatively homogenous ZnO NR deposition when combined with stirring. Based on our research and tests, we successfully fabricated uniform and large-scale ZnO NRs (8×10 cm^2) by optimizing the solution flow and electric field distribution.

2. Materials and Methods

In the deposition process, we immersed all samples in an aqueous solution containing 7 mM zinc nitrate (Zn(NO$_3$)$_2$.6H$_2$O) and 5 mM ammonium nitrate (NH$_4$NO$_3$) (purity > 99.99%, purchased from Alfa Aesar, Haverhill, MA, USA), subjecting them to a potential of -1.4 V in potentiostatic mode. This electrochemical procedure employed a three-electrode configuration, incorporating a Pt counter electrode and a pseudo-reference electrode (IviumStat. H, Ivium, Eindhove, The Netherlands), both immersed in a large water bath to ensure optimal temperature control. The deposition temperature was meticulously maintained at 75 °C. To systematically explore the impact of the solution flow on the homogeneity of the ZnO nanorods (NRs), we deposited samples 1–3 onto fluorine-doped SnO$_2$ (FTO) substrates. These substrates were strategically paired with a spacious counter electrode (12×12 cm^2 Pt foil) under various conditions, including stirring at the center, no stirring, and stirring at the bottom, strategically positioned between the counter and working electrodes within the electrochemical cell. Additionally, we prepared the ZnO NRs in sample 4 utilizing a smaller counter electrode (5×6 cm^2 Pt foil, centrally clipped onto 10×10 cm^2 polycarbonate plates) without employing stirring. This approach was designed to isolate the influence of the electrical field, allowing for a comparative analysis against the results obtained using larger counter electrodes. It is noteworthy that these samples were deposited using a single-contact working electrode (top side). Subsequently, we cultivated relatively homogeneous ZnO NRs in sample 5 on FTO substrates that were pre-coated with a ~30 nm undoped ZnO film (i-ZnO/FTO). This innovative process involved utilizing both sides (left and right) of the working electrode and the smaller counter electrode, with consistent stirring. It is essential to highlight that all the substrates used in this study were 8×10 cm^2 in size. Prior to the deposition process, each substrate underwent ultrasonic cleaning in acetone and ethanol, followed a thorough rinsing in Milli-Q water for 5 minutes at each stage. Subsequent to this meticulous cleaning procedure, the substrates were dried using nitrogen gas to ensure a pristine surface for the subsequent deposition steps.

Assessment of the homogeneity across all the samples was meticulously carried out employing a state-of-the-art Philips MagiX Pro wavelength-dispersive X-ray fluorescence (XRF) spectrometer (PANalytical B.V., Alemlo, The Netherlands). This cutting-edge instrument facilitated a detailed analysis by converting the detected Zn content in the ZnO nanorod (NR) region into the film thickness. This innovative approach enabled us to derive the homogeneity data by meticulously comparing the thickness variations across different regions within each sample. To delve even further into the structural characteristics, the morphology of the relatively homogeneous sample was subjected to rigorous scrutiny utilizing a Zeiss LEO 1530 scanning electron microscope (SEM, Zeiss, Oberkochen, Germany) equipped with a Gemini 4000 column. This advanced microscopic examination not only provided valuable insights into the structural uniformity but also offered a closer look at the surface characteristics of the ZnO nanorods. Such detailed observations are instrumental in comprehending the intricacies of the nanostructures and optimizing their performance in optoelectronic applications. Moving beyond structural analysis, we conducted room-temperature photoluminescence (PL) measurements for the aforementioned sample. These were meticulously recorded utilizing a He-Cd laser, employing an excitation wavelength of 325 nm. The resulting PL spectra serve as a crucial source of information concerning the optical properties and emission characteristics of the ZnO nanorods under ambient conditions. This spectral analysis is indispensable for understanding the electronic transitions within the material, shedding light on its photonic behavior. In addition to the PL studies, we ventured into an assessment of the reflectance spectra within the 400–800 nm wavelength range. This was achieved using a UV–vis–NIR spectrophotometer (UV 3600 Plus, Japan Shimadzu, Kyoto, Japan), coupled with an integrating sphere. The acquired reflectance spectra furnish valuable insights into the interaction of light with the ZnO-nanorod-coated substrates. Analyzing the reflectance characteristics is pivotal for gaining a comprehensive understanding of the optical performance and antireflective properties exhibited by the deposited ZnO nanorods in this specific spectral range. This multifaceted approach contributes to a holistic exploration of the material's optical behavior, enriching the scientific discourse and paving the way for advancements in semiconductor optoelectronics.

3. Results and Discussion

Table 1 provides a comprehensive overview of the X-ray fluorescence (XRF) results for samples 1–3. The segmentation into Top (T), Middle (M), and Bottom (B) regions, along with the Left (L), Middle (M), and Right (R) regions within each sample, facilitates a detailed analysis of the spatial variations in ZnO content. An intriguing observation emerges from samples 1 and 2, where a discernible reduction in ZnO presence is noted in the central regions. This intriguing phenomenon suggests a potential disparity in the density or dimensions of the ZnO nanorods (NRs) within these areas, particularly when ZnO NRs are deposited under conditions involving central stirring or the absence of stirring. Our hypothesis is rooted in the notion that the solution flow dynamics contribute to the observed variations. Specifically, we postulate that the influx of reaction ions onto the substrate is less pronounced in the central region compared to the surrounding areas. This discrepancy in the ion concentration is likely a consequence of the solution's intricate flow patterns during the deposition process. According to the following equations that were suggested by Izaki et al. [44,45],

$$Zn(NO_3)_2 <=> Zn^{2+} + 2NO_3^{-} \tag{1}$$

$$NO_3^{-} + 2e + H_2O \rightarrow 2OH^{-} + NO_2^{-} \tag{2}$$

$$Zn^{2+} + 2OH^- <=> Zn(OH)_2 \qquad (3)$$

$$Zn(OH)_2 \rightarrow ZnO + H_2O \qquad (4)$$

Table 1. Normalized XRF results of samples 1–3. Converting Zn content into ZnO thickness in testing NR region.

Sample	TL	TM	TR	ML	MM	MR	BL	BM	BR
1	1.46	1.1	1.54	1.21	1	1.30	1.47	1.06	1.43
2	1.41	1.15	1.42	1.30	1	1.36	1.30	1.07	1.34
3	1.29	1.11	1.30	1.18	1	1.13	1.16	0.95	1.16

These reactions provide valuable insights into the complex processes occurring during ZnO NR deposition. Notably, we draw attention to the fact that the growth rate of NRs in the central region is anticipated to be slower. This deceleration can be attributed to the lower availability of reacting species reaching the substrate simultaneously. Since these reactions primarily unfold on the substrate's surface during deposition, it is plausible that the solution in the central region readily exchanges with a more reacted solution than in the surrounding areas. To illustrate this concept, we present depositing schematic diagrams (Figure 1a,b), visually demonstrating the slower NR deposition rate in the central region under the conditions of central stirring or the absence of stirring. These schematics serve as a valuable tool for comprehending the dynamic interactions influencing ZnO NR growth across different regions within the samples. The spatial variations in the ZnO content observed in samples 1 and 2 align with our hypothesis of differential reaction ion availability being influenced by the solution flow dynamics. This nuanced understanding contributes to the optimization of ZnO NR deposition processes, offering insights that can enhance the uniformity and quality of nanorod structures.

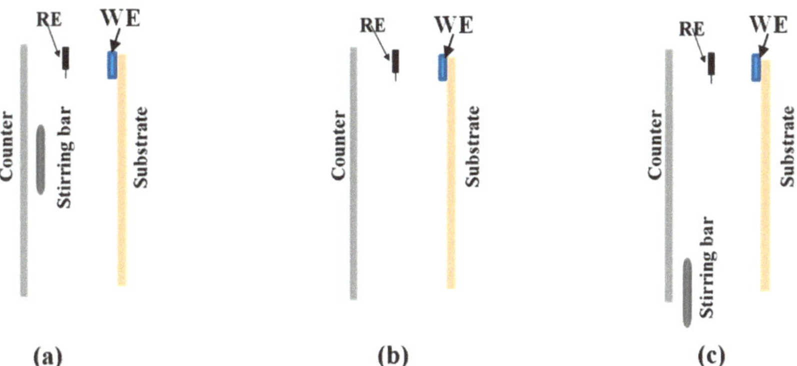

Figure 1. Schematic diagram of samples 1–3 deposited under stirring in the middle, without stirring, and stirring away from the center, which correspond to pictures (**a–c**), respectively. (RE: pseudo-reference electrode; WE: working electrode).

In the deposition process of sample 3, a unique approach was implemented, with stirring taking place away from the center and the stirring bar strategically positioned at the bottom of the counter electrode, as illustrated in Figure 1c. Subsequent to the deposition, an in-depth X-ray fluorescence (XRF) analysis of sample 3 uncovered a compelling observation—a discernibly thinner ZnO layer in the Bottom (BM) region, as detailed in Table 1. This finding serves as a critical piece of evidence, further substantiating the hypothesis that the dynamic nature of the solution flow exerts a substantial influence on the homogeneity of the ZnO nanorod (NR) deposition. Upon a meticulous comparison of the XRF results between sample 1 and sample 3, a nuanced inference comes to light. It becomes evident that the solution in close proximity to the substrate, particularly facing the stirring center, exhibits a heightened susceptibility to exchange with the surrounding reacted solution, facilitated by the stirring action. This intricate interplay results in a fascinating helical structure-like contraction of the solution flow toward the stirring bar, a phenomenon visualized in Figure 2—a schematic diagram illustrating the dynamic solution flow under stirring conditions. The consequence of this dynamic behavior extends to local growth rate differences, particularly near the surface of the large-scale substrate. The fluctuation in the concentration of the reacting species within both the central and surrounding regions amplifies the heterogeneity in the growth rates of the ZnO nanorods. This localized growth rate difference is particularly pronounced in the vicinity of the stirring bar, contributing to the observed variations in the ZnO layer's thickness in the BM region of sample 3. In essence, the stirring-induced solution flow dynamics play a pivotal role in dictating the homogeneity of the ZnO NR deposition, leading to nuanced localized growth rate differences. This newfound understanding holds significant implications for optimizing large-scale ZnO NR fabrication processes, enhancing the precision and uniformity of nanorod structures.

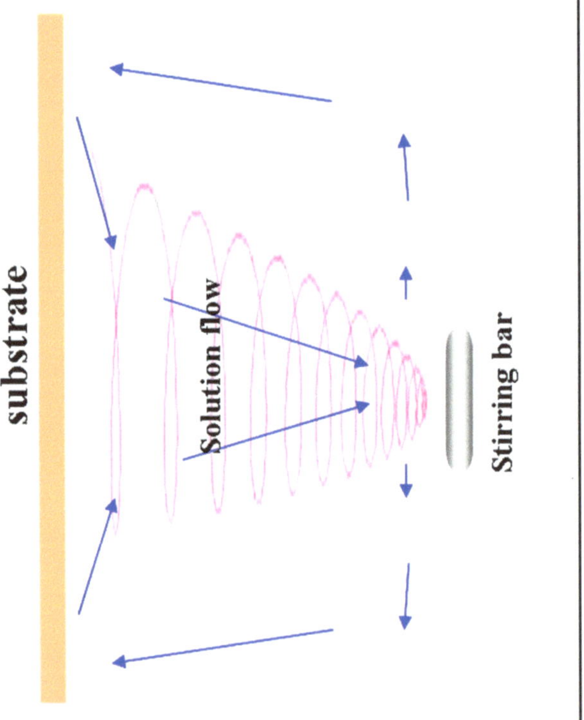

Figure 2. The schematic diagram of solution flow that was stirred with stirring bar located at center. (The direction of the arrow indicates the direction of liquid flow and exchange.)

Table 2 provides a detailed comparison of the X-ray fluorescence (XRF) results for samples 2 and 4, both deposited without stirring, utilizing large and small counters, respectively. Unlike sample 2, where a higher ZnO content was observed in the surrounding regions, sample 4 exhibited a more uniform ZnO distribution, showcasing a closer match between the central and surrounding areas. This notable uniformity is attributed to the variations in the electrical field distribution, influenced by the counter size. In previous studies, researchers like M.-M. Zhang and R.G. Reddy have simulated the electrical field distribution during electrodeposition in ionic liquids, demonstrating that the main electrical gradient exists around the edges of electrodes. Their modeling results indicate that the current density near the edges of the electrode is generally more intensive than elsewhere within the cross section [46]. Applying this concept to our experiments, the larger counter likely created a stronger electrical field around the edge of the working electrode, promoting enhanced ZnO deposition in the surrounding regions. Conversely, in the case of the smaller counter, the electrical field may have been weakened in the surrounding region near the substrate, effectively balancing the transportation of the reacting ions toward the substrate. Consequently, for sample 4, the growth rate of the nanorods (NRs) in the center approximates that of the surrounding regions. This intriguing observation underscores the importance of the counter size in controlling the electrical field and, subsequently, the homogeneity of ZnO NR deposition.

Table 2. Normalized XRF results of samples 2 and 4. Converting Zn content into ZnO thickness in testing NR region.

Sample	TL	TM	TR	ML	MM	MR	BL	BM	BR
2	1.41	1.15	1.42	1.30	1	1.36	1.30	1.07	1.34
4	1.02	0.93	1.22	0.99	1	1.12	0.87	0.96	0.96

Based on the aforementioned details, sample 5 was meticulously prepared on i-ZnO/FTO with stirring conditions, utilizing Cu foils as a dual-sided working electrode to mitigate the observed ZnO gradient tendency in sample 4. To address the potential reduction from top to bottom, both sides of the Cu foils were employed during the nanorod (NR) electrodeposition. The normalized X-ray fluorescence (XRF) results in Table 3 reveal a less pronounced gradient from top to bottom in sample 5. This suggests that the observed gradient in sample 4, attributed to potential reduction, is mitigated or at least not significantly pronounced. The refined experimental setup emphasizes the role of the solution flow in influencing the homogeneity of ZnO NR deposition, offering valuable insights for optimizing large-scale fabrication processes.

Table 3. Normalized XRF results of samples 5. Converting Zn content into ZnO thickness in testing NR region.

Sample	TL	TM	TR	ML	MM	MR	BL	BM	BR
5	1.16	1.11	1.12	1.06	1	1.11	0.99	0.95	1.02

By analyzing the XRF data, the average thickness of the ZnO, measured in atom units (a.u.) after the normalized XRF results, is determined to be 1.06, with a standard deviation (σ) of 0.07 a.u., as illustrated in Figure 3. The SEM results for the TR, MM, and BL regions are displayed in Figure 4. Top-view images indicate comparable homogeneity across different regions, and the enlarged insets reveal the hexagonal structure of the ZnO NRs. Cross-sectional micrographs are presented at the bottom of Figure 4, showing that the diameter of the NRs is 82 ± 5 nm. The density of these NRs is calculated to be $(5.9 \pm 0.2) \times 10^9$ cm^{-2}. The length of the NRs in the MM and BL regions is 480 ± 20 nm, while it is slightly longer in the TR region, measuring 510 ± 20 nm, aligning with the differences observed in the XRF results. Furthermore, the room-temperature photoluminescence (PL) spectra reveal a dominant and intense near-band-edge (NBE) emission at approximately 377 nm, consistent across different ZnO NR regions, as shown in Figure 5. Additionally, a weak broad defect emission in the visible spectrum and a near-infrared peak at approximately 750 nm are observed, as referenced in [47,48]. The optical quality of these large-scale as-grown ZnO NRs is noteworthy, and the PL results also indicate the homogeneity of the sample. These findings from sample 5 suggest that we can achieve relatively homogeneous NRs using the electrodeposition process.

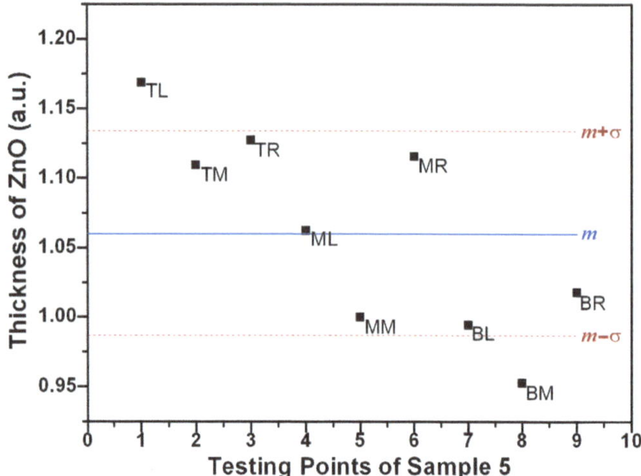

Figure 3. Normalized XRF results of sample 5 with an average thickness of ZnO m = 1.06 (a.u.) and a standard deviation σ = 0.07 (a.u.).

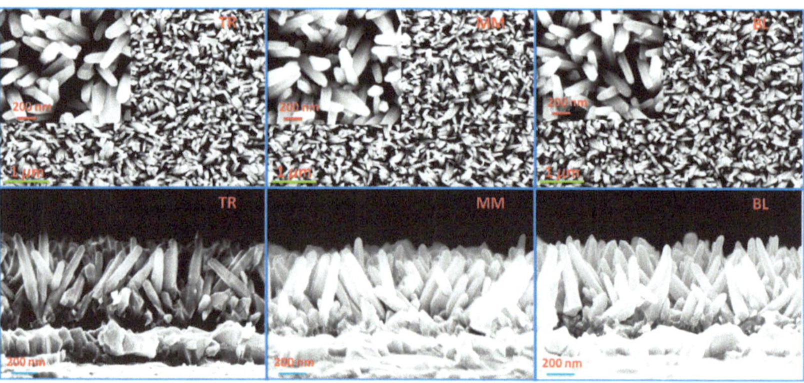

Figure 4. Top-view and cross-section SEM micrographs of ZnO nanorods in the TR, MM, and BL sections of sample 5. Enlarged images are shown in the insets.

To enhance our understanding of the impact of ZnO NRs as an antireflective coating (ARC) and to augment its light-trapping capabilities, we conducted an analysis of the wavelength-dependent reflectance of the sample (referred to as sample 5) both before and after the application of the ZnO NRs. This investigation spanned three distinct areas, as depicted in Figure 6. After the deposition of the ZnO NRs, a notable change was observed when compared with the FTO substrate. Specifically, the interference fringes, initially present, vanished post the growth of the ZnO NRs. Additionally, there was a significant reduction in the reflectance across a broad wavelength spectrum ranging from 400 to 800 nm. For comparison, we calculated the parameter of the averaged reflectance (Rave) without and with the ZnO NR ARCs, including the FTO sample and the TR, MM, and BL regions of sample 5, respectively, with the Air Mass 1.5 terrestrial global spectrum in the wavelength range of 400–800 nm, as given by [25]:

$$R_{ave} = \frac{\int F(\lambda)R(\lambda)d\lambda}{\int F(\lambda)d\lambda}$$

where $R(\lambda)$ is the surface reflectance in the wavelength range of 400–800 nm, and $F(\lambda)$ is the photon flux at Air Mass 1.5. The calculated solar average reflectance for the FTO and the three fabricated NR sections are 6.887%, 2.434%, 2.598%, and 2.267%, respectively. The average reflectance with the AM1.5 solar spectrum decreases significantly from 6.887% to 2.267–2.434%, so we can notice that the use of ZnO NRs as an antireflective layer on an FTO surface can significantly increase its light absorption.

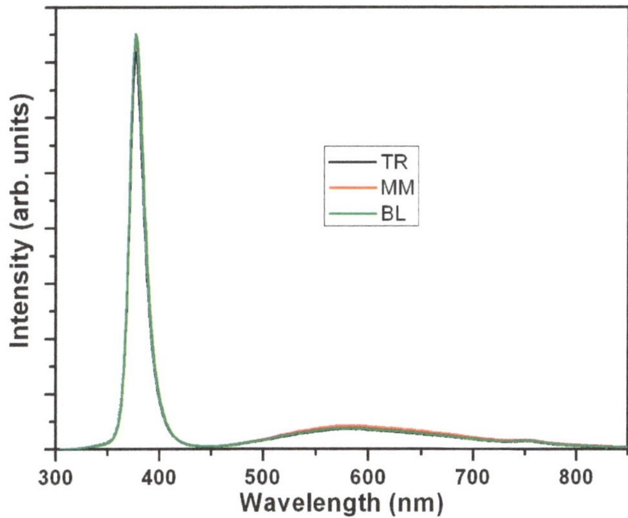

Figure 5. Room-temperature PL spectra of ZnO NRs in the TR, MM, and BL regions of sample 5.

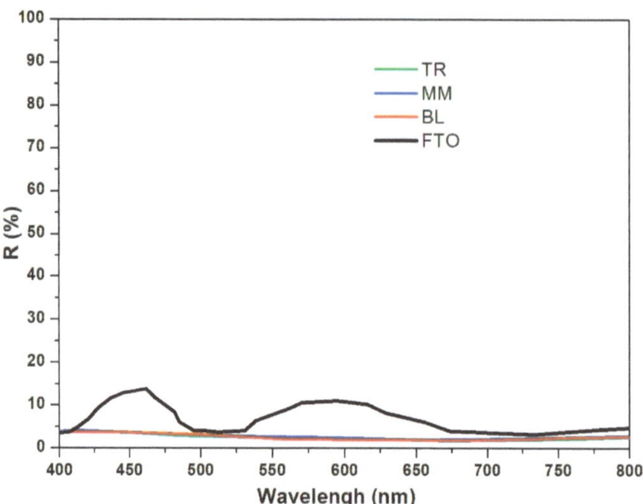

Figure 6. Reflectance spectra of FTO substrate without ZnO nanorods, and with ZnO nanorods in the TR, MM, and BL sections of sample 5.

This effect remained consistent across various sections of sample 5, namely the TR, MM, and BL sections, maintaining a similar spectral shape in these regions. Compared to FTO, the pronounced decrease in reflectance across this broad range can be attributed to the so-called 'moth-eye effect', which arises due to a continuously varying refractive index, leading to broadband suppression of reflection within the nanorod structure [49]. The theoretical analysis of this effect is elaborated upon and usually conducted using rigorous coupled-wave analysis (RCWA) [50]. The uniform antireflective property in different regions of the sample, demonstrated by the significant moth-eye effect in different regions of sample 5, also indicates the achieved homogeneity of the large-scale ZnO NR ARC.

Based on our experimental results and analysis, two primary factors influencing the homogeneity of large-scale zinc oxide nanorod (ZnO NR) electrodeposition are the solution flow and the electrical field. By appropriately adjusting the solution flow and the distribution of the electrical field, we can enhance the homogeneity of large-scale nanorods. This study further demonstrates that ZnO NRs are effective as an antireflective layer, offering potential for cost-effective, large-scale production and fabrication. Such advancements pave the way for increased industrial adoption and eventual commercialization, particularly in the realm of large-scale solar cells. The homogeneous, large-scale deposition of ZnO NRs as an antireflective coating (ARC) on solar cell surfaces significantly reduces sunlight reflection. This, in turn, enhances solar light absorption, boosting the photovoltaic conversion efficiency of these cells. Moreover, the ability of ZnO NRs to improve solar energy absorption makes them suitable for various heat collection applications, such as solar water heaters, greenhouses, etc. The outcomes of this research offer tangible benefits for advancing renewable clean energy resources, conserving conventional energy, and reducing carbon dioxide emissions. These findings underscore the potential of ZnO NRs in the field of green energy and highlight their role in promoting more sustainable energy practices.

4. Conclusions

We investigated the electrodeposition of ZnO NRs under varying solution flow states and using two sizes of Pt counter electrodes. Using experiments, we demonstrated that both the solution flow and the electric field significantly influence the homogeneity of large-scale ZnO NR electrodeposition. Utilizing a small counter electrode and stirring in the center, we successfully grew relatively homogeneous ZnO NRs on an 8×10 cm^2 i-

ZnO/FTO substrate. The photoluminescence (PL) results exhibited outstanding properties. The reflectance of the ZnO NRs, as examined in their reflectance spectra, was broadly suppressed across the wavelength region of 400–800 nm. We believe that by controlling the solution flow and adjusting the electric field distribution, larger and more homogeneous ZnO NRs with enhanced economic benefits can be achieved in future work. The optimized electrodeposited large-scale ZnO NRs show promise for applications in solar cells and heat collection systems, improving their light utilization efficiency. Additionally, this technology holds potential for LED applications, where ZnO NR ARCs fabricated on LED surfaces could improve the light output flux efficiency. Therefore, these prospective applications underscore the significance of this technology in energy conservation and emission reduction, making it worthy of further study and development.

Author Contributions: Conceptualization, Y.Z. and K.L.; methodology, K.L. and J.T.; software, X.H.; validation, X.W. and X.Z.; formal analysis, Y.H. and J.T.; investigation, Y.Z. and X.H.; resources, X.X. and Z.W.; data curation, Y.H., X.T., and X.Z.; writing—original draft, Y.Z. and K.L.; writing—review and editing, K.L. and F.L.; visualization, X.T., H.Z., and X.W.; supervision, X.X. and H.Z.; project administration F.L.; funding acquisition, Y.Z. and Z.W. All authors have read and agreed to the published version of the manuscript.

Funding: This research was funded by the National Natural Science Foundation of China (12074045, 62027820); the Natural Science Foundation of Jilin Province (No. 20210101408JC, 20230101352JC); the Introduced Innovative Research Team Program of Zhongshan Institute of Changchun University of Science and Technology (No. CXTD2023004).

Institutional Review Board Statement: Not applicable.

Informed Consent Statement: Not applicable.

Data Availability Statement: Data are contained within the article.

Conflicts of Interest: Author Xida Xing was employed by the company Shandong North Optical Electronic Co., Ltd. Authors Xiao Zhao and Haibin Zhu were employed by the company Northeast Industry Group Co., Ltd. The remaining authors declare that the research was conducted in the absence of any commercial or financial relationships that could be construed as a potential conflict of interest.

References

1. Green, M.A.; Blakers, A.W.; Shi, J.; Keller, E.M.; Wenham, S.R. High-efficiency silicon solar cells. *IEEE Trans. Electron. Dev.* **1984**, *31*, 679–683. [CrossRef]
2. Shockley, W.; Queisser, H.J. Detailed balance limit of efficiency of p-n junction solar cells. *J. Appl. Phys.* **1961**, *32*, 510. [CrossRef]
3. Gao, Z.L.; Lin, G.L.; Chen, Y.C.; Zheng, Y.P.; Sang, N.; Li, Y.F.; Chen, L.; Li, M.C. Moth-eye nanostructure PDMS films for reducing reflection and retaining flexibility in ultra-thin c-Si solar cells. *Solar Energy* **2020**, *205*, 275–281. [CrossRef]
4. Look, D.C. Recent advances in ZnO material and device. *Mater. Sci. Eng. B* **2001**, *80*, 383. [CrossRef]
5. Desai, A.V.; Haque, M.A. Mechanical properties of ZnO nanowires. *Sens. Actuators A Phys.* **2007**, *134*, 169–176. [CrossRef]
6. Özgür, Ü.; Alivov, Y.A.; Liu, C.; Teke, A.; Reshchikov, M.A.; Doğan, S.; Avrutin, V.; Cho, S.; Morkoc, H.-J. A comprehensive review of ZnO materials and devices. *J. Appl. Phys.* **2005**, *98*, 041301. [CrossRef]
7. Peng, Y.; Que, M.; Lee, H.E.; Bao, R.; Wang, X.; Lu, J.; Yuan, Z.; Li, X.; Tao, J.; Sun, J. Achieving high-resolution pressure mapping via flexible GaN/ZnO nanowire LEDs array by piezo-phototronic effect. *Nano Energy* **2019**, *58*, 633–640. [CrossRef]
8. Zhao, X.; Li, Q.; Xu, L.; Zhang, Z.; Kang, Z.; Liao, Q.; Zhang, Y. Interface engineering in 1D ZnO-based heterostructures for photoelectrical devices. *Adv. Funct. Mater.* **2022**, *32*, 2106887. [CrossRef]
9. Chen, T.; Gao, X.; Zhang, J.Y.; Xu, J.L.; Wang, S.D. Ultrasensitive ZnO nanowire photodetectors with a polymer electret interlayer for minimizing dark current. *Adv. Opt. Mater.* **2020**, *8*, 1901289. [CrossRef]
10. Kumaresan, Y.; Min, G.; Dahiya, A.S.; Ejaz, A.; Shakthivel, D.; Dahiya, R. Kirigami and Mogul-Patterned Ultra-Stretchable High-Performance ZnO Nanowires-Based Photodetector. *Adv. Mater. Technol.* **2022**, *7*, 2100804. [CrossRef]
11. Gullu, H.; Yildiz, D.; Kocyigit, A.; Yıldırım, M. Electrical properties of Al/PCBM: ZnO/p-Si heterojunction for photodiode application. *J. Alloys Compd.* **2020**, *827*, 154279. [CrossRef]
12. Choi, M.S.; Kim, M.Y.; Mirzaei, A.; Kim, H.-S.; Kim, S.-i.; Baek, S.-H.; Chun, D.W.; Jin, C.; Lee, K.H. Selective, sensitive, and stable NO$_2$ gas sensor based on porous ZnO nanosheets. *Appl. Surf. Sci.* **2021**, *568*, 150910. [CrossRef]
13. Wibowo, A.; Marsudi, M.A.; Amal, M.I.; Ananda, M.B.; Stephanie, R.; Ardy, H.; Diguna, L.J. ZnO nanostructured materials for emerging solar cell applications. *RSC Adv.* **2020**, *10*, 42838–42859. [CrossRef]

14. Senol, S.; Ozugurlu, E.; Arda, L. Synthesis, structure and optical properties of (Mn/Cu) co-doped ZnO nanoparticles. *J. Alloys Compd.* **2020**, *822*, 153514. [CrossRef]
15. Tosun, M.; Senol, S.; Arda, L. Effect of Mn/Cu co-doping on the structural, optical and photocatalytic properties of ZnO nanorods. *J. Mol. Struct.* **2020**, *1212*, 128071. [CrossRef]
16. Hendrickson, J.R.; Vangala, S.; Nader, N.; Leedy, K.; Guo, J.; Cleary, J.W. Plasmon resonance and perfect light absorption in subwavelength trench arrays etched in gallium-doped zinc oxide film. *Appl. Phys. Lett.* **2015**, *107*, 191906. [CrossRef]
17. Wilson, S.; Hutley, M. The optical properties of 'moth eye' antireflection surfaces. *Opt. Acta* **1982**, *29*, 993–1009. [CrossRef]
18. Aé, L.; Kieven, D.; Chen, J.; Klenk, R.; Rissom, T.; Tang, Y.; Lux-Steiner, M.C. ZnO nanorod arrays as an antireflective coating for Cu(In, Ga)Se2 thin film solar cells. *Prog. Photovolt. Res. Appl.* **2010**, *18*, 209–213. [CrossRef]
19. Lee, Y.J.; Ruby, D.S.; Peters, D.W.; McKenzie, B.B.; Hsu, J.W.P. ZnO nanostructures as efficient antireflection layers in solar cells. *Nano Lett.* **2008**, *8*, 1501–1505. [CrossRef]
20. Leem, J.W.; Joo, D.H.; Yu, J.S. Biomimetic parabola-shaped AZO subwavelength grating structures for efficient antireflection of Si-based solar cells. *Sol. Energy Mater. Sol. Cells* **2011**, *95*, 2221–2227. [CrossRef]
21. Liu, B.; Wang, C.; Bazri, S.; Badruddin, I.A.; Orooji, Y.; Saeidi, S.; Wongwises, S.; Mahian, O. Optical properties and thermal stability evaluation of solar absorbers enhanced by nanostructured selective coating films. *Powder Technol.* **2021**, *377*, 939–957. [CrossRef]
22. Tamulevičius, T.; Laurikėnas, P.; Juodėnas, M.; Mardosaitė, R.; Abakevičienė, B.; Pereyra, C.J.; Račkauskas, S. Antireflection Coatings Based on Randomly Oriented ZnO Nanowires. *Sol. RRL* **2023**, *7*, 2201056. [CrossRef]
23. Chen, J.; Ye, H.; Aé, L.; Tang, Y.; Kieven, D.; Rissom, T.; Neuendorf, J.; Lux-Steiner, M.C. Tapered aluminum-doped vertical zinc oxide nanorod arrays as light coupling layer for solar energy applications. *Sol. Energy Mater. Sol. Cells* **2011**, *95*, 1437–1440. [CrossRef]
24. Zhao, Y.; Tang, Y.; Han, Z. Low-temperature rapid syntheses of high-quality ZnO nanostructure arrays induced by ammonium salt. *Chem. Phys. Lett.* **2017**, *668*, 47–55. [CrossRef]
25. Qu, Y.; Huang, X.; Li, Y.; Lin, G.; Guo, B.; Song, D.; Cheng, Q. Chemical bath deposition produced ZnO nanorod arrays as an antireflective layer in the polycrystalline Si solar cells. *J. Alloys Compd.* **2017**, *698*, 719–724. [CrossRef]
26. Huang, F.; Guo, B.; Li, S.; Fu, J.; Zhang, L.; Lin, G.; Yang, Q.; Cheng, Q. Plasma-produced ZnO nanorod arrays as an antireflective layer in c-Si solar cells. *J. Mater. Sci.* **2019**, *54*, 4011–4023. [CrossRef]
27. Bai, A.; Tang, Y.; Chen, J. Efficient photon capturing in Cu (In, Ga) Se$_2$ thin film solar cells with ZnO nanorod arrays as an antireflective coating. *Chem. Phys. Lett.* **2015**, *636*, 134–140. [CrossRef]
28. Navarrete, E.; Güell, F.; Martinez-Alanis, P.R.; Llobet, E. Chemical vapour deposited ZnO nanowires for detecting ethanol and NO$_2$. *J. Alloys Compd.* **2022**, *890*, 161923. [CrossRef]
29. Swathi, S.; Babu, E.S.; Yuvakkumar, R.; Ravi, G.; Chinnathambi, A.; Alharbi, S.A.; Velauthapillai, D. Branched and unbranched ZnO nanorods grown via chemical vapor deposition for photoelectrochemical water-splitting applications. *Ceram. Int.* **2021**, *47*, 9785–9790. [CrossRef]
30. Huang, M.H.; Wu, Y.; Feick, H.; Tran, N.; Weber, E.; Yang, P. Catalytic growth of zinc oxide nanowires by vapor transport. *Adv. Mater.* **2001**, *13*, 113–116. [CrossRef]
31. Zheng, J.; Jiang, Q.; Lian, J. Synthesis and optical properties of flower-like ZnO nanorods by thermal evaporation method. *Appl. Surf. Sci.* **2011**, *257*, 5083–5087. [CrossRef]
32. Yao, Y.F.; Chou, K.P.; Lin, H.H.; Chen, C.C.; Kiang, Y.W.; Yang, C. Polarity control in growing highly Ga-doped ZnO nanowires with the vapor–liquid–solid process. *ACS Appl. Mater. Interfaces* **2018**, *10*, 40764–40772. [CrossRef]
33. Poornajar, M.; Marashi, P.; Fatmehsari, D.H.; Esfahani, M.K. Synthesis of ZnO nanorods via chemical bath deposition method: The effects of physicochemical factors. *Ceram. Int.* **2016**, *42*, 173–184. [CrossRef]
34. Vessalli, B.A.; Zito, C.A.; Perfecto, T.M.; Volanti, D.P.; Mazon, T. ZnO nanorods/graphene oxide sheets prepared by chemical bath deposition for volatile organic compounds detection. *J. Alloys Compd.* **2017**, *696*, 996–1003. [CrossRef]
35. Pourshaban, E.; Abdizadeh, H.; Golobostanfard, M.R. A close correlation between nucleation sites, growth and final properties of ZnO nanorod arrays: Sol-gel assisted chemical bath deposition process. *Ceram. Int.* **2016**, *42*, 14721–14729. [CrossRef]
36. Tang, Y.; Chen, J.; Greiner, D.; Aé, L.; Baier, R.; Lehmann, J.; Sadewasser, S.; Lux-Steiner, M.C. Fast growth of high work function and high-quality ZnO nanorods from an aqueous solution. *J. Phys. Chem. C* **2011**, *115*, 5239–5243. [CrossRef]
37. Venditti, I.; Barbero, N.; Russo, M.V.; Di Carlo, A.; Decker, F.; Fratoddi, I.; Barolo, C.; Dini, D. Electrodeposited ZnO with squaraine sentisizers as photoactive anode of DSCs. *Mater. Res. Express* **2014**, *1*, 015040. [CrossRef]
38. Bonomo, M.; Naponiello, G.; Venditti, I.; Zardetto, V.; Di Carlo, A.; Dini, D. Electrochemical and photoelectrochemical properties of screen-printed nickel oxide thin films obtained from precursor pastes with different compositions. *J. Electrochem. Soc.* **2017**, *164*, H137–H147. [CrossRef]
39. Jakšić, M.M. Hydrodynamic effects on the macromorphology of electrodeposited zinc and flow visualization: A flat plate electrode in a channel flow cell. *J. Electroanal. Chem. Interfacial Electrochem.* **1988**, *249*, 35–62. [CrossRef]
40. ShakeriHosseinabad, F.; Daemi, S.R.; Momodu, D.; Brett, D.J.; Shearing, P.R.; Roberts, E.P. Influence of flow field design on zinc deposition and performance in a zinc-iodide flow battery. *ACS Appl. Mater. Interfaces* **2021**, *13*, 41563–41572. [CrossRef]
41. Naybour, R. The effect of electrolyte flow on the morphology of zinc electrodeposited from aqueous alkaline solution containing zincate ions. *J. Electrochem. Soc.* **1969**, *116*, 520. [CrossRef]

42. Naybour, R. Morphologies of zinc electrodeposited from zinc-saturated aqueous alkaline solution. *Electrochim. Acta* **1968**, *13*, 763–769. [CrossRef]
43. Lloyd, J.; Sparrow, E.; Eckert, E. Local natural convection mass transfer measurements. *J. Electrochem. Soc.* **1972**, *119*, 702. [CrossRef]
44. Izaki, M.; Omi, T. Electrolyte optimization for cathodic growth of zinc oxide films. *J. Electrochem. Soc.* **1996**, *143*, L53. [CrossRef]
45. Abderrahmane, B.; Djamila, A.; Chaabia, N.; Fodil, R. Improvement of ZnO nanorods photoelectrochemical, optical, structural and morphological characterizations by cerium ions doping. *J. Alloys Compd.* **2020**, *829*, 154498. [CrossRef]
46. Zhang, M.; Reddy, R. Electrical field and current density distribution modeling of aluminum electrodeposition in ionic liquid electrolytes. *ECS Trans.* **2006**, *1*, 47. [CrossRef]
47. Cross, R.; De Souza, M.; Narayanan, E.S. A low temperature combination method for the production of ZnO nanowires. *Nanotechnology* **2005**, *16*, 2188. [CrossRef] [PubMed]
48. Das, A.; Ghosh, S.; Mahapatra, A.D.; Kabiraj, D.; Basak, D. Highly enhanced ultraviolet to visible room temperature photoluminescence emission ratio in Al implanted ZnO nanorods. *Appl. Surf. Sci.* **2019**, *495*, 143615. [CrossRef]
49. Chopra, K.L.; Ranjan, S. *Thin Film Solar Cells*; Springer: New York, NY, USA, 1983; pp. 515–520.
50. Moharam, M.G.; Gaylord, T.K. Rigorous coupled-wave analysis of metallic surface-relief gratings. *J. Opt. Soc. Am. A* **1986**, *3*, 1780–1787. [CrossRef]

Disclaimer/Publisher's Note: The statements, opinions and data contained in all publications are solely those of the individual author(s) and contributor(s) and not of MDPI and/or the editor(s). MDPI and/or the editor(s) disclaim responsibility for any injury to people or property resulting from any ideas, methods, instructions or products referred to in the content.

materials

Article

Ultra-Smooth Polishing of Single-Crystal Silicon Carbide by Pulsed-Ion-Beam Sputtering of Quantum-Dot Sacrificial Layers

Dongyang Qiao [1,2,3], Feng Shi [1,2,3,*], Ye Tian [1,2,3], Wanli Zhang [1,2,3], Lingbo Xie [1,2,3], Shuangpeng Guo [1,2,3], Ci Song [1,2,3] and Guipeng Tie [1,2,3,4]

1   College of Intelligence Science and Technology, National University of Defense Technology, Changsha 410073, China; dyqiao@nudt.edu.cn (D.Q.); tianyecomeon@sina.cn (Y.T.); zhangwanli17@nudt.edu.cn (W.Z.); lingbotse@163.com (L.X.); g18231033287@163.com (S.G.); songci@nudt.edu.cn (C.S.); tieguipeng@163.com (G.T.)
2   Hunan Key Laboratory of Ultra-Precision Machining Technology, Changsha 410073, China
3   Laboratory of Science and Technology on Integrated Logistics Support, National University of Defense Technology, 109 Deya Road, Changsha 410073, China
4   Precision Optical Manufacturing and Testing Center, Shanghai Institute of Optics and Fine Mechanics, Chinese Academy of Sciences, Shanghai 201800, China
*   Correspondence: shifeng@nudt.edu.cn; Tel.: +86-135-7485-9732

Abstract: Single-crystal silicon carbide has excellent electrical, mechanical, and chemical properties. However, due to its high hardness material properties, achieving high-precision manufacturing of single-crystal silicon carbide with an ultra-smooth surface is difficult. In this work, quantum dots were introduced as a sacrificial layer in polishing for pulsed-ion-beam sputtering of single-crystal SiC. The surface of single-crystal silicon carbide with a quantum-dot sacrificial layer was sputtered using a pulsed-ion beam and compared with the surface of single-crystal silicon carbide sputtered directly. The surface roughness evolution of single-crystal silicon carbide etched using a pulsed ion beam was studied, and the mechanism of sacrificial layer sputtering was analyzed theoretically. The results show that direct sputtering of single-crystal silicon carbide will deteriorate the surface quality. On the contrary, the surface roughness of single-crystal silicon carbide with a quantum-dot sacrificial layer added using pulsed-ion-beam sputtering was effectively suppressed, the surface shape accuracy of the Ø120 mm sample was converged to 7.63 nm RMS, and the roughness was reduced to 0.21 nm RMS. Therefore, the single-crystal silicon carbide with the quantum-dot sacrificial layer added via pulsed-ion-beam sputtering can effectively reduce the micro-morphology roughness phenomenon caused by ion-beam sputtering, and it is expected to realize the manufacture of a high-precision ultra-smooth surface of single-crystal silicon carbide.

Keywords: single-crystal SiC; quantum dots; ultra-smooth polishing; pulsed-ion-beam; sacrificial layer

check for updates

Citation: Qiao, D.; Shi, F.; Tian, Y.; Zhang, W.; Xie, L.; Guo, S.; Song, C.; Tie, G. Ultra-Smooth Polishing of Single-Crystal Silicon Carbide by Pulsed-Ion-Beam Sputtering of Quantum-Dot Sacrificial Layers. *Materials* **2024**, *17*, 157. https:// doi.org/10.3390/ma17010157

Academic Editor: Heesun Yang

Received: 30 November 2023
Revised: 23 December 2023
Accepted: 26 December 2023
Published: 27 December 2023

Copyright: © 2023 by the authors. Licensee MDPI, Basel, Switzerland. This article is an open access article distributed under the terms and conditions of the Creative Commons Attribution (CC BY) license (https:// creativecommons.org/licenses/by/ 4.0/).

1. Introduction

The single-crystal silicon carbide (SiC) is a highly promising third-generation semiconductor material with excellent physical and mechanical properties such as large bandgap, high thermal conductivity, high specific stiffness, low coefficient of thermal expansion, and good abrasion resistance [1,2]. It has been widely used in high frequency, high temperature, radiation resistance, photoelectric, and other fields. A smooth surface with low damage is an inevitable requirement for the application of single-crystal silicon carbide. However, due to its high brittleness, hardness, and low fracture toughness, defects such as rough surfaces and microcracks are inevitably formed during the surface processing of SiC [3,4]. Achieving high-precision single-crystal silicon carbide manufacturing with an ultra-smooth surface is still a great challenge.

Polishing is the last step of single-crystal silicon carbide processing, and its surface quality directly affects the performance of semiconductor materials [5]. The chemical mechanical polishing (CMP) method is currently a typical method for preparing SiC [6]. However, due to the high chemical inertness of the surface of single-crystal silicon carbide, its etching efficiency is low. In addition, the process consumes a large amount of slurry and introduces impurity pollution [7], and the CMP method does not have a polishing ability for optical elements of free-form surfaces. Ion-beam sputtering (IBS) is based on the physical sputtering effect, using ions with a certain energy to bombard the surface of the optical element to achieve atomic-level material removal, which is less affected by the chemical properties and hardness of the element. IBS can polish complex free-form surfaces without inducing subsurface damage [8]. In order to improve the machining accuracy of optical components of high-hardness materials, F. Shi et al. proposed a pulsed-ion-beam machining technology with sub-nanometer machining accuracy based on traditional ion-beam machining [9–11]. In light of the evolution of the surface morphology of ion-beam sputtering elements, researchers have conducted many experimental and theoretical studies. Allen [12] studied the surface roughness evolution of ion-beam polishing fused silica, and the results showed that the surface roughness value increased with an increase in the removal depth. At the same time, experiments have also shown that ion-beam sputtering can effectively reduce surface roughness [13,14]. Bradley and Harper established the linear evolution theory (BH model) of surface micro-topography based on the Sigmund sputtering theory. They pointed out that the local etching rate is related to the local curvature, and the energy deposited in the local pits is more than that in the bulge, so the etching rate of the pits is greater than that of the bulge, resulting in the roughening of the surface micro-topography [15]. At the same time, the thermally induced surface diffusion effect and surface porosity mechanism make the ion sputtering have a smoothing effect on the surface [16,17]. Due to the uncertainty of the ion beam smoothing the surface of optical elements, the IOM Institute [18,19] proposed a sacrificial layer-assisted polishing method. A material layer such as photoresist, silicon, and SiO_2 is uniformly covered on the initial surface by coating or sputtering deposition, and then the material is smoothed directly using ion-beam sputtering until an ultra-smooth surface is obtained.

At first, quantum dots were introduced into the machining process as a subsurface damage detection method. K.L.M Williams et al. [20] used a nano-scale quantum dot solution to label the subsurface defects formed in the grinding and polishing process. The depth of the subsurface defect layer was calibrated via fluorescence labeling. M. Chen et al. [21] showed that improving the size distribution of HgSe quantum dots could exponentially increase their mobility, and it is necessary to improve the size distribution when using in-band photodetectors. Benjamin T et al. proved that CdSe nanosheet quantum wells had a narrow, polarized intersubband and absorption characteristics under light excitation or external bias [22]. Xin Tang et al. fabricated functional quasi-3D nanophotonic structures directly into colloidal quantum dot (CQD) films using the one-step imprinting method, and the diffraction gratings, double-layer wire-grid polarizers, and resonant metal mesh long-pass filters were imprinted on the CQD films [23]. These studies show that quantum dots exhibit large specific surface areas with excellent optical, electrical, and thermal properties due to the quantum confinement effect and boundary effect and can be used as sacrificial layer materials to introduce the polishing process of optical elements.

This paper aimed to study the roughness evolution of single-crystal SiC surfaces using pulsed-ion-beam sputtering and to realize the ultra-smooth machining of single-crystal SiC surfaces. Through experimental research and theoretical analysis, we found that adding a quantum dot coating as a sacrificial layer on single-crystal silicon carbide could hinder the roughening phenomenon caused by different sputtering characteristics of traditional ion beam bombardment of dual-phase materials and achieve high-precision modification of single-crystal silicon carbide while obtaining a higher surface quality. We anticipate that this method will apply to industrial-scale ultra-smooth polishing of SiC.

2. Experimental Setup

2.1. Material

This study used two 4H-SiC lenses provided by TankeBlue Semiconductor Co. Ltd., Beijing, China, 120 mm in diameter and 10 mm in thickness. All experiments were carried out on the most commonly used Si (0001) surface of electronic devices, and the samples are shown in Figure 1b,c. First, all samples were ultrasonically cleaned with deionized water (conductivity < 0.5 us/cm) for 30 min, then dehydrated and dried with anhydrous ethanol. Deionized water and anhydrous ethanol were obtained from Aladdin Scientific Corp. In this study, we used a commercially available (Mesolight Co. Ltd., Suzhou, China) water-soluble CdSe/ZnS core–shell structured quantum dot solution with a luminescence wavelength of 544 nm $\pm$ 10 nm (Figure 1a).

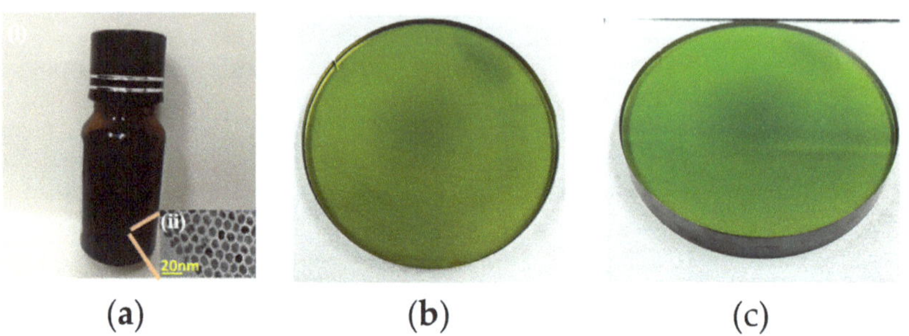

(a)   (b)   (c)

Figure 1. Materials used in the experiment. (**a**) CdSe/ZnS quantum dot solution. (i) Reagent bottle containing quantum dot solution. (ii) Transmission electron micrograph of a quantum dot. (**b**) Single-crystal silicon carbide sample 1. (**c**) Single-crystal silicon carbide sample 2.

2.2. Pulsed-Ion-Beam Etching Equipment

The process of pulsed-ion-beam etching of single-crystal silicon carbide was carried out on the ion-beam etching machine shown in Figure 2a. Argon (Ar) is ionized into Ar^+ in the ion source, forming a plasma in the cavity. The three-grid ion optical system accelerated and focused the ion beam, which was then energized using the accelerated electric field to form material removal on the surface of the workpiece.

As shown in Figure 2b, a highly controllable high-voltage pulse power supply was connected to the screen grid of the ion optical system. The high-voltage pulse power supply supplied power to the screen grid, and the extraction method of the ion beam was changed from continuous extraction to pulsed extraction. The continuous ion beam was also intercepted as a pulsed ion beam. The self-developed pulsed ion-beam processing equipment used argon as the etching gas, and its pulse frequency was adjusted in the range of 1–500 Hz. The pulse ion-beam removal resolution can reach 0.07 nm in a single shot, achieving the optical mirror's atomic level removal accuracy.

2.3. Method

The experiment was carried out in a self-developed IBE system. The working pressure was 2.5×10^{-3} Pa, the Ar^+ ion energy was 800 eV, the beam density was 10–25 mA, and the working temperature was 70 °C.

The surface of sample 1 was left untreated. A layer of quantum dots was coated on the surface of single-crystal silicon carbide sample 2 using the spin-coating method, in which a quantum dot solution was dropped on the surface of the substrate. Then, the substrate was rotated to allow the quantum dots to cover the surface uniformly by centrifugal force, and then the etching and polishing research was carried out under the same etching parameters. The same pulse frequency was maintained during the pulse-ion-beam etching and polishing process to ensure the stability of the removal function. The etching process is shown in Figure 3. In order to observe the evolution of surface quality, the white light interferometer

(Figure 4a) was used to observe the surface morphology of single-crystal silicon carbide. A commercial Zygo New View 700 s white light interferometer was used to detect the medium and high frequencies of the micron scale on the wafer surface. The lens was equipped with magnifications of 10× and 50×, and the resolution of the data was 1.5 μm and 7.5 μm, respectively. The detection range was 468 μm × 351 μm. The VeriFire Asphere laser wavefront interferometer by Zygo was used to detect low-frequency surface errors, as shown in Figure 4b. The maximum aperture of the interferometer standard lens was 150 mm, and the maximum resolution was 1024 × 1024 pixels.

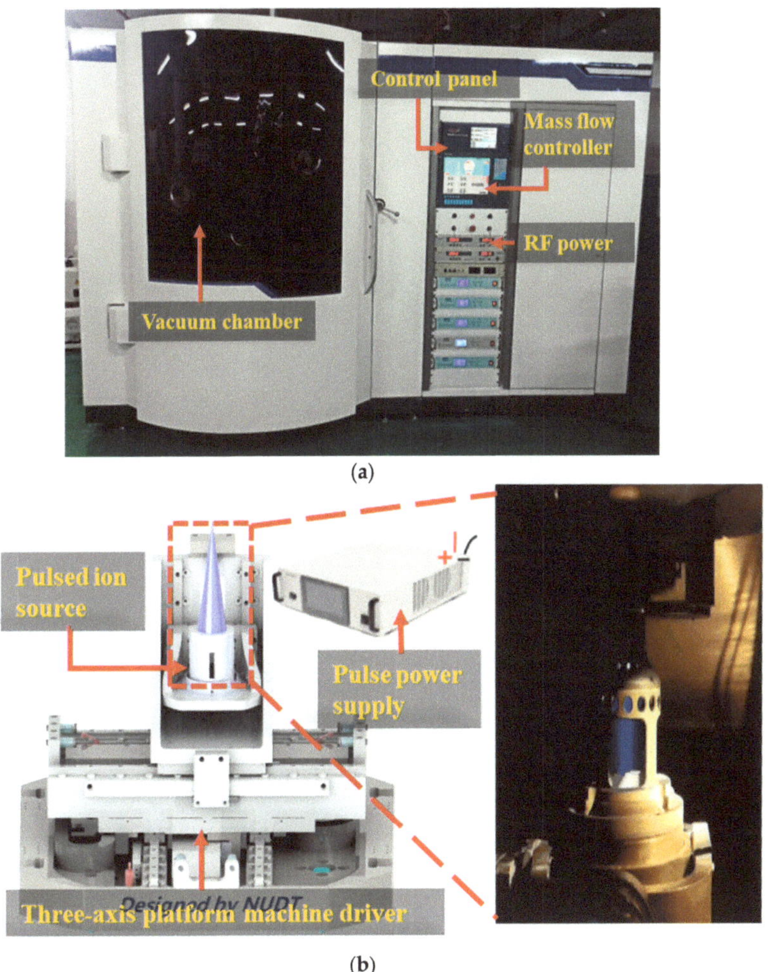

Figure 2. Physical diagram and schematic diagram of the pulsed-ion-beam etching equipment. (**a**) NUDT-IBE700 equipment. (**b**) Pulsed ion source model schematic diagram and pulsed ion source device physical diagram.

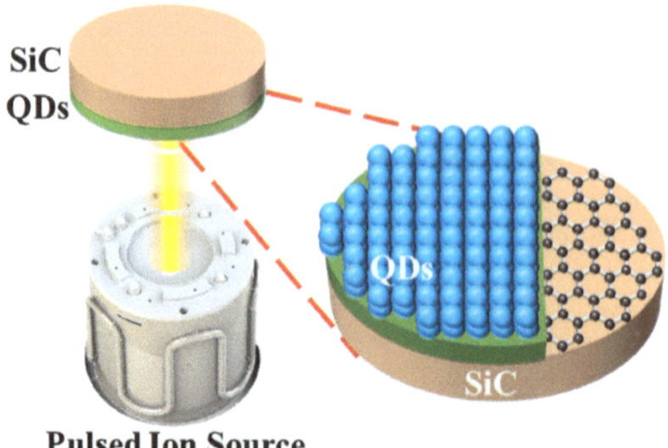

Figure 3. Quantum-dot sacrificial layer pulsed-ion-beam etching method.

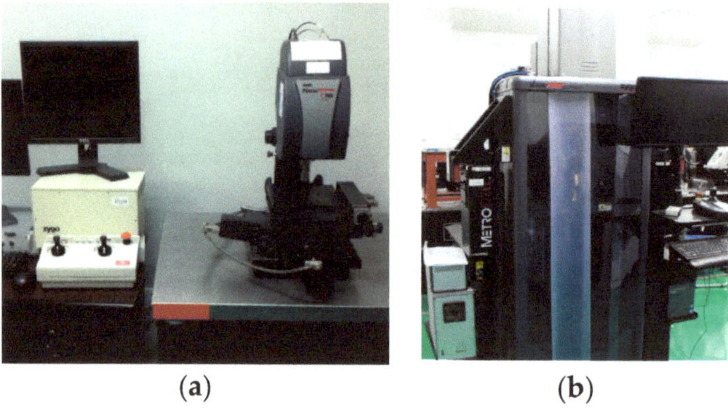

(a)   (b)

Figure 4. Measurement equipment. (**a**) Zygo white light interferometer. (**b**) Zygo wavefront interferometer.

3. Results and Discussion

Figure 5 shows the change in surface roughness of single-crystal silicon carbide without a quantum-dot sacrificial layer under common etching parameters (ion energy of 800 eV, beam density of 20 mA, and duty cycle of 50%). Two points were taken on the surface of the original sample to measure the initial roughness. PV (peak to valley) is the difference between the highest and lowest part of the surface, and RMS (root mean square) is the root mean square value. The original surface roughness was 0.25 nm RMS (Figure 5a) and 0.30 nm RMS (Figure 5b). After ion-beam etching for 30 nm, the surface roughness of the same area became 0.38 nm RMS (Figure 5c) and 0.35 nm RMS (Figure 5d). The surface roughness of single-crystal silicon carbide increased, and the surface quality deteriorated. According to the results of the PSD curve (Figure 6), the error of medium- and high-frequency bands was worse than that of the initial surface after ion-beam sputtering on the surface of single-crystal silicon carbide, which shows that there is a rough effect on the surface of single-crystal silicon carbide directly bombarded by the ion beam.

Under the same etching parameters (ion energy of 800 eV, beam current size of 20 mA, and duty cycle of 50%), 30 nm was etched on the surface of single-crystal silicon carbide with a sacrificial layer. In contrast, the experimental results in Figure 7 show that the surface quality was effectively improved by adding quantum dots as sacrificial layers after pulsed-ion-beam removal. The original surface roughness of the sample was 0.27 nm RMS (Figure 7a) and 0.29 nm RMS (Figure 7b). After ion-beam etching, the surface roughness of the same area became 0.21 nm RMS (Figure 7c) and 0.22 nm RMS (Figure 7d). According to the PSD curve (Figure 8), the results showed that the full-band curve of the surface after pulsed-ion-beam sputtering was lower than that of the initial surface. After adding the quantum-dot sacrificial layer to the surface of single-crystal silicon carbide, pulsed-ion-beam sputtering could improve the surface quality.

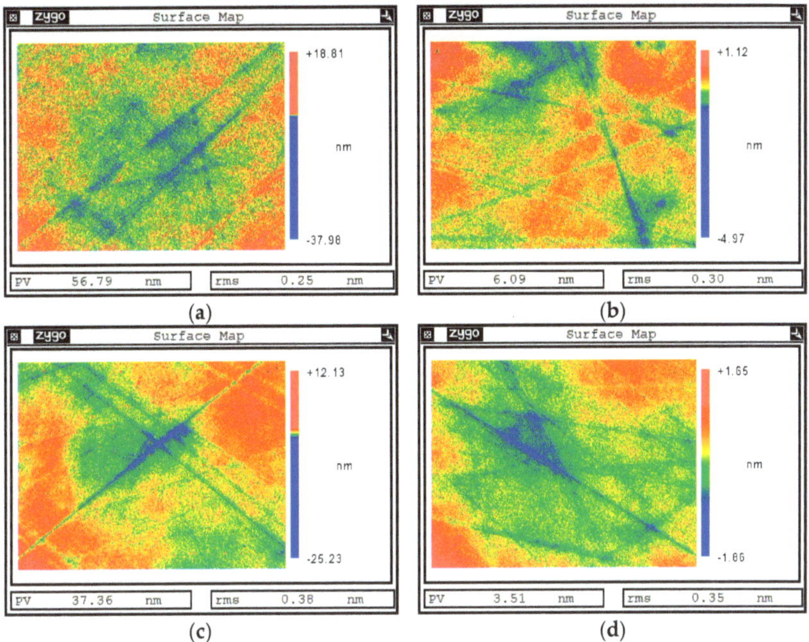

Figure 5. Detection results of white light interferometer on the surface of sample 1. (**a**) Surface area a of the sample before sputtering. (**b**) Surface area b of the sample before sputtering. (**c**) Surface area a of the sample after sputtering. (**d**) Surface area b of the sample after sputtering.

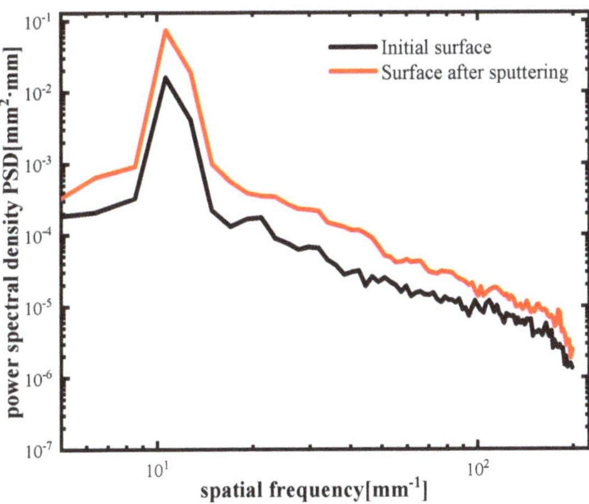

Figure 6. The PSD curves of the initial surface without quantum dots and after IBE polishing.

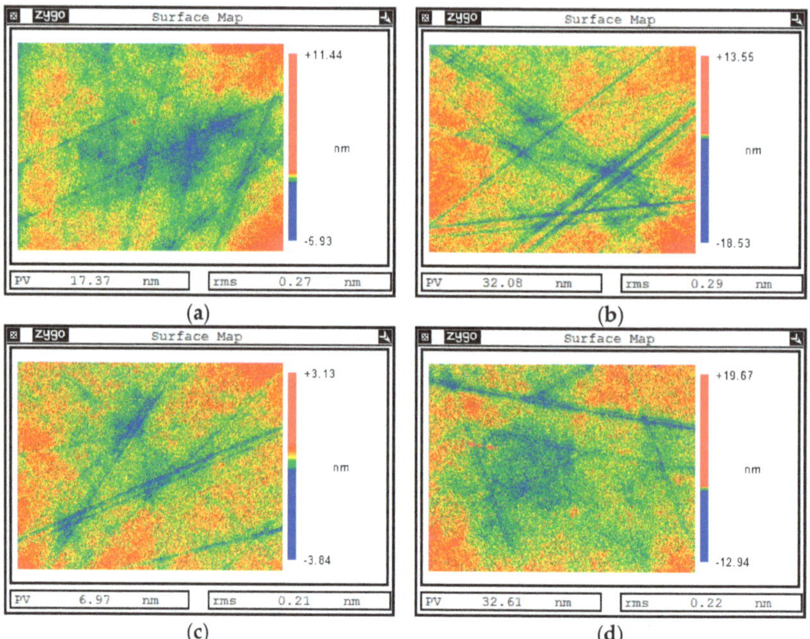

Figure 7. Surface white light interferometer test results of sample 2 with a quantum-dot sacrificial layer. (**a**) Surface area a of the sample before sputtering. (**b**) Surface area b of the sample before sputtering. (**c**) Surface area a of the sample after sputtering. (**d**) Surface area b of the sample after sputtering.

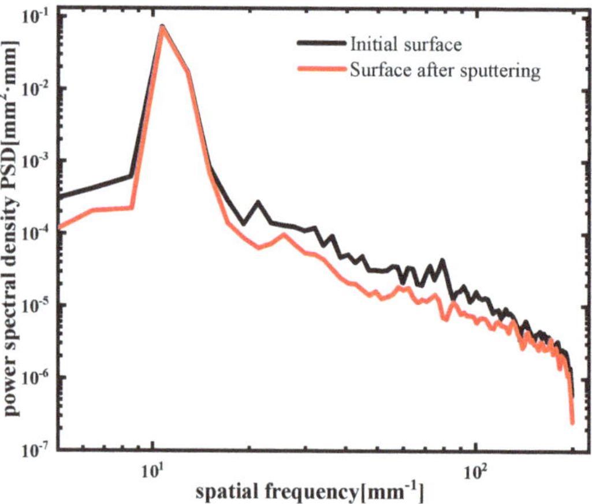

Figure 8. The PSD curves of the initial surface with quantum dots and after IBE polishing.

The roughness and smoothing effect interaction dominates the morphology change with bombardment time in ion sputtering. For any microscopic morphology $m(x, y, t)$ combining partial differential equations and statistical methods, the change in surface morphology $m(q, t)$ and in frequency space with time can be expressed as:

$$\frac{\partial m(q, t)}{\partial t} = -m(q, t)M(q) + \eta(q, t) \tag{1}$$

For the statistical linear equation of Equation (1), Moselerp [24] and Spiller et al. [25] derived the relevant solutions and obtained the power density function:

$$PSDm(q,t) = PSD(q,t=0)exp(2M(q)t) + A\frac{1 - exp(2M(q)t)}{M(q)} \tag{2}$$

$M(q)$ depends on the processing parameters of ion-beam sputtering. The q^n component in the polynomial represents different sputtering roughness and smoothing processes. t is the introduced sacrificial layer parameter, namely:

$$M(q) = (M_{2s} + M_{2B})q^2 - t \times M_{4f}q^4 \tag{3}$$

Therefore, it can be seen that the evolution of surface micro-morphology is the result of the combined action of surface roughness and smoothness, and different processing conditions determine the development trend of surface morphology. If the effect of viscous flow and elastic diffusion is dominant, $M(q)$ is negative, and the surface develops in a smooth direction. Otherwise, $M(q)$ is positive, and the surface will become rougher. The root mean square roughness Rq of the microstructure can be obtained as follows:

$$Rq = 2\pi \int_0^\infty PSD(q,t)qdq \tag{4}$$

According to the above theoretical research, the evolution process of the ion-beam etching of single-crystal silicon carbide was analyzed, and the change in surface roughness under ion beam bombardment was obtained.

The process of ion-beam sputtering of the single-crystal silicon carbide surface has both surface smoothing and rough effects. The experimental results in Figure 5 also verified this theory. When the effect of viscous flow and elastic diffusion on the surface of sputtered single-crystal silicon carbide cannot eliminate the influence of the roughness effect, the roughness effect plays a leading role, the surface quality will deteriorate, and the roughness will increase, as shown in Figure 5c,d. RMS increased to 0.38 nm and 0.35 nm. The PSD curve in Figure 6 also shows that the error in the surface's middle- and high-frequency bands after ion-beam sputtering deteriorated.

The quantum-dot sacrificial layer added to the surface of single-crystal silicon carbide made the $t \times M_{4f}$ part negative, while the front positive value was very small compared to the change in t. It can be seen that $M(q)$ always remained negative, and the surface of the single-crystal silicon carbide always developed in a smooth direction. The results in Figure 7 also show that the roughness of the surface was reduced after adding the quantum dot solution as the sacrificial layer, and the minimum RMS could reach 0.21 nm, which realizes the ultra-smooth surface manufacturing of single-crystal silicon carbide. The PSD curve in Figure 8 also shows that the surface's middle- and high-frequency bands greatly improved after adding a quantum-dot sacrificial layer via ion-beam sputtering on the surface.

Figure 9 demonstrates the pulsed-ion-beam sputtering of a single-crystal SiC surface with a sacrificial layer of quantum dots. Quantum dots were added to the rough initial surface, as shown in Figure 9b; the added quantum dots filled the surface scratches and pits and made the surface flat. Then, the surface was etched using a pulsed ion beam to remove the added quantum-dot sacrificial layer, as shown in Figure 9c. As the pits were filled, the difference in the energy deposition and etching rate at various ion-beam sputtering pits was suppressed. Thus, the surface was uniformly removed, and after the quantum-dot sacrificial layer was completely removed, an ultra-smooth surface was obtained, as shown in Figure 9d.

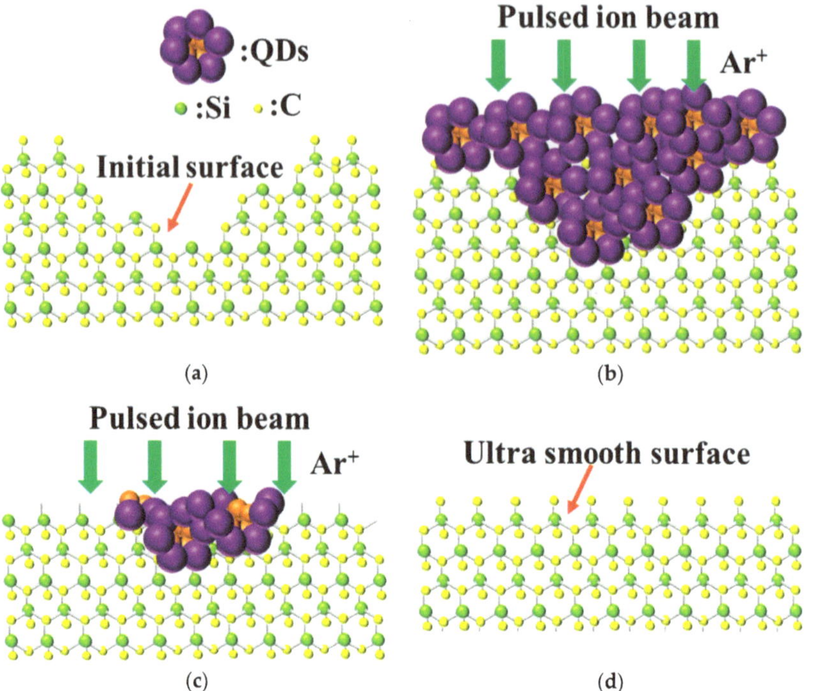

Figure 9. Ion-beam sputtering removal with a sacrificial layer. (**a**) Initial surface of single-crystal silicon carbide. (**b**) Surface after adding the quantum-dot sacrificial layer. (**c**) Polishing process. (**d**) Ultra-smooth surface of single-crystal silicon carbide.

The quantum-dot sacrificial layer was coated on the surface of the single-crystal silicon carbide with an aperture of 120 mm, and the whole surface was modified by the pulsed ion beam. After polishing, the surface shape accuracy converged to 7.63 nm RMS (Figure 10), and the roughness was reduced to 0.21 nm RMS, which realizes the high-precision ultra-smooth polishing of single-crystal silicon carbide.

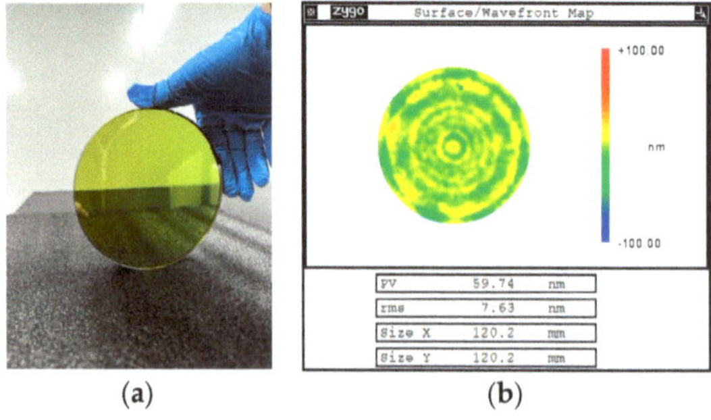

Figure 10. Pulsed-ion-beam sputtering of quantum-dot sacrificial layer single-crystal silicon carbide processing examples. (**a**) Physical drawing of the processing sample. (**b**) Surface shape results after processing.

4. Conclusions

This paper proposed a pulsed-ion-beam ultra-smooth polishing method for single-crystal SiC by introducing quantum dots as sacrificial layers. A water-soluble CdSe/ZnS core–shell quantum dot solution was used to coat the surface of the single-crystal silicon carbide, and then pulsed-ion-beam sputtering etching was carried out at an ion energy of 800 eV and a beam density of 20 mA. An ultra-smooth surface with a roughness of 0.21 nm RMS was fabricated. The main conclusions of this paper are as follows:

(1) Pulsed-ion-beam etching based on the material removal mechanism by physical sputtering can be used as a high-precision and low-damage process to achieve ultra-smooth polishing of high-hardness single-crystal silicon carbide;

(2) Due to energy deposition and thermal diffusion, the surface roughness of single-crystal silicon carbide surfaces with different initial morphology can be increased by ion-beam sputtering, and the surface quality can deteriorate. This phenomenon can be attributed to the coexistence of the smoothing and roughening effects in ion-beam sputtering surface material;

(3) The introduction of quantum dots as a sacrificial layer can change the energy deposition distribution and etching rate to increase the surface smoothing effect of pulsed-ion-beam sputtering and realize the ultra-smooth surface polishing of the single-crystal silicon carbide surface. The experimental results showed that the surface shape accuracy of the quantum-dot sacrificial layer after pulsed-ion-beam sputtering etching was 7.63 nm RMS, and an ultra-smooth surface of single-crystal silicon carbide with a roughness of 0.21 nm RMS was realized.

In summary, the introduction of quantum dots as a sacrificial layer in the process of pulsed-ion-beam etching to polish single-crystal silicon carbide can improve the surface quality, which provides theoretical and technical support for the acquisition of an ultra-smooth surface of single-crystal silicon carbide, and also provides a new idea for the ultra-smooth polishing of high-hardness optical components.

Author Contributions: Conceptualization, F.S.; Methodology, Y.T. and C.S.; Validation, D.Q., L.X. and S.G.; Formal analysis, W.Z. and L.X.; Investigation, D.Q.; Resources, F.S., Y.T. and G.T.; Data curation, W.Z. and S.G.; Writing—original draft, D.Q.; Writing—review & editing, F.S. and C.S.; Supervision, C.S.; Project administration, Y.T. and G.T.; Funding acquisition, C.S. All authors have read and agreed to the published version of the manuscript.

Funding: This project was supported by the National Natural Science Foundation of China (62175259) and the National Key R&D Program of China (No. 2021YFB3701504 and No. 2021YFC2202403).

Institutional Review Board Statement: Not applicable.

Informed Consent Statement: Not applicable.

Data Availability Statement: The data presented in this study are available on request from the corresponding author. The data are not publicly available because the data are also part of an ongoing study.

Acknowledgments: This work was also supported by the School of Optics and Photonics at the Beijing Institute of Technology.

Conflicts of Interest: The authors declare no conflict of interest.

References

1. Zhao, L.; Zhang, J.; Pfetzing, J.; Alam, M.; Hartmaier, A. Depth-sensing ductile and brittle deformation in 3C-SiC under Berkovich nanoindentation. *Mater. Des.* **2020**, *197*, 109223. [CrossRef]
2. Li, H.; Cui, C.; Bian, S.; Lu, J.; Xu, X.; Arteaga, O. Double-sided and single-sided polished 6H-SiC wafers with subsurface damage layer studied by Mueller matrix ellipsometry. *J. Appl. Phys.* **2020**, *128*, 235304. [CrossRef]
3. Wu, M.; Huang, H.; Luo, Q.; Wu, Y. A novel approach to obtain near damage-free surface/subsurface in machining of single crystal 4H-SiC substrate using pure metal mediated friction. *Appl. Surf. Sci.* **2022**, *588*, 152963. [CrossRef]
4. Deng, H.; Endo, K.; Yamamura, K. Competition between surface modification and abrasive polishing: A method of controlling the surface atomic structure of 4H-SiC (0001). *Sci. Rep.* **2015**, *5*, 8947. [CrossRef]
5. Ji, J.; Yamamura, K.; Deng, H. Plasma-assisted polishing for atomic surface fabrication of single crystal SiC. *Acta Phys. Sin.* **2021**, *70*, 068102. [CrossRef]

6.	Hu, Y.; Shi, D.; Hu, Y.; Zhao, H.W.; Sun, X.D. Investigation on the material removal and surface generation of a single crystal SiC wafer by ultrasonic chemical mechanical polishing combined with ultrasonic lapping. *Materials* **2018**, *11*, 2022. [CrossRef] [PubMed]
7.	Speed, D.; Westerhoff, P.; Sierra-Alvarez, R.; Draper, R.; Pantano, P.; Aravamudhan, S.; Chen, K.L.; Hristovski, K.; Herckes, P.; Bi, X.; et al. Physical, chemical, and in vitro toxicological characterization of nanoparticles in chemical mechanical planarization suspensions used in the semiconductor industry: Towards environmental health and safety assessments. *Environ. Sci. Nano* **2015**, *2*, 227–244. [CrossRef]
8.	Drueding, T.; Bifano, T.; Fawcett, S. Contouring algorithm for ion figuring. *Precis. Eng.* **1995**, *17*, 10–21. [CrossRef]
9.	Xie, L.; Tian, Y.; Shi, F.; Song, C.; Tie, G.; Zhou, G.; Shao, J.; Liu, S. Nonlinear Effects of Pulsed Ion Beam in Ultra-High Resolution Material Removal. *Micromachines* **2022**, *13*, 1097. [CrossRef]
10.	Zhou, G.; Tian, Y.; Shi, F.; Song, C.; Tie, G.; Zhou, G.; Xie, L.; Shao, J.; Wu, Z. Low-Energy Pulsed Ion Beam Technology with Ultra-High Material Removal Resolution and Widely Adjustable Removal Efficiency. *Micromachines* **2021**, *12*, 1370. [CrossRef]
11.	Xie, L.; Tian, Y.; Shi, F.; Zhou, G.; Guo, S.; Zhu, Z.; Song, C.; Tie, G. Figuring Method of High Convergence Ratio for Pulsed Ion Beams Based on Frequency-Domain Parameter Control. *Micromachines* **2022**, *13*, 1159. [CrossRef] [PubMed]
12.	Allen, L.N. Progress in ion figuring large optics. In *Laser-Induced Damage in Optical Materials: 1994*; SPIE: Bellingham, WA, USA, 1995; pp. 237–247. [CrossRef]
13.	Keller, A.; Facsko, S.; Moller, W. The morphology of amorphous SiO_2 surfaces during low energy ion sputtering. *J. Phys. Condens. Matter* **2009**, *21*, 495305. [CrossRef] [PubMed]
14.	Hans, S.; Parida, B.K.; Pachchigar, V.; Augustine, S.; Saini, M.; Sooraj, K.; Ranjan, M. Temperature influence on the formation of triangular features superimposed on nanoripples produced by low-energy ion beam. *Surf. Interfaces* **2022**, *28*, 101619. [CrossRef]
15.	Bradley, R.M.; Harper, J. Theory of ripple topography induced by ion bombardment. *J. Vac. Sci. Technol.* **1988**, *6*, 2390–2395. [CrossRef]
16.	Ozaydin, G.; Ludwig, K.F.; Zhou, H.; Zhou, L.; Headrick, R.L. Transition behavior of surface morphology evolution of Si(100) during low-energy normal-incidence Ar^+ ion bombardment. *J. Appl. Phys.* **2008**, *103*, 4713. [CrossRef]
17.	Anzenberg, E.; Perkinson, J.C.; Madi, C.S.; Aziz, M.J.; Ludwig, K.F. Nanoscale surface pattern formation kinetics on germanium irradiated by Kr^+ ions. *Phys. Rev. B* **2012**, *86*, 59–73. [CrossRef]
18.	Arnold, T.; Böhm, G.; Fechner, R.; Meister, J.; Nickel, A.; Frost, F.; Hänsel, T.; Schindler, A. Ultra-precision surface finishing by ion beam and plasma jet techniques—Status and outlook. *Nucl. Instrum. Methods Phys. Res. Sect. A* **2010**, *616*, 147–156. [CrossRef]
19.	Fechner, R.; Flamm, D.; Frank, W.; Schindler, A.; Frost, F.; Ziberi, B. Ion beam assisted smoothing of optical surfaces. *Appl. Phys. A* **2004**, *78*, 651–654. [CrossRef]
20.	Williams, W.; Mullany, B.; Parker, W.; Moyer, P.; Randles, M. Using quantum dots to evaluate subsurface damage depths and formation mechanisms in glass. *CIRP Ann. Manuf. Technol.* **2010**, *59*, 569–572. [CrossRef]
21.	Chen, M.; Shen, G.; Guyot-Sionnest, P. Size Distribution Effects on Mobility and Intraband Gap of HgSe Quantum Dots. *J. Phys. Chem. C* **2020**, *124*, 16216–16221. [CrossRef]
22.	Chen, M.; Shen, G.; Guyot-Sionnest, P. Polarized near-infrared intersubband absorptions in CdSe colloidal quantum wells. *Nat. Commun.* **2019**, *10*, 4511. [CrossRef]
23.	Tang, X.; Chen, M.; Ackerman, M.M.; Melnychuk, C.; Guyot-Sionnest, P. Direct Imprinting of Quasi-3D Nanophotonic Structures into Colloidal Quantum-Dot Devices. *Adv. Mater.* **2020**, *32*, e1906590. [CrossRef] [PubMed]
24.	Moseler, M.; Rattunde, O.; Nordiek, J.; Haberland, H. On the origin of surface smoothing by energetic cluster impact: Molecular dynamics simulation and mesoscopic modeling. *Nucl. Instrum. Methods Phys. Res. Sect. B* **2000**, *164*, 522–536. [CrossRef]
25.	Spiller, E.; Stearns, D.; Krumrey, M. Multilayer X-ray mirrors: Interfacial roughness, scattering, and image quality. *J. Appl. Phys.* **1993**, *74*, 107–118. [CrossRef]

Disclaimer/Publisher's Note: The statements, opinions and data contained in all publications are solely those of the individual author(s) and contributor(s) and not of MDPI and/or the editor(s). MDPI and/or the editor(s) disclaim responsibility for any injury to people or property resulting from any ideas, methods, instructions or products referred to in the content.

Article

Type-I CdSe@CdS@ZnS Heterostructured Nanocrystals with Long Fluorescence Lifetime

Yuzhe Wang [†], Yueqi Zhong [†], Jiangzhi Zi and Zichao Lian *[ID]

School of Materials and Chemistry, University of Shanghai for Science and Technology, Shanghai 200093, China; 212142401@st.usst.edu.cn (Y.W.); 203612315@st.usst.edu.cn (Y.Z.); jiangzhizi@usst.edu.cn (J.Z.)
* Correspondence: zichaolian@usst.edu.cn
† These authors contributed equally to this work.

Abstract: Conventional single-component quantum dots (QDs) suffer from low recombination rates of photogenerated electrons and holes, which hinders their ability to meet the requirements for LED and laser applications. Therefore, it is urgent to design multicomponent heterojunction nanocrystals with these properties. Herein, we used CdSe quantum dot nanocrystals as a typical model, which were synthesized by means of a colloidal chemistry method at high temperatures. Then, CdS with a wide band gap was used to encapsulate the CdSe QDs, forming a CdSe@CdS core@shell heterojunction. Finally, the CdSe@CdS core@shell was modified through the growth of the ZnS shell to obtain CdSe@CdS@ZnS heterojunction nanocrystal hybrids. The morphologies, phases, structures and performance characteristics of CdSe@CdS@ZnS were evaluated using various analytical techniques, including transmission electron microscopy, X-ray diffraction, UV-vis absorption spectroscopy, fluorescence spectroscopy and time-resolved transient photoluminescence spectroscopy. The results show that the energy band structure is transformed from type II to type I after the ZnS growth. The photoluminescence lifetime increases from 41.4 ns to 88.8 ns and the photoluminescence quantum efficiency reaches 17.05% compared with that of pristine CdSe QDs. This paper provides a fundamental study and a new route for studying light-emitting devices and biological imaging based on multicomponent QDs.

Keywords: CdSe quantum dots; charge separation; type-I heterostructured semiconductor; luminescence quantum efficiency and lifetime

Citation: Wang, Y.; Zhong, Y.; Zi, J.; Lian, Z. Type-I CdSe@CdS@ZnS Heterostructured Nanocrystals with Long Fluorescence Lifetime. *Materials* **2023**, *16*, 7007. https://doi.org/10.3390/ma16217007

Received: 4 October 2023
Revised: 29 October 2023
Accepted: 30 October 2023
Published: 1 November 2023

Copyright: © 2023 by the authors. Licensee MDPI, Basel, Switzerland. This article is an open access article distributed under the terms and conditions of the Creative Commons Attribution (CC BY) license (https://creativecommons.org/licenses/by/4.0/).

1. Introduction

In recent years, colloidal nanocrystals have aroused extensive attention in the fields of basic research and practical applications due to their excellent optical and photoelectrical properties [1–5]. For example, they have broad application prospects in the fields of fluorescent probes [6], lasers [7], solar cells [8,9], bioimaging [10,11] and luminescent devices [12–14]. Cadmium selenide (CdSe) quantum dots (QDs) have been studied extensively because of their quantum size effects and because their luminous wavelength can be extended to the entire visible region. QDs can be synthesized via surface modification of organic molecules not only to improve their luminescent properties and stability but also to obtain a narrow size distribution [15]. However, the presence of many dangling bonds on the surface of the material may lead to defects, which can greatly reduce the stability and luminescent properties of the material [16]. In order to solve this problem, many researchers adopt a method of surface modification of inorganic materials to construct type-I semiconductor band structures by generating a passivation effect of core@shell heterojunction on the surface of materials [17,18]. For instance, Zhu et al. [19] synthesized CdSe/ZnS type I core–shell quantum dots by choosing wide-band-gap ZnS as the shell material and developed a method to optimize the charge separation rate and lifetime by controlling the thickness of the shell material. Toufanian et al. [20] prepared type I InP/ZnSe/ZnS quantum dots and overcame the inherent brightness mismatch seen in QDs through concerted materials design of heterostructured core/shell InP-based QDs. Panda et al. [21] alloyed CdSe/CdS quantum dots at elevated temperatures to prepare high-quality CdZnSe quantum dots and obtained very high environmental stability. Research has also found that heterostructures can effectively resist their own chemical degradation and photocorrosion, thereby improving their photoluminescent quantum efficiency. However, type-II semiconductor heterostructures can reduce the photoluminescent quantum efficiency of materials by promoting the separation of photo-generated electrons and holes. Therefore, this method is not conducive to enhancing the luminescent properties of materials [22]. However, there are few reports on the construction of multi-component structures from type II to type I to enhance the luminescence properties of materials [23,24].

In this study, we first constructed type-II heterojunction with CdSe quantum dots as the core and CdS as the shell. Then, we used ZnS with a wide band gap for modification and constructed type-I heterojunction, which enhanced the fluorescence quantum efficiency and luminescence lifetime of the material. After CdSe was coated with CdS, the UV-vis absorption spectrum and fluorescence emission spectrum of CdSe@CdS showed a red shift, the fluorescence lifetime increased from 41.4 ns to 59.0 ns, but the fluorescence quantum efficiency decreased from 15.89% to 6.32%. Additionally, the effects of different proportions of ZnS on the surface modification of CdSe@CdS were studied. The UV-vis absorption spectra and fluorescence emission spectra of CdSe@CdS@ZnS showed a red shift after the modification of ZnS. When the optimal ratio of Zn:S = 0.6:1.2, the fluorescence lifetime of the material increased from 59.0 ns to 88.8 ns and the fluorescence quantum efficiency increased from 6.32% to 17.05%, showing good luminescence properties.

2. Results and Discussion

2.1. Characterization of Synthetic Materials

We synthesized novel CdSe@CdS@ZnS heterojunction nanocrystals (NCs) and studied their luminescent properties and the lifetime of photogenerated electrons and holes. The whole synthetic processes are shown in Figure 1a. First, CdSe quantum dots were synthesized via the thermally injected wet chemical method using CdO as the cadmium source and a selenium precursor as the selenium source [25]. As shown in Figure 1b, the representative transmission electron microscope (TEM) image shows that the particle size of CdSe QDs was 4.3 ± 0.1 nm (Figure 1e). The CdSe@CdS core@shell structure was obtained by slowly injecting a Cd-and-S-mixed precursor solution. As shown in Figure 1c, its particle size distribution was 6.8 ± 0.1 nm (Figure 1f). Compared with CdSe QDs, the size of CdSe@CdS nanocrystals was increased by 2.5 nm. It could be concluded that CdS was coated on the surface of CdSe, corresponding to the thickness of CdS of about 1.25 nm. Then, we used CdSe@CdS as the core, zinc acetate as the zinc source, and a sulfur–trioctylphosphine (S-TOP) solution as the sulfur source, which were slowly injected into the reaction solution to obtain CdSe@CdS@ZnS heterojunction NCs. The TEM image is shown in Figure 1d, and the size distribution shows that its size was 8.8 ± 0.1 nm (Figure 1g) corresponding to the thickness of ZnS of about 1.0 nm. As the shells were loaded, the size of the materials gradually increased. The HRTEM image in Figure 1h shows the high crystallinity of the CdSe@CdS@ZnS, which revealed the existence of the CdSe, CdS and ZnS phases. The different lattice spacings of 0.32 nm, 0.33 nm and 0.31 nm corresponded to wurtzite CdSe (w-CdSe) (101), zinc-blende CdS (w-CdSe) (111) and zinc-blende ZnS (zb-ZnS) (111), respectively. The HRTEM (Figure 1h), high-angle annular dark-field (HAADF) scanning TEM (STEM) and STEM-energy dispersive X-ray spectrometry (EDS) mapping (Figure 1i–n) of CdSe@CdS@ZnS showed that CdSe was coated with CdS and ZnS.

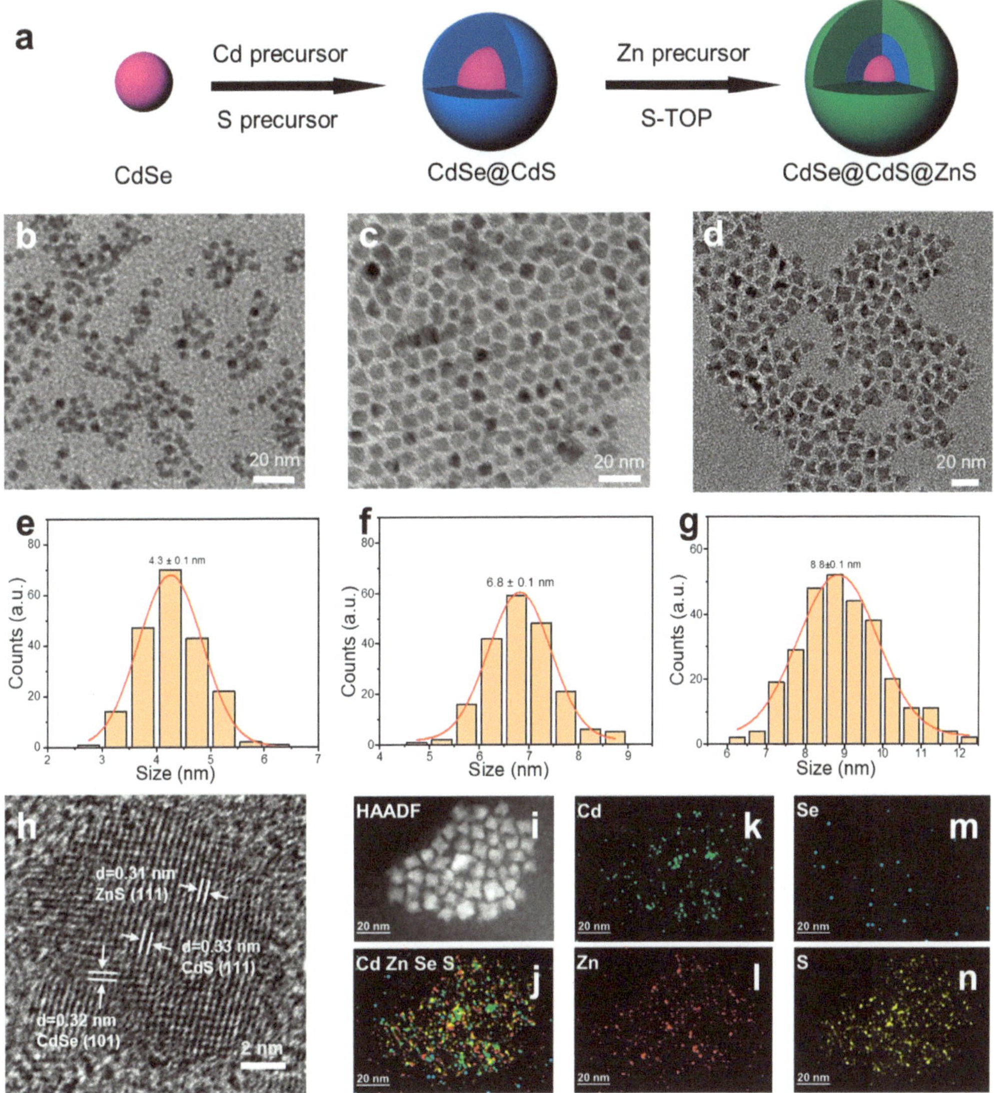

Figure 1. (**a**) The scheme of the synthetic processes of CdSe@CdS@ZnS. Representative TEM images of (**b**) CdSe QDs; (**c**) CdSe@CdS; and (**d**) CdSe@CdS@ZnS. Size distribution diagrams of (**e**) CdSe QDs; (**f**) CdSe@CdS; and (**g**) CdSe@CdS@ZnS. (**h**) HRTEM image. (**i–n**) HAADF-STEM-EDS elemental mapping images of CdSe@CdS@ZnS.

The crystal structures of the nanomaterials were characterized via X-ray diffraction (XRD). As shown in Figure 2, the synthesized CdSe QDs were w-CdSe (Joint Committee on Powder Diffraction Standards (JCPDS) no. 08-0459) [26]. When CdS was coated on CdSe to form a CdSe@CdS core@shell structure, only the diffraction peak of zb-CdS (JCPDS no. 10-0454) showed in the material, which might be due to the thickness of the CdS coating on the surface masking the diffraction peak of CdSe [27,28]. The diffraction peak of CdSe@CdS@ZnS had a red shift compared with CdSe@CdS, which was attributed to the heterogeneous epitaxial growth of zb-ZnS (JCPDS no. 05-0566) [29,30]. In summary, it could be deduced that the synthesized nanomaterials were composed of the three NCs.

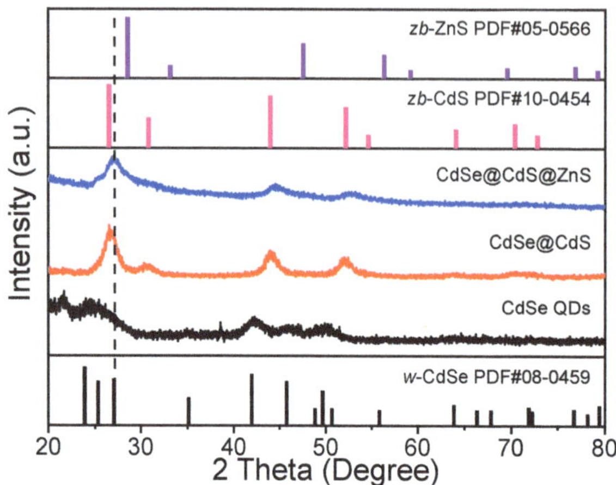

Figure 2. XRD patterns of CdSe QDs, CdSe@CdS and CdSe@CdS@ZnS.

2.2. Optical Properties of Materials

We used a UV-vis spectrophotometer and a fluorescence spectrometer to observe the changes in the optical properties of the materials. First, as shown in Figure 3a, multiple exciton peaks could be seen in the absorption spectrum of CdSe QDs, which were caused by the quantum size effects [25]. The first exciton absorption peak of CdSe QDs was located at 593 nm. When the CdS shell was grown on CdSe QDs, a red shift of the first exciton absorption peak was observed. The slight absorption spectrum of the CdSe@CdS core@shell structure at about 640 nm was attributed to the low-intensity absorption band of CdSe. However, this phenomenon was not observed in CdSe@CdS@ZnS, indicating a change in the charge-transfer mode between it. The photoluminescence (PL) spectra are shown in Figure 3b, which indicate that CdSe modified with CdS exhibited two fluorescence peaks: 509 nm for the outer layer of CdS and 641 nm for the core of CdSe. Due to the quantum size effects, the characteristic diffraction peaks were red-shifted compared to pure CdSe QDs [25]. After being modified with ZnS, the fluorescence spectrum of CdSe@CdS@ZnS still exhibited these two fluorescence peaks, but the fluorescence intensity was improved. This indicated that ZnS could effectively restrict the photogenerated electrons and holes within the interior of CdSe@CdS, thus greatly improving the luminescent properties of the materials [31].

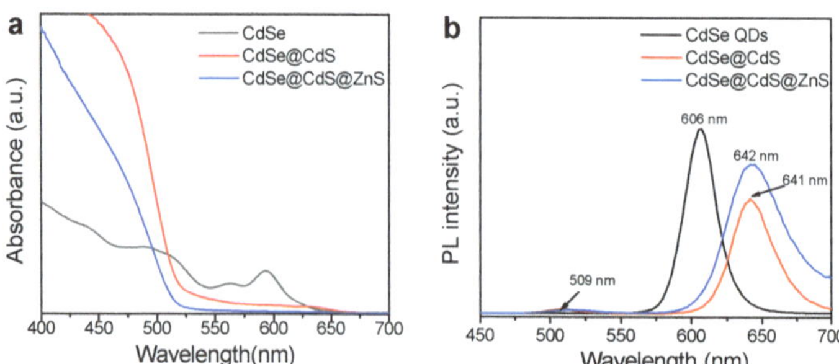

Figure 3. UV-vis absorbance spectra (**a**) and photoluminescence (PL) spectra (**b**) of CdSe QDs, CdSe@CdS and CdSe@CdS@ZnS.

In addition, the band gaps of CdSe, CdS and ZnS were calculated according to the UV-Vis absorption spectra (Figure 3a and Figure S1). Since CdSe, CdS and ZnS are all direct-gap semiconductors, the band gap can be determined via linear extrapolation from the absorption shoulder to $(Ah\nu)^{1/2}$ to hν. According to the Tauc diagram of Figure 4a–c, the band-gap energy (E_g) of CdSe, CdS and ZnS is 1.94 eV [32], 2.12 eV [33–35] and 3.62 eV [5], respectively. The band positions of CdSe, CdS and ZnS can be calculated using the following empirical formulae [36,37]:

$$E_{VB} = X - E_c + 1/2E_g,\qquad(1)$$

$$E_{CB} = E_{VB} - E_g.\qquad(2)$$

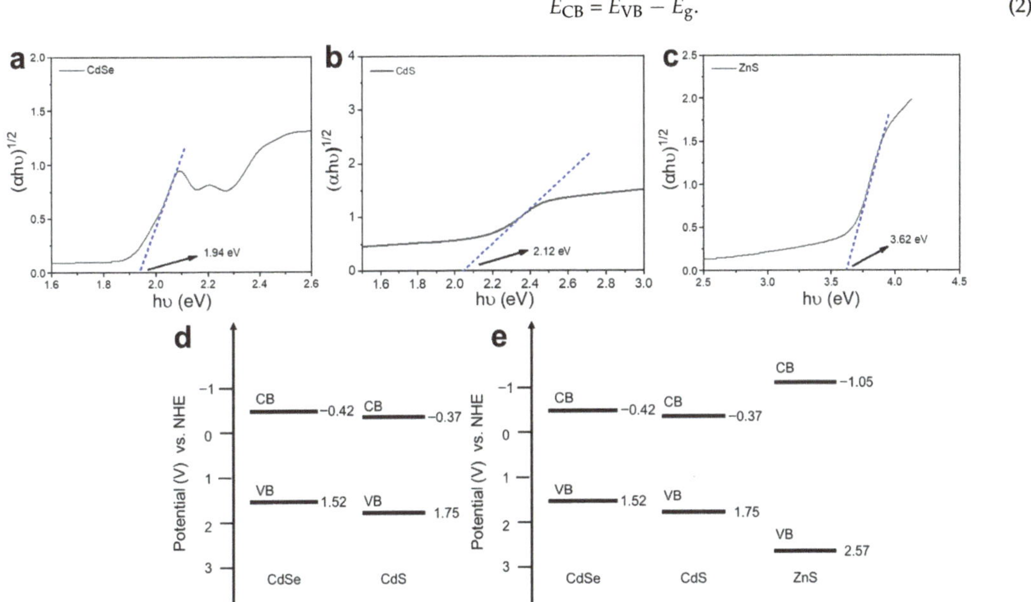

Figure 4. Tauc plots to estimate bandgaps of (**a**) CdSe QDs, (**b**) CdS NPs and (**c**) ZnS NPs. The band structure of (**d**) CdSe@CdS and (**e**) CdSe@CdS@ZnS.

In the empirical formulae, X is the geometric mean of the absolute electronegativity of each atom in the semiconductor [38], E_c is the energy of free electrons on the hydrogen scale (E_c = 4.5 eV), E_g is the band gap of the semiconductor, E_{VB} is the valence band potential, and E_{CB} is the conduction band potential [39]. The X values for CdSe, CdS and ZnS were calculated to be 5.05, 5.19 and 5.26, respectively. According to these parameters, the energy value of the valence band (VB) was calculated to be 1.52 eV, 1.75 eV and 2.57 eV, respectively, and the energy value of the conduction band (CB) was calculated to be -0.42 eV, -0.37 eV and -1.05 eV, respectively. The specific values of the band positions are shown in Table S1. Combined with the above results, the band structure of CdSe@CdS and CdSe@CdS@ZnS is shown in Figure 4d. It could be concluded that CdSe@CdS constituted a type-II band structure. After it was modified by ZnS, CdSe@CdS@ZnS constituted a type-I band structure.

In addition, we also studied the influences of different synthetic reaction times on the optical properties of CdSe@CdS. The analysis was performed by taking equal amounts of the samples (1 mL) from the three-necked flask at different reaction times and immediately cooled them to room temperature with hexane. As shown in Figure 5a,b, the obtained samples were tested using the UV-vis absorption spectra and fluorescence spectra. We found that, with an extension of reaction time, the UV-vis absorption peaks of CdSe gradually weakened until they disappeared, and the fluorescence peaks of the CdSe phase and the CdS phase showed a red shift, indicating that the thickness of the CdS shell had gradually increased [40].

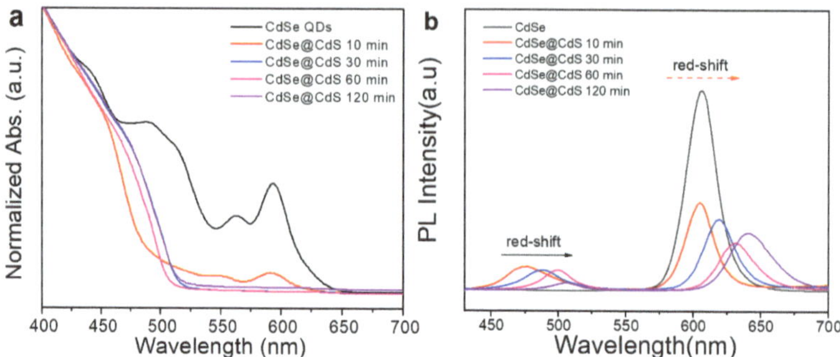

Figure 5. UV-vis absorbance spectra (**a**) and photoluminescence (PL) spectra (**b**) of CdSe@CdS at different synthetic reaction times.

At the same time, we also examined the influence of adjusting the amount of Zn on the optical properties of CdSe@CdS@ZnS while keeping the same molar ratio of Zn to S. Figure 6a,b show the absorption and fluorescence spectra of the multicomponent heterostructures. The results show that the fluorescence peaks of the CdSe phase and CdS phase were still maintained in the multicomponent heterojunctions, indicating that the fluorescence peaks of the materials were not significantly affected by changing the amount of ZnS. The limiting effects of ZnS on the photogenic charge in CdSe@CdS were further proved.

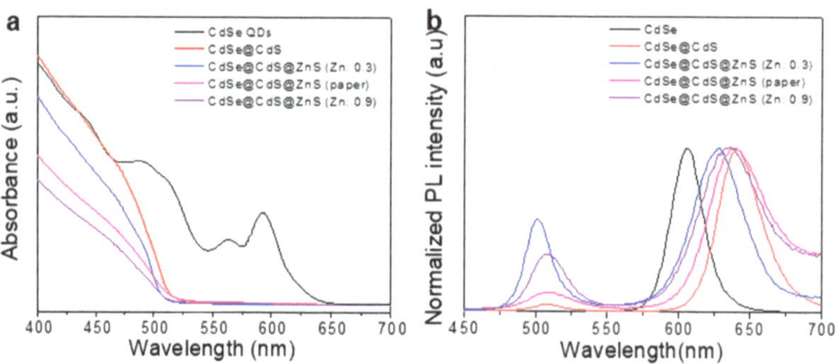

Figure 6. UV-vis absorbance spectra (**a**) and photoluminescence (PL) spectra (**b**) of CdSe@CdS@ZnS with different amounts of Zn.

2.3. Fluorescence Lifetime of Materials

To further verify the excellent optical properties of CdSe@CdS@ZnS, the fluorescence lifetime of the materials was characterized using a time-resolved transient fluorescence spectrometer. Figure 7 shows the time-resolved PL kinetic profiles of the CdSe QDs, CdSe@CdS and CdSe@CdS@ZnS. In Table S2, the average fluorescence lifetime of single CdSe QDs is 41.4 ns, while the fluorescence quantum efficiency is 15.89%. Because CdSe @ CdS constitutes a type-II band structure, the spatial separation of photogenerated charges leads to a decrease in quantum efficiency [41]. Therefore, after CdS coating to form a CdSe @ CdS core–shell structure, its fluorescence lifetime increased to 59.0 ns, but its fluorescence quantum efficiency decreased to 6.32%. After it was modified by ZnS, CdSe@CdS@ZnS constituted a type-I band structure, and the fluorescence lifetime reached 88.8 ns, which was about 2 times higher than that of CdSe QDs and higher than the previous report (Table S3). The quantum efficiency of CdSe@CdS@ZnS was increased to 17.05%. To further confirm that coating wide-bandgap semiconductors to form a type-I semiconductor heterojunction could enhance the luminescent quantum efficiency and extend the lifetime [42], we synthesized CdSe@CdS@ZnSe for comparison. As shown in Figure S2a,b, the particle size of CdSe@CdS@ZnSe was 17.2 ± 0.2 nm, and the diffraction peaks of zinc-blende ZnSe (zb-ZnSe, JCPDS no. 37-1463) existed in the material (Figure S2c). Moreover, the fluorescence peaks of the CdS phase and CdSe phase still appeared at corresponding positions in CdSe@CdS@ZnSe (Figure S2d). As shown in Figure 7 and Figure S2 and Table S4, the fluorescence quantum efficiency of CdSe@CdS@ZnSe is less than 1% and the luminescence lifetime is 11.5 ns. Therefore, ZnSe and CdS also constituted a type-II band structure (Figure S3). All photogenerated electrons transfer to the conduction band of CdS, while the holes transfer to the valence bands of CdSe and ZnSe.

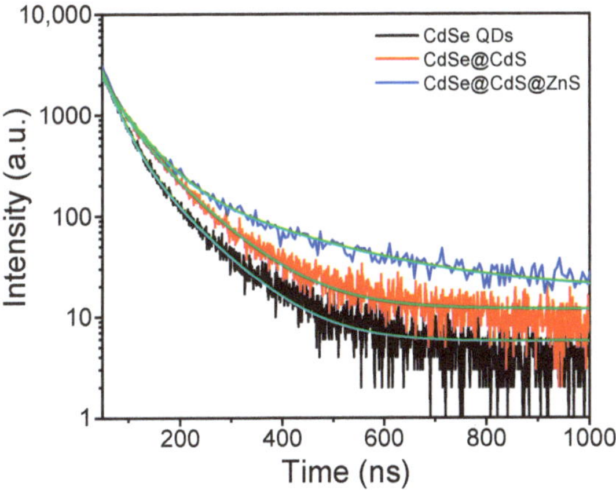

Figure 7. Time-resolved PL at the main peak for CdSe QDs, CdSe@CdS and CdSe@CdS@ZnS. Best-fit lines obtained using biexponential fitting are shown in green.

2.4. Charge-Transfer Mechanism between Heterojunctions

Based on the above results, we proposed the conversion of type II to type I in multicomponent heterojunctions, which could greatly improve the fluorescence quantum efficiency and fluorescence lifetime of the materials. As shown in Figure 8, photogenerated electrons were transferred from the conduction band of CdSe to that of CdS in CdSe@CdS, while photogenerated holes were transferred from the valence band of CdS to that of CdSe, forming a semiconductor heterojunction of type II. This was consistent with the results of the electron-and-hole separation states formed at the fluorescence peak of 641 nm [25]. After the modification of CdSe@CdS by wide-band-gap ZnS, the type-II heterojunction structure was transformed into type-I heterojunction structure, and photogenerated electrons and holes were confined inside the CdSe@CdS@ZnS, which eliminated the surface defect states, thus improving the fluorescence quantum efficiency and prolonging the fluorescence lifetime of the materials [40].

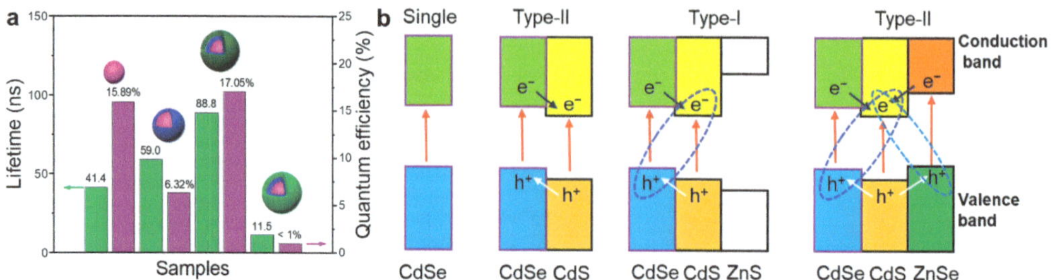

Figure 8. (**a**) The PL lifetime and quantum efficiency of CdSe QDs, CdSe@CdS, CdSe@CdS@ZnS and CdSe@CdS@ZnSe. (**b**) The energetic band positions of all the samples as comparisons, and photoinduced carrier transfer in the type-II and type-I band alignments.

3. Experimental Section

3.1. Materials

The materials included tetramethylammonium hydroxide pentahydrate (97%+, Adamas-beta), cadmium acetate dihydrate (Cd(Ac)$_2$·2H$_2$O, AR, Shanghai Aladdin Biochemical Technology Co., Ltd., Shanghai, China), zinc acetate dihydrate (Zn(Ac)$_2$·2H$_2$O, 99.99% metals basis, Shanghai Aladdin Biochemical Technology Co., Ltd., Shanghai, China), cadmium oxide (CdO, AR, Shanghai Aladdin Biochemical Technology Co., Ltd.), stearic acid (99%, Adamas-beta), octade cylamine (80%+, Adamas-beta), selenium (Se, ≥99.999% metals basis, Shanghai Aladdin Biochemical Technology Co., Ltd.), sublimed sulfur (S, AR, Shanghai Aladdin Biochemical Technology Co., Ltd.), trioctylphosphine (TOP, 90%, Shanghai Aladdin Biochemical Technology Co., Ltd.), trioctylphosphine oxide (TOPO, 90%, Aldrich), Tri-n-Butylphosphine (TBP, 98%+, Adamas-beta), oleic acid (OAc, AR, Shanghai Aladdin Biochemical Technology Co., Ltd.), oleylamine (OAm, C18: 80–90%, Shanghai Aladdin Biochemical Technology Co., Ltd.), 1-octadecene (ODE, 90%, Sigma-Aldrich, St. Louis, MO, USA), octanoic acid (Hoc, 99%, Adamas-beta), hexane (AR, General-reagent, Belmont, NC, USA), trichloromethane (CHCl$_3$, AR, Shanghai Hushi Chemical Co., Ltd., Shanghai, China), methyl alcohol (MeOH, 99.8%, Adamas-beta), and ethanol absolute (AR, General-reagent).

3.2. Methods

3.2.1. Synthesis of the Cadmium Oleate (Cd(Ol)$_2$)

First, 20 mmol (3.6246 g) of tetramethylammonium hydroxide pentahydrate and 6.4 mL of OAc were added to 50 mL of MeOH. The mixture was vigorously stirred to form homogeneous solution A. Then, homogeneous solution B was formed by adding 10 mmol (2.6653 g) of Cd(Ac)$_2$·2H$_2$O to 50 mL of MeOH and stirring vigorously. Then, solution B was slowly added to solution A under stirring and a milky white cadmium oleate precipitate was immediately produced. After all the mixture was added, the solution was strongly stirred for 20 min. Finally, the precipitate was centrifuged with methanol about 3 times and then dried in a vacuum oven at 40 °C. After drying, the product was stored in a vial.

3.2.2. Preparation of Selenium Precursor

Se powder at 10 mmol (0.7896 g) was dissolved in 2.36 g (2.91 mL) TBP to prepare 0.1 mol/L TBP-Se solution, and then it was diluted with 6.85 g (8.68 mL) ODE to obtain selenium precursor.

3.2.3. Preparation of Mixed Precursor

First, 0.3 mmol (0.2277 g) of Cd(Ol)$_2$ and 0.6 mmol (0.0192 g) of S powder were added to a vial (5 mL). Subsequently, a total of 3 mL of mixed precursor solution was prepared by adding 1.5 mmol (0.5 mL) of OAm, 1.5 mmol (0.24 mL) of HOc and 2.26 mL of ODE to the above vial. The mixed solution was ultrasonically dispersed at 50 °C and stored after complete dissolution.

3.2.4. Synthesis of CdSe Core Nanocrystals

In this process, 0.2 mmol (0.0256 g) of CdO and 0.8 mmol (0.2277 g) of stearic acid were added to a three-necked flask with 10 mL of ODE. The flask was then mounted on a heating magnetic stirrer. The mixture was degassed at room temperature for 30 min and then heated at 270 °C for 1 h under a nitrogen atmosphere until the liquid turned light yellow. It was cooled to room temperature, and 0.5 g of TOPO and 1.5 g of stearic amine were added to the above solution. Then, it was heated to 60 °C and stirred to dissolve the solid completely. It was degassed again until the solution had no bubbles. Then, it was pumped with N_2 and stirred for 10 minutes, before heating to 290 °C. At this temperature, 1 mL of the TBP-Se solution was injected rapidly. After the injection, the temperature was reduced to 250 °C for 5 min and then cooled to room temperature. The precipitate was centrifuged 1–2 times with 5 mL of hexane and 15 mL of ethanol and dispersed in 5 mL of chloroform for further characterization.

3.2.5. Synthesis of CdSe@CdS Core@Shell Structure

CdSe core nanocrystals in 500 µL of chloroform solution and 7.5 mL of ODE were placed in a three-necked flask. The flask was then mounted on a heating magnetic stirrer. The mixture was degassed at 120 °C for 30 min until the solution had no bubbles. Then, it was heated to 230 °C at a rate of 18 °C/min under a nitrogen atmosphere. At this temperature, 3 mL of mixed precursor started to be injected at a rate of 1.5 mL/h. At the same time, the temperature was increased to 250 °C at the same rate. When the reaction time was 10, 30, 60 and 120 min, a small sample was extracted from the mixture for testing. Finally, the solution was cooled to room temperature and centrifuged 1–2 times with 5 mL of hexane and 15 mL of ethanol. This was then dispersed in 5 mL of $CHCl_3$ for further characterizations.

3.2.6. Synthesis of CdSe@CdS@ZnS

An x mmol (x = 0.3, 0.6, 0.9 mmol) of sulfur powder and 2 mL of TOP were put into a 5 mL test flask for pre-ultrasonic dispersion and dissolution to obtain the precursor. Then y mmol $Zn(Ac)_2 \cdot 2H_2O$ (x:y = 1:2, y = 0.6, 1.2, 1.8 mmol) was added to a three-necked flask with 2 mL of OAc and 6 mL of ODE. The flask was mounted on a heating magnetic stirrer. The mixture was degassed at 120 °C for 30 min and then heated at 250 °C for 1 h under a nitrogen atmosphere to ensure complete dissolution of the solid. After that, the solution was cooled to 70 °C, and 1 mL of CdSe@CdS chloroform solution was quickly injected into the flask and degassed for 30 min, followed by heating at 250 °C under a nitrogen atmosphere. Then, the precursor was slowly injected into the above solution at a rate of 0.1 mL/min and reacted for another 40 min after the injection processes. The solution was cooled to room temperature and centrifuged 1–2 times with 5 mL of hexane and 15 mL of ethanol. Finally, this was dispersed in 5 mL of $CHCl_3$ for further characterizations.

3.2.7. Characterizations

The morphological structures of the samples were measured using scanning electron microscopy (FESEM, Hitachi S4800), transmission electron microscopy (TEM, Hitachi HT7820) and high-resolution transmission electron microscopy (HRTEM, Philips CM100). To determine the crystal phases, X-ray diffraction (XRD) patterns were obtained using a Rigaku Dmax-3C with Cu Kα irradiation (λ = 1.5406 Å). The UV-vis absorption spectra were acquired using a Shimadzu 1900i spectrophotometer. Steady-state fluorescence spectroscopic measurements were performed using a fluorescence spectrophotometer (Hitachi F-7000). Fluorescence quantum efficiency and fluorescence lifetime were measured using a time-resolved transient absorption spectrometer and a spectrometer with an integrating sphere (Edinburgh Instrument, EI FLS1000).

4. Conclusions

In this study, CdSe@CdS@ZnS nanocrystals with multi-component heterostructures were constructed utilizing the band structures between semiconductors to transition from a type-II to a type-I structure, resulting in improved luminescence quantum efficiency and increased fluorescence lifetime. Compared with CdSe quantum dots, the luminescence quantum efficiency of the CdSe@CdS core@shell structure of the type-II heterojunction decreased by 60%, while the fluorescence lifetime increased by 42.5%. The luminescence quantum efficiency of the type-I heterojunction CdSe@CdS@ZnS was 17.05%, and the fluorescence lifetime was 88.8 ns. Compared with nanocrystalline CdSe QDs, its fluorescence lifetime was increased by about 2 times. This study provided a new method and concept for future research on the construction and charge transfer of multicomponent semiconductor nanocrystals for enhanced PL properties.

Supplementary Materials: The following supporting information can be downloaded at https://www.mdpi.com/article/10.3390/ma16217007/s1. Figure S1: UV-vis absorbance spectra of (a) CdS and (b) ZnS; Figure S2: (a) Representative TEM images CdSe@CdS@ZnSe; (b) Size distribution diagrams of CdSe@CdS@ZnSe; (c) XRD patterns of CdSe QDs, CdSe@CdS and CdSe@CdS@ZnSe; (d) UV-vis absorbance spectra and photoluminescence (PL) spectra of CdSe@CdS@ZnSe; Figure S3: Time-resolved PL at the main peak for CdSe@CdS@ZnSe. Best fit in green line using biexponential; Figure S4: Representative TEM images of (a) CdS; (b) ZnS; Size distribution diagrams of (c) CdS; (d) ZnS; Table S1: X, E_g, E_{CB} and E_{VB} of CdSe, CdS and ZnS; Table S2: Time-resolved PL kinetics of CdSe QDs, CdSe@CdS and CdSe@CdS@ZnS upon the excitation of 400 nm. Fitting procedures are described by biexponential function analysis; Table S3: Comparison of the reported photoluminescence lifetimes of different core-shell materials; Table S4: Time-resolved PL kinetics of CdSe@CdS@ZnSe upon the excitation of 400 nm. Fitting procedures are described by biexponential function analysis [43–47].

Author Contributions: Conceptualization, Z.L.; Validation, J.Z.; Formal analysis, Y.Z.; Investigation, Y.W., Y.Z. and J.Z.; Data curation, Y.W. and J.Z.; Writing—original draft, Y.W. and Z.L.; Writing—review & editing, J.Z. and Z.L.; Supervision, Z.L.; Project administration, Z.L.; Funding acquisition, Z.L. All authors have read and agreed to the published version of the manuscript.

Funding: This research was funded by the National Natural Science Foundation of China (22109097) and the Natural Science Foundation of Shanghai (20ZR1472000).

Institutional Review Board Statement: Not applicable.

Informed Consent Statement: Not applicable.

Data Availability Statement: The data presented in this study are available from the corresponding author upon request.

Acknowledgments: The authors are grateful for the financial support received for the project.

Conflicts of Interest: The authors declare no conflict of interest.

References

1. Zhang, J.; Li, J.; Ye, Z.; Cui, J.; Peng, X. Hot-Electron-Induced Photochemical Properties of CdSe/ZnSe Core/Shell Quantum Dots under an Ambient Environment. *J. Am. Chem. Soc.* **2023**, *145*, 13938–13949. [CrossRef] [PubMed]
2. Lian, Z.; Wu, F.; Zhong, Y.; Zi, J.; Li, Z.; Wang, X.; Nakagawa, T.; Li, H.; Sakamoto, M. Tuning Plasmonic p–n Junction for Efficient Infrared–Light–Responsive Hydrogen Evolution. *Appl. Catal. B Environ.* **2022**, *318*, 121860. [CrossRef]
3. Lian, Z.; Kobayashi, Y.; Vequizo, J.J.M.; Ranasinghe, C.S.K.; Yamakata, A.; Nagai, T.; Kimoto, K.; Kobayashi, K.; Tanaka, K.; Teranishi, T.; et al. Harnessing Infrared Solar Energy with Plasmonic Energy Upconversion. *Nat. Sustain.* **2022**, *5*, 1092–1099. [CrossRef]
4. Zhong, Y.; Zi, J.; Wu, F.; Li, Z.; Luan, X.; Gao, F.; Lian, Z. Defect–Mediated Electron Transfer in Pt–CuInS$_2$/CdS Heterostructured Nanocrystals for Enhanced Photocatalytic H$_2$ Evolution. *ACS Appl. Nano Mater.* **2022**, *5*, 7704–7713. [CrossRef]
5. Zhong, Y.; Li, M.; Luan, X.; Gao, F.; Wu, H.; Zi, J.; Lian, Z. Ultrathin ZnIn$_2$S$_4$/ZnSe Heteronanosheets with Modulated S–Scheme Enable High Efficiency of Visible–Light–Responsive Photocatalytic Hydrogen Evolution. *Appl. Catal. B Environ.* **2023**, *335*, 122859. [CrossRef]

6. Yang, Y.; Xie, Y.; Zhang, F. Second near–Infrared Window Fluorescence Nanoprobes for Deep–Tissue in Vivo Multiplexed Bioimaging. *Adv. Drug Deliv. Rev.* **2023**, *193*, 114697. [CrossRef]

7. Whitworth, G.L.; Dalmases, M.; Taghipour, N.; Konstantatos, G. Solution–Processed Pbs Quantum Dot Infrared Laser with Room–Temperature Tuneable Emission in the Optical Telecommunications Window. *Nat. Photonics* **2021**, *15*, 738–742. [CrossRef] [PubMed]

8. Sfaelou, S.; Kontos, A.G.; Givalou, L.; Falaras, P.; Lianos, P. Study of the Stability of Quantum Dot Sensitized Solar Cells. *Catal. Today* **2014**, *230*, 221–226. [CrossRef]

9. Sfaelou, S.; Kontos, A.G.; Falaras, P.; Lianos, P. Micro-Raman, Photoluminescence and Photocurrent Studies on the Photostability of Quantum Dot Sensitized Photoanodes. *J. Photoch. Photobio. A* **2014**, *275*, 127–133. [CrossRef]

10. Piao, Z.Y.; Yang, D.; Cui, Z.J.; He, H.Y.; Mei, S.L.; Lu, H.X.; Fu, Z.Z.; Wang, L.; Zhang, W.L.; Guo, R.Q. Recent Advances in Metal Chalcogenide Quantum Dots: From Material Design to Biomedical Applications. *Adv. Funct. Mater.* **2022**, *32*, 2207662. [CrossRef]

11. Lu, H.; Li, W.; Dong, H.; Wei, M. Graphene Quantum Dots for Optical Bioimaging. *Small* **2019**, *15*, e1902136. [CrossRef] [PubMed]

12. Ye, S.; Yan, W.; Zhao, M.; Peng, X.; Song, J.; Qu, J. Low–Saturation–Intensity, High–Photostability, and High–Resolution STED Nanoscopy Assisted by CsPbBr$_3$ Quantum Dots. *Adv. Mater.* **2018**, *30*, e1800167. [CrossRef] [PubMed]

13. Panfil, Y.E.; Oded, M.; Banin, U. Colloidal Quantum Nanostructures: Emerging Materials for Display Applications. *Angew. Chem. Int. Ed.* **2018**, *57*, 4274–4295. [CrossRef] [PubMed]

14. Shen, Z.; Zhao, S.; Song, D.; Xu, Z.; Qiao, B.; Song, P.; Bai, Q.; Cao, J.; Zhang, G.; Swelm, W. Improving the Quality and Luminescence Performance of All–Inorganic Perovskite Nanomaterials for Light–Emitting Devices by Surface Engineering. *Small* **2020**, *16*, e1907089. [CrossRef] [PubMed]

15. Wang, H.; Guo, Y.; Hao, H.; Bian, H.; Aubin, H.; Wei, Y.; Li, H.; Liu, T.; Degiron, A.; Wang, H. Bright CdSe/CdS Quantum Dot Light–Emitting Diodes with Modulated Carrier Dynamics Via the Local Kirchhoff Law. *ACS Appl. Mater. Interfaces* **2021**, *13*, 56476–56484. [CrossRef] [PubMed]

16. Sobhanan, J.; Rival, J.V.; Anas, A.; Sidharth Shibu, E.; Takano, Y.; Biju, V. Luminescent Quantum Dots: Synthesis, Optical Properties, Bioimaging and Toxicity. *Adv. Drug. Deliv. Rev.* **2023**, *197*, 114830. [CrossRef]

17. Liu, L.; Li, H.; Liu, Z.; Xie, Y.H. The Conversion of CuInS$_2$/ZnS Core/Shell Structure from Type I to Quasi–Type II and the Shell Thickness–Dependent Solar Cell Performance. *J. Colloid Interface Sci.* **2019**, *546*, 276–284. [CrossRef]

18. Yan, C.; Wen, J.; Lin, P.; Sun, Z. A Tunneling Dielectric Layer Free Floating Gate Nonvolatile Memory Employing Type–I Core–Shell Quantum Dots as Discrete Charge–Trapping/Tunneling Centers. *Small* **2022**, *15*, e1804156. [CrossRef] [PubMed]

19. Zhu, H.; Song, N.; Lian, T. Controlling Charge Separation and Recombination Rates in CdSe/ZnS Type I Core−Shell Quantum Dots by Shell Thicknesses. *J. Am. Chem. Soc.* **2010**, *132*, 15038–15045. [CrossRef]

20. Toufanian, R.; Chern, M.; Kong, V.H.; Dennis, A.M. Engineering Brightness-Matched Indium Phosphide Quantum Dots. *Chem. Mater.* **2021**, *33*, 1964–1975. [CrossRef]

21. Panda, S.K.; Hickey, S.G.; Waurisch, C.; Eychmüller, A. Gradated alloyed CdZnSe nanocrystals with high luminescence quantum yields and stability for optoelectronic and biological applications. *J. Mater. Chem.* **2011**, *21*, 11550–11555. [CrossRef]

22. Altintas, Y.; Quliyeva, U.; Gungor, K.; Erdem, O.; Kelestemur, Y.; Mutlugun, E.; Kovalenko, M.V.; Demir, H.V. Highly Stable, near–Unity Efficiency Atomically Flat Semiconductor Nanocrystals of CdSe/ZnS Hetero–Nanoplatelets Enabled by ZnS–Shell Hot–Injection Growth. *Small* **2019**, *15*, e1804854. [CrossRef] [PubMed]

23. Niu, J.Z.; Shen, H.; Zhou, C.; Xu, W.; Li, X.; Wang, H.; Lou, S.; Du, Z.; Li, L.S. Controlled synthesis of high quality type-II/type-I CdS/ZnSe/ZnS core/shell1/shell2 nanocrystals. *Dalton Trans.* **2010**, *39*, 3308–3314. [CrossRef] [PubMed]

24. Blackman, B.; Battaglia, D.; Peng, X. Bright and Water-Soluble Near IR-Emitting CdSe/CdTe/ZnSe Type-II/Type-I Nanocrystals, Tuning the Efficiency and Stability by Growth. *Chem. Mater.* **2008**, *20*, 4847–4853. [CrossRef]

25. Wang, Y.; Pu, C.; Lei, H.; Qin, H.; Peng, X. CdSe@CdS Dot@Platelet Nanocrystals: Controlled Epitaxy, Monoexponential Decay of Two–Dimensional Exciton, and Nonblinking Photoluminescence of Single Nanocrystal. *J. Am. Chem. Soc.* **2019**, *141*, 17617–17628. [CrossRef]

26. Park, Y.S.; Lim, J.; Klimov, V.I. Asymmetrically Strained Quantum Dots with Non–Fluctuating Single–Dot Emission Spectra and Subthermal Room–Temperature Linewidths. *Nat. Mater.* **2019**, *18*, 249–255. [CrossRef]

27. Nan, W.; Niu, Y.; Qin, H.; Cui, F.; Yang, Y.; Lai, R.; Lin, W.; Peng, X. Crystal Structure Control of Zinc-Blende CdSe/CdS Core/Shell Nanocrystals: Synthesis and Structure–Dependent Optical Properties. *J. Am. Chem. Soc.* **2012**, *134*, 19685–19693. [CrossRef]

28. Peng, X.; Schlamp, M.C.; Kadavanich, A.V.; Alivisatos, A.P. Epitaxial Growth of Highly Luminescent CdSe/CdS Core/Shell Nanocrystals with Photostability and Electronic Accessibility. *J. Am. Chem. Soc.* **1997**, *119*, 7019–7029. [CrossRef]

29. Soni, U.; Pal, A.; Singh, S.; Mittal, M.; Yadav, S.; Elangovan, R.; Sapra, S. Simultaneous Type–I/Type–II Emission from CdSe/CdS/ZnSe Nano-Heterostructures. *ACS Nano* **2014**, *8*, 113–123. [CrossRef]

30. Li, J.; Zheng, H.; Zheng, Z.; Rong, H.; Zeng, Z.; Zeng, H. Synthesis of CdSe and CdSe/ZnS Quantum Dots with Tunable Crystal Structure and Photoluminescent Properties. *Nanomaterials* **2022**, *12*, 2969. [CrossRef]

31. Fernández-Delgado, N.; Herrera, M.; Tavabi, A.H.; Luysberg, M.; Dunin-Borkowski, R.E.; Rodriguez-Cantó, P.J.; Abargues, R.; Martínez-Pastor, J.P.; Molina, S.I. Structural and Chemical Characterization of CdSe–ZnS Core-Shell Quantum Dots. *Appl. Surf. Sci.* **2018**, *457*, 93–97. [CrossRef]

32. Fan, Z.; Liu, Y.; Guo, X.; Jin, Z. Construct Organic/Inorganic Heterojunction Photocatalyst of Benzene-Ring-Grafted g-C$_3$N$_4$/CdSe for Photocatalytic H$_2$ Evolution. *Int. J. Hydrogen Energy* **2023**, *48*, 19137–19152. [CrossRef]

33. Dang, Y.; Luo, L.; Wang, W.; Hu, W.; Wen, X.; Lin, K.; Ma, B. Improving the Photocatalytic H_2 Evolution of CdS by Adjusting the (002) Crystal Facet. *J. Phys. Chem. C* **2022**, *126*, 1346–1355. [CrossRef]
34. Chi, L.-X.; Clarisse, N.D.; Yang, S.; Bao, W.-L.; Xiao, M.; Zhao, Y.-P.; Wang, J.-T.; Chen, Q.; Zhang, Z.-H. Boosting Photocatalytic Hydrogen Production Based on Amino Acid Derived Zn-MOF/CdS Composite Photocatalysts. *J. Solid State Chem.* **2023**, *324*, 124117. [CrossRef]
35. Zhu, C.-Y.; Shen, M.-T.; Qi, M.-J.; Zhao, Y.-Y.; Xu, Z.; Li, P.; Ru, J.; Gao, W.; Zhang, X.-M. Constructed CdS/Mn-MOF Heterostructure for Promoting Photocatalytic Degradation of Rhodamine B. *Dyes Pigments* **2023**, *219*, 111607. [CrossRef]
36. Zheng, X.; Han, H.; Ye, X.; Meng, S.; Zhao, S.; Wang, X.; Chen, S. Fabrication of Z-Scheme WO_3/$KNbO_3$ Photocatalyst with Enhanced Separation of Charge Carriers. *Chem. Res. Chinese U* **2020**, *36*, 901–907. [CrossRef]
37. Liu, Y.; Dai, F.; Zhao, R.; Huai, X.; Han, J.; Wang, L. Aqueous Synthesis of Core/Shell/Shell CdSe/CdS/ZnS Quantum Dots for Photocatalytic Hydrogen Generation. *J. Mater. Sci.* **2019**, *54*, 8571–8580. [CrossRef]
38. Lin, X.; Xing, J.; Wang, W.; Shan, Z.; Xu, F.; Huang, F. Photocatalytic Activities of Heterojunction Semiconductors Bi_2O_3/$BaTiO_3$: A Strategy for the Design of Efficient Combined Photocatalysts. *J. Phys. Chem. C* **2007**, *111*, 18288–18293. [CrossRef]
39. Nguyen, A.T.; Lin, W.-H.; Lu, Y.-H.; Chiou, Y.-D.; Hsu, Y.-J. First Demonstration of Rainbow Photocatalysts Using Ternary Cd1-xZnxSe Nanorods of Varying Compositions. *Appl. Catal. A Gen.* **2014**, *476*, 140–147. [CrossRef]
40. Xie, R.; Kolb, U.; Li, J.; Basche, T.; Mews, A. Synthesis and Characterization of Highly Luminescent CdSe–Core CdS/$Zn_{0.5}Cd_{0.5}$S/ZnS Multishell Nanocrystals. *J. Am. Chem. Soc.* **2005**, *127*, 7480–7488. [CrossRef] [PubMed]
41. Jain, S.; Bharti, S.; Bhullar, G.K.; Tripathi, S.K. I-III-VI Core/Shell QDs: Synthesis, Characterizations and Applications. *J. Lumin.* **2020**, *219*, 116912. [CrossRef]
42. Vasudevan, D.; Gaddam, R.R.; Trinchi, A.; Cole, I. Core–Shell Quantum Dots: Properties and Applications. *J. Alloys Compd.* **2015**, *636*, 395–404. [CrossRef]
43. Zi, J.; Zhong, Y.; Li, Z.; Wu, F.; Yang, W.; Lian, Z. Type-I CdSe/ZnS Heteronanoplatelets Exhibit Enhanced Photocatalytic Hydrogen Evolution by Interfacial Trap-Mediated Hole Transfer. *J. Phys. Chem. C* **2021**, *125*, 23945–23951. [CrossRef]
44. Wang, X.; Qu, L.; Zhang, J.; Peng, X.; Xiao, M. Surface-Related Emission in Highly Luminescent CdSe Quantum Dots. *Nano Lett.* **2003**, *3*, 1103–1106. [CrossRef]
45. Drijvers, E.; De Roo, J.; Geiregat, P.; Fehér, K.; Hens, Z.; Aubert, T. Revisited Wurtzite CdSe Synthesis: A Gateway for the Versatile Flash Synthesis of Multishell Quantum Dots and Rods. *Chem. Mater.* **2016**, *28*, 7311–7323. [CrossRef]
46. Tao, C.-L.; Ma, J.; Wei, C.; Xu, D.; Xie, Z.; Jiang, Z.; Ge, F.; Zhang, H.; Xie, M.; Ye, Z.; et al. Scalable Synthesis of High-Quality Core/Shell Quantum Dots with Suppressed Blinking. *Adv. Opt. Mater.* **2023**, *11*, 2300533. [CrossRef]
47. Wang, K.; Tao, Y.; Tang, Z.; Benetti, D.; Vidal, F.; Zhao, H.; Rosei, F.; Sun, X. Heterostructured core/gradient multi-shell quantum dots for high-performance and durable photoelectrochemical hydrogen generation. *Nano Energy* **2022**, *100*, 107524. [CrossRef]

Disclaimer/Publisher's Note: The statements, opinions and data contained in all publications are solely those of the individual author(s) and contributor(s) and not of MDPI and/or the editor(s). MDPI and/or the editor(s) disclaim responsibility for any injury to people or property resulting from any ideas, methods, instructions or products referred to in the content.

Article

CH$_3$NH$_3$PbI$_3$/Au/Mg$_{0.2}$Zn$_{0.8}$O Heterojunction Self-Powered Photodetectors with Suppressed Dark Current and Enhanced Detectivity

Meijiao Wang [1,2]**, Man Zhao** [1,]*** and Dayong Jiang** [1,3,]***

[1] School of Materials Science and Engineering, Changchun University of Science and Technology, Changchun 130022, China; 2019200119@mails.cust.edu.cn
[2] School of Optoelectronic Engineering, Changchun College of Electronic Technology, Changchun 130022, China
[3] Engineering Research Center of Optoelectronic Functional Materials, Ministry of Education, Changchun 130022, China
* Correspondence: 2017800059@cust.edu.cn (M.Z.); 2006800026@cust.edu.cn (D.J.)

Abstract: Interface engineering of the hole transport layer in CH$_3$NH$_3$PbI$_3$ photodetectors has resulted in significantly increased carrier accumulation and dark current as well as energy band mismatch, thus achieving the goal of high-power conversion efficiency. However, the reported heterojunction perovskite photodetectors exhibit high dark currents and low responsivities. Herein, heterojunction self-powered photodetectors, composed of p-type CH$_3$NH$_3$PbI$_3$ and n-type Mg$_{0.2}$Zn$_{0.8}$O, are prepared through the spin coating and magnetron sputtering. The obtained heterojunctions exhibit a high responsivity of 0.58 A/W, and the EQE of the CH$_3$NH$_3$PbI$_3$/Au/Mg$_{0.2}$Zn$_{0.8}$O heterojunction self-powered photodetectors is 10.23 times that of the CH$_3$NH$_3$PbI$_3$/Au photodetectors and 84.51 times that of the Mg$_{0.2}$ZnO$_{0.8}$/Au photodetectors. The built-in electric field of the p-n heterojunction significantly suppresses the dark current and improves the responsivity. Remarkably, in the self-supply voltage detection mode, the heterojunction achieves a high responsivity of up to 1.1 mA/W. The dark current of the CH$_3$NH$_3$PbI$_3$/Au/Mg$_{0.2}$Zn$_{0.8}$O heterojunction self-powered photodetectors is less than 1.4×10^{-1} pA at 0 V, which is more than 10 times lower than that of the CH$_3$NH$_3$PbI$_3$ photodetectors. The best value of the detectivity is as high as 4.7×10^{12} Jones. Furthermore, the heterojunction self-powered photodetectors exhibit a uniform photodetection response over a wide spectral range from 200 to 850 nm. This work provides guidance for achieving a low dark current and high detectivity for perovskite photodetectors.

Keywords: CH$_3$NH$_3$PbI$_3$ photodetectors; heterojunction; self-powered photodetectors; Mg$_{0.2}$Zn$_{0.8}$O

Citation: Wang, M.; Zhao, M.; Jiang, D. CH$_3$NH$_3$PbI$_3$/Au/Mg$_{0.2}$Zn$_{0.8}$O Heterojunction Self-Powered Photodetectors with Suppressed Dark Current and Enhanced Detectivity. *Materials* **2023**, *16*, 4330. https://doi.org/10.3390/ma16124330

Academic Editor: Matteo Cantoni

Received: 7 May 2023
Revised: 30 May 2023
Accepted: 7 June 2023
Published: 12 June 2023

Copyright: © 2023 by the authors. Licensee MDPI, Basel, Switzerland. This article is an open access article distributed under the terms and conditions of the Creative Commons Attribution (CC BY) license (https://creativecommons.org/licenses/by/4.0/).

1. Introduction

With the rapid development of two-dimensional (2D) materials, such as graphene, transition metal dichlorides (TMDs), and black phosphorus, 2D perovskites have emerged, which combine the excellent properties of 2D materials and perovskites, namely the good solution processability, molecular-scale self-assembly, and film formation of the former alongside the direct tunable band gap, high carrier mobility, and high absorption coefficient of the latter [1–7]. Among these, CH$_3$NH$_3$PbI$_3$ possesses excellent absorption and has been applied to photodetectors (PDs); in particular, self-powered PDs (SPPDs) have drawn considerable research interest [8]. SPPDs can detect light without the need for any power supply; furthermore, they can be miniaturized and integrated into nanodevices with remote wireless control. However, due to the complex production process, existing perovskite SPPDs show a high dark current and a low detection rate, which greatly limits their wide application. For example, a common approach to realizing perovskite SPPDs is to construct p-i-n with vertical multilayer structures, which require complex and rigorous processes [9–12]. They are important candidates in the field of renewable photovoltaics and photoelectric devices [13–16]. Since $MAPbI_3/TiO_2$ PDs were reported in 2014, various PDs have been successfully fabricated using $MAPbI_3$. However, it has been shown that perovskite SPPDs exhibit high dark currents and low responsivities [17]. Furthermore, perovskite heterojunction PDs are being quickly developed as well [18–21]. The built-in electric field formed at the junction interface can act as a driving force to separate the photogenerated electron-hole (e-h) pairs in the depletion region, drive the transport of the separated photogenerated carriers, and then generate the photocurrent in the PDs without an external power supply. Herein, heterojunction SPPDs are designed.

Due to their advantages of lightweight and easy processing, perovskites have attracted more and more attention, and they have shown great application potential in various fields in recent years. In particular, wideband organic photoelectric detection (OPD) has been successfully applied in many important fields, such as astronomical exploration, remote sensing, and infrared imaging. The combination of MgZnO and the advantages of organic polymers can improve the performance of PDs, so organic PDs show fascinating characteristics. Perovskite heterojunction PDs performance can be improved by using metal oxide dense films, such as MgZnO or ZnO, which are usually used as an electron-transport layer to transport electrons and holes [22–25]. MgZnO and ZnO with wide band gaps exhibit a large ultraviolet (UV)-light absorption coefficient and high carrier mobility. It is well known that intrinsic point defects, such as oxygen vacancies and metal interstitials, have an important impact on the electronic properties of metal oxides. It has been widely accepted that the presence of oxygen vacancies in $Mg_{0.2}Zn_{0.8}O$ can increase the charge density of $Mg_{0.2}Zn_{0.8}O$ to form a native n-type semiconductor [26–29]. Thus, it is urgently required to fabricate PDs with a simple alternative method that can also endow them with self-powered functionality. The most common method for fabricating SPPDs relies on the E_b created by heterojunctions. To date, $Mg_{0.2}Zn_{0.8}O$ has been rarely reported for use as an electron-transport layer in heterojunction SPPDs. $Mg_{0.2}Zn_{0.8}O$ has a wide band gap and acts as the n-type layer. $CH_3NH_3PbI_3/Au/Mg_{0.2}Zn_{0.8}O$ heterojunction SPPDs with excellent comprehensive properties have been successfully prepared. Therefore, it is very important to develop $CH_3NH_3PbI_3/Au/Mg_{0.2}Zn_{0.8}O$ heterojunction SPPDs with a low dark current and high detectivity to improve their performance.

Herein, high-quality $CH_3NH_3PbI_3$ perovskites were prepared as hole-transport layers through the one-step spin coating method and were then used in $Mg_{0.2}Zn_{0.8}O$ PDs prepared by magnetron sputtering. These devices possess light responsiveness in a broad range, from UV to near-infrared (NIR), high responsivity, and low dark current. The $CH_3NH_3PbI_3/Au/Mg_{0.2}Zn_{0.8}O$ heterojunction SPPDs exhibit an excellent responsivity of up to 0.58 A/W. In the self-supply voltage detection mode, it achieves a high responsivity of up to 1.1 mA/W, and the dark current is less than 1.4×10^{-1} pA at 0 V. The best value of the detectivity is as high as 4.7×10^{12} Jones. This work presents a new route for designing SPPDs with a low dark current and high detectivity.

2. Experimental Section

2.1. Materials

All the reagents used in the experiments were of analytic grade and used without further purification. The conjugated polymers PbI_2 and CH_3NH_3I were purchased from Xi'an Polymer Light Technology Corp., Xi'an, China.

2.2. PDs Preparation Process

2.2.1. Preparation of the $CH_3NH_3PbI_3/Au$ PDs

Firstly, Au thin films were prepared on a 2×3 cm^2 polyethylene terephthalate (PET) using radio-frequency (RF) magnetron sputtering (JZ-RF600A). Subsequently, the metal semiconductor metal (MSM) structure was realized through the following steps: gluing, exposure, development, etching, and resist removal. The electrode width and spacing were both 5 μm, and the electrode length was 500 μm. We masked the PET film with 5 μm-wide channels with polyimide tape. In the next step, using a vacuum spin coater (VTC-200), the $CH_3NH_3PbI_3$ layer was deposited on the PET substrate. Briefly, 10 mg of CH_3NH_3I and 460 mg of PbI_2 were dissolved in 1 mL of N, N-dimethylformamide (DMF) and stirred at 70 °C for 2 h to obtain a homogeneous $CH_3NH_3PbI_3$ precursor solution. Then, the above precursor solution was spin coated on the PET substrate at an initial speed of 500 r/min for 10 s and then at a speed of 3000 r/min for 50 s. Subsequently, PET with a spin coating was heated in a vacuum oven at 70 °C for 3 min to remove the residual DMF. After cooling, the polyimide tape was removed, and the $CH_3NH_3PbI_3$ layer was obtained upon annealing at 90 °C for 10 min. Finally, the $CH_3NH_3PbI_3$/Au PDs were obtained.

2.2.2. Preparation of the $Au/Mg_{0.2}Zn_{0.8}O$ PDs

$Mg_{0.2}Zn_{0.8}O$ thin films were prepared on a 2×3 cm^2 PET using RF magnetron sputtering (JZ-RF600A). The vacuum chamber was initially evacuated to 5×10^{-4} Pa, and O_2 and Ar were introduced into the chamber in a flow ratio of 10:40. PET was washed with acetone, absolute ethanol, and deionized water for 10 min. The $Mg_{0.2}Zn_{0.8}O$ films were sputtered on the PET substrates at a total pressure of 4 Pa and an RF power of 150 W for 30 min to obtain the $Mg_{0.2}Zn_{0.8}O$ thin films, as shown in Figure 1a,b.

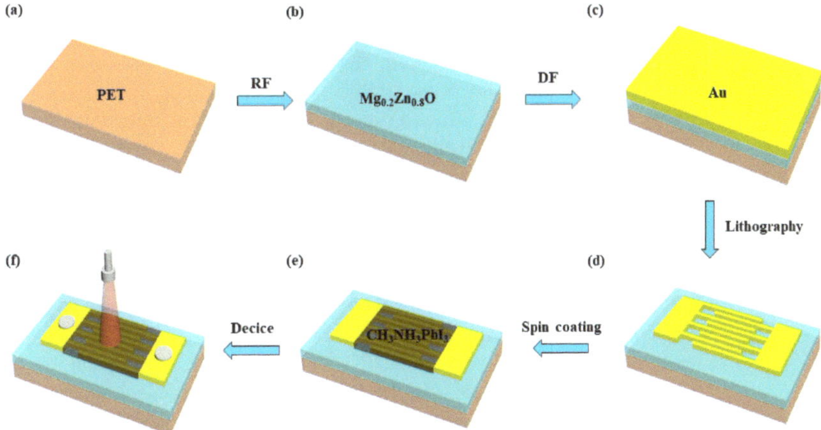

Figure 1. Preparation process of the $CH_3NH_3PbI_3$/Au/$Mg_{0.2}Zn_{0.8}O$ flexible PDs. (**a**) PET substrate. (**b**) Preparation of $Mg_{0.2}Zn_{0.8}O$/Au thin films. (**c**) Preparation of Au thin films. (**d**) Photolithography. (**e**) Preparation of $CH_3NH_3PbI_3$ thin films. (**f**) Preparation of $CH_3NH_3PbI_3$/Au/$Mg_{0.2}Zn_{0.8}O$ SPPDs.

Next, Au was RF sputtered on the $Mg_{0.2}Zn_{0.8}O$ thin films to realize the MSM structure, as shown in Figure 1c,d; the process included gluing, exposure, development, etching, and resist removal. The electrode width and spacing were both 5 μm, and the electrode length was 500 μm. The obtained $Au/Mg_{0.2}Zn_{0.8}O$ UV PDs are shown in Figure 1d.

2.2.3. Preparation of the $CH_3NH_3PbI_3$/Au/$Mg_{0.2}Zn_{0.8}O$ Heterojunction SPPDs

Firstly, we masked the $Mg_{0.2}Zn_{0.8}O$ film with 5 μm-wide channels with the polyimide tape. In the second step, using a vacuum spin coater (VTC-200), the $CH_3NH_3PbI_3$ layer was deposited on the $Mg_{0.2}Zn_{0.8}O$ thin film. 10 mg of CH_3NH_3I and 460 mg of PbI_2 were dissolved in 1 mL of DMF and stirred at 70 °C for 2 h to obtain a homogeneous $CH_3NH_3PbI_3$ precursor solution. Then, the above precursor solution was spin coated on the $Mg_{0.2}Zn_{0.8}O$ thin film at an initial speed of 500 r/min for 10 s and then at a speed of 3000 r/min for 50 s. Then, a $Mg_{0.2}Zn_{0.8}O$ thin film with spin coating was heated in a vacuum oven at 70 °C for 3 min to remove the residual DMF. After cooling, the polyimide tape was removed, and a $CH_3NH_3PbI_3$ layer was obtained upon annealing at 90 °C for

10 min. Finally, the $CH_3NH_3PbI_3/Au/Mg_{0.2}Zn_{0.8}O$ heterojunction SPPDs were obtained, as shown in Figure 1e,f.

2.3. Device Characterization

The crystal structures of $CH_3NH_3PbI_3/Au$ PDs and $CH_3NH_3PbI_3/Au/Mg_{0.2}Zn_{0.8}O$ heterojunction SPPDs were characterized using a Rigaku Ultima VI X-ray diffractometer (XRD). The morphology was characterized by scanning electron microscopy (SEM) using a JEOL JSM-7600F microscope. The UV-visible (Vis)-NIR absorption spectra of the $CH_3NH_3PbI_3/Au$ PDs and $CH_3NH_3PbI_3/Au/Mg_{0.2}Zn_{0.8}O$ heterojunction SPPDs were measured using a PerkinElmer Lambda 950 spectrophotometer. The dark and photocurrent-voltage (I-V) curves of the $CH_3NH_3PbI_3/Au$ PDs and $CH_3NH_3PbI_3/Au/Mg_{0.2}Zn_{0.8}O$ heterojunction SPPDs were measured using an Agilent 16442A test fixture. The responsivity spectra of the $CH_3NH_3PbI_3/Au$ PDs and $CH_3NH_3PbI_3/Au/Mg_{0.2}Zn_{0.8}O$ heterojunction SPPDs were measured using a Zolix DR800-CUST testing system.

3. Results and Discussion

As shown in Figure 2a, the XRD pattern of the $CH_3NH_3PbI_3/Au$ PDs exhibits diffraction peaks at 2θ = 14.1°, 28.4°, 31.9°, and 40.8°, which are associated with the (110), (220), (310), and (224) planes, respectively; the strongest diffraction peaks are (110) and (220), which shows that the materials grow preferentially along the (110) direction, which is consistent with previous studies [30–32]. The still-remaining diffraction peak at 12.65° suggests the level of the PbI_2 impurity phase [33]. The XRD pattern of the $CH_3NH_3PbI_3/Au/Mg_{0.2}Zn_{0.8}O$ heterojunction SPPDs shows a peak at 34.51°, which corresponds to the (002) planes of $Mg_{0.2}Zn_{0.8}O$ [34,35]. The other peaks are coincident with those of $CH_3NH_3PbI_3$, indicating the diffraction peak of $Mg_{0.2}Zn_{0.8}O$ film. The $Mg_{0.2}Zn_{0.8}O$ thin film has no effect on the crystallinity of the $CH_3NH_3PbI_3$ layer.

Figure 2b shows the normalized absorption spectra of the $Mg_{0.2}Zn_{0.8}O/Au$ PDs, $CH_3NH_3PbI_3/Au$ PDs, and $CH_3NH_3PbI_3/Au/Mg_{0.2}Zn_{0.8}O$ heterojunction SPPDs. For the $Mg_{0.2}Zn_{0.8}O/Au$ PDs, there is an absorption peak at a wavelength of 330 nm. A broad absorption band from 330 to 780 nm is observed for the $CH_3NH_3PbI_3/Au$ PDs. It is worth noting that the absorption of the $CH_3NH_3PbI_3/Au/Mg_{0.2}Zn_{0.8}O$ heterojunction SPPDs is in the range from 330 to 780 nm, and the absorption performance is better than that of the $CH_3NH_3PbI_3/Au$ PDs. Overall, the excellent light absorption properties make $CH_3NH_3PbI_3/Au/Mg_{0.2}Zn_{0.8}O$ heterojunction SPPDs promising candidates for high-performance PDs. It can be seen from Figure 2c,d that the cuboid-shaped $CH_3NH_3PbI_3$ grains are uniformly distributed on the substrate. Through the double-layer film of $Mg_{0.2}Zn_{0.8}O$ and $CH_3NH_3PbI_3$, it can be seen that the upper layer of the $CH_3NH_3PbI_3$ polycrystalline film is more dense; at the same time, the crystal grains of $CH_3NH_3PbI_3$ can be enlarged. Larger $CH_3NH_3PbI_3$ grains mean fewer perovskite grain boundaries. This result suggests that the $Mg_{0.2}Zn_{0.8}O$ film can passivate the surface defects of the perovskite and that a uniform and flat $CH_3NH_3PbI_3$ layer is formed.

Figure 3a shows the responsivity (R) spectra of the $CH_3NH_3PbI_3/Au/Mg_{0.2}Zn_{0.8}O$ heterojunction SPPDs in the UV-Vis-NIR range. It can be seen that in the range of 250–850 nm, the R value is significantly higher than that of the $Mg_{0.2}Zn_{0.8}O/Au$ PDs or $CH_3NH_3PbI_3/Au$ PDs under 1 V. For the $CH_3NH_3PbI_3/Au/Mg_{0.2}Zn_{0.8}O$ heterojunction SPPDs, the highest R value is up to 0.58 A/W, which is 11.09 times higher than that of the best $CH_3NH_3PbI_3/Au$ PDs and is 161 times higher than that of the best $Mg_{0.2}Zn_{0.8}O/Au$ PDs. It can be seen that the $CH_3NH_3PbI_3/Au/Mg_{0.2}Zn_{0.8}O$ PDs have a higher R than the $Mg_{0.2}Zn_{0.8}O/Au$ PDs and the $CH_3NH_3PbI_3/Au$ PDs alone.

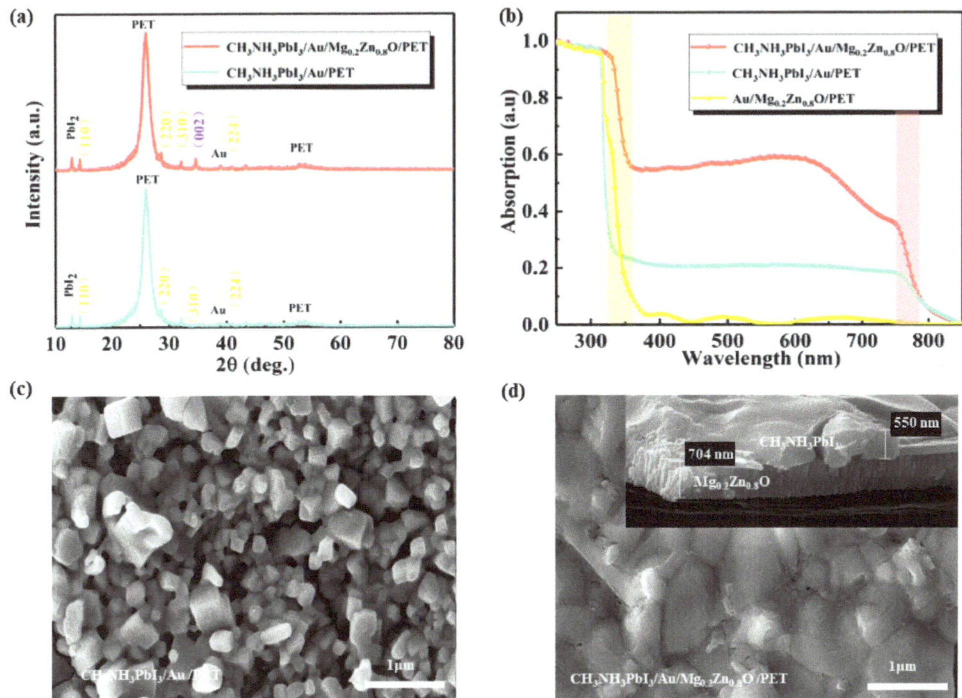

Figure 2. (**a**) XRD spectra of the $CH_3NH_3PbI_3/Au$ PDs and $CH_3NH_3PbI_3/Au/Mg_{0.2}Zn_{0.8}O$ SP-PDs. (**b**) Normalized absorption spectra of the $Mg_{0.2}Zn_{0.8}O/Au$ PDs, $CH_3NH_3PbI_3/Au$ PDs, and $CH_3NH_3PbI_3/Au/Mg_{0.2}Zn_{0.8}O$ SPPDs. (**c**,**d**) SEM spectra of the $CH_3NH_3PbI_3$ PDs and $CH_3NH_3PbI_3/Au/Mg_{0.2}Zn_{0.8}O$ SPPDs.

Figure 3b shows a comparison of the external quantum efficiency (EQE) measured under 1 V. The EQE curve is similar to the absorbance curve of the $CH_3NH_3PbI_3/Au$ PDs. The EQE is one of the most important performance parameters of perovskite PDs; the EQE represents the number of electron-hole pairs generated for a single incident photon. The EQE of the two groups of devices was calculated using formula (1). In the case of a certain bias, the EQE of the $CH_3NH_3PbI_3/Au/Mg_{0.2}Zn_{0.8}O$ heterojunction SPPDs is 10.23 times that of the $CH_3NH_3PbI_3/Au$ PDs and 84.51 times that of the $Mg_{0.2}ZnO_{0.8}/Au$ PDs. The EQE is defined as:

$$EQE(\lambda) = \frac{R \times h \times c}{q \times \lambda} \tag{1}$$

The I-V curves of the $CH_3NH_3PbI_3/Au$ PDs and $CH_3NH_3PbI_3/Au/Mg_{0.2}Zn_{0.8}O$ heterojunction SPPDs under dark and light conditions are shown in Figure 3c,d. The light-dark current ratio of the $CH_3NH_3PbI_3/Au/Mg_{0.2}Zn_{0.8}O$ heterojunction SPPDs is 100 times that of the $CH_3NH_3PbI_3/Au$ PDs. Additionally, the nonlinear I-V curves indicate that Schottky metal-semiconductor contacts were formed. By comparing Figure 3c,d, it is found that this is mainly attributed to the fact that the heterojunction inhibits the rise of the dark current. From Figure 3d, it can be seen that the dark current of the $CH_3NH_3PbI_3/Au/Mg_{0.2}Zn_{0.8}O$ heterojunction SPPDs is very low, and the light-dark current ratio is about 10^3. The dark current of the $CH_3NH_3PbI_3/Au/Mg_{0.2}Zn_{0.8}O$ heterojunction SPPDs is less than 1.4×10^{-1} pA at 0 V, which is more than 10 times less than that of the $CH_3NH_3PbI_3$ PDs. Furthermore, the dark current is very low, which can effectively reduce the lowest detectable optical power and enhance the capability of detecting weak light [36]. When the $CH_3NH_3PbI_3$ thin film is spin coated on $Mg_{0.2}Zn_{0.8}O$, the dark current of the device decreases, the photocurrent increases, and the photoresponsivity gain is enhanced for the heterojunction. There are few carrier-donating defects in the bilayer, and the interfacial charge transfer reduces the carrier concentration in the dissipative region. The higher the light current-dark current ratio, the better the device detection performance. At the same time, the recombination of the electron pairs in the $Mg_{0.2}Zn_{0.8}O$ thin film is reduced in the gap between the $CH_3NH_3PbI_3$ grains, which is beneficial for the hole separation and transport in the $CH_3NH_3PbI_3/Au/Mg_{0.2}Zn_{0.8}O$ heterojunction thin film; this enables the realization of a high response and a very low dark current.

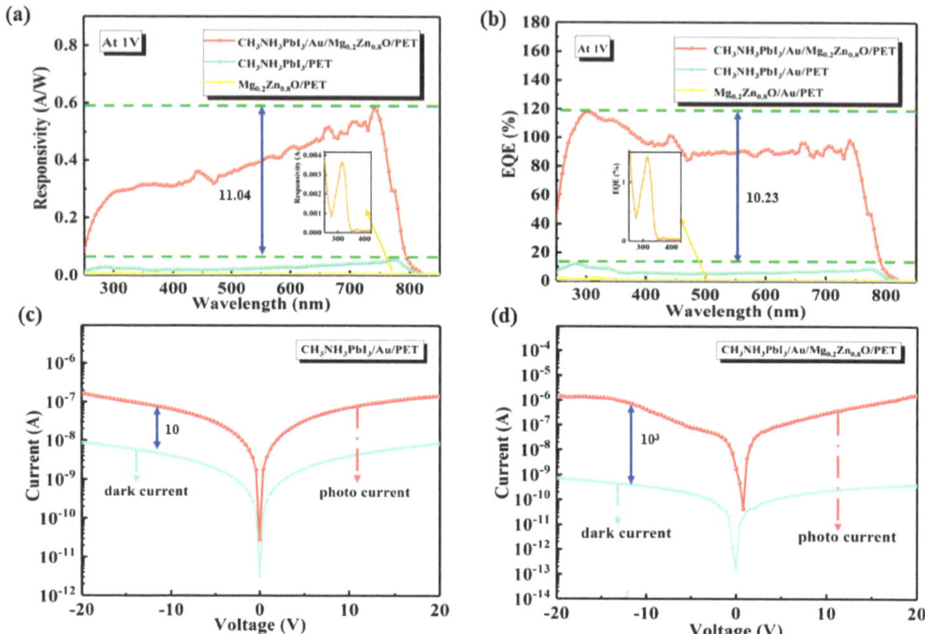

Figure 3. (a) Responsivity spectra of the $CH_3NH_3PbI_3/Au$ PDs, $Mg_{0.2}Zn_{0.8}O$ PDs, and $CH_3NH_3PbI_3/Au/Mg_{0.2}Zn_{0.8}O$ heterojunction SPPDs under 1 V. (b) EQE of the $CH_3NH_3PbI_3/Au$ PDs, $Mg_{0.2}Zn_{0.8}O$ PDs, and $CH_3NH_3PbI_3/Au/Mg_{0.2}Zn_{0.8}O$ heterojunction SPPDs under 1 V. (c,d) I-V curves of the $CH_3NH_3PbI_3/Au$ PDs and $CH_3NH_3PbI_3/Au/Mg_{0.2}Zn_{0.8}O$ heterojunction SPPDs under dark and light conditions.

Figure 4a,b show the I-V curves for different wavelengths (namely 330, 550, and 760 nm). Clearly, the photocurrent density increases gradually with the incident light wavelength. The main reason is that the dark current is ultimately limited by the recombination current, which is an inherent property of semiconductor materials and heterojunctions. The built-in electric field of the $CH_3NH_3PbI_3/Au/Mg_{0.2}Zn_{0.8}O$ heterojunction with the increased Fermi level of $Mg_{0.2}Zn_{0.8}O$ provides a strong driving force to separate and transfer the photogenerated carriers, and the depletion layer becomes wider, which decreases the recombination of the carriers and then reduces the dark current. This endows the heterojunction devices with a high photoresponse performance under external bias. The high detection rate of the perovskite PDs is mainly due to their very low dark current under reverse bias. Such a small dark current explains the reason for its high photodetection and also shows that perovskite-based PDs can exhibit very good detection capability. Moreover, the device shows an apparent photovoltaic behavior under illumination, and the offset voltage is 0 V, which exhibits a self-powered characteristic behavior, as shown in Figure 4c. In the self-supply voltage-detection mode, the device achieves a high responsivity of up to

1.1 mA/W. A large number of electron-hole pairs are generated in the film due to the internal electric field at the surface. The space charges separate and drift in opposite directions, generating photocurrent at the electrode to produce zero bias.

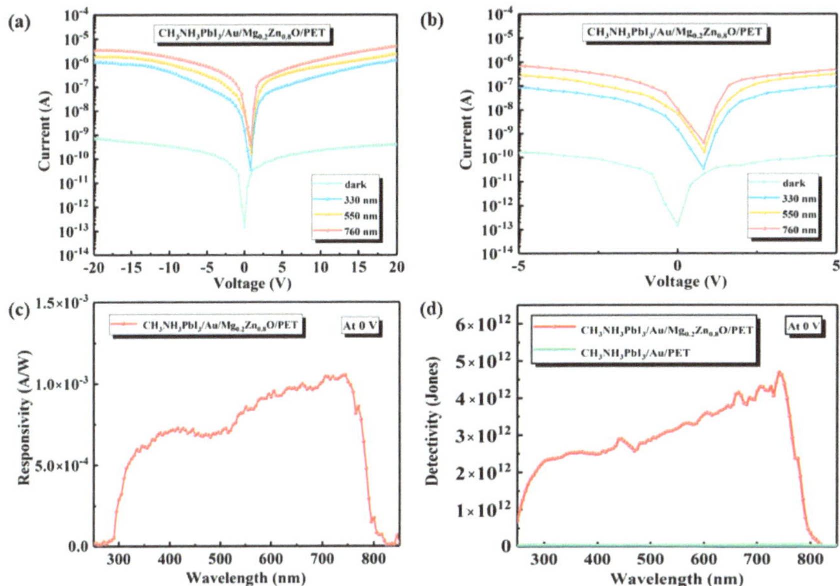

Figure 4. (**a**,**b**) I-V curves of the $CH_3NH_3PbI_3/Au/Mg_{0.2}Zn_{0.8}O$ heterojunction PDs for different wavelengths. (**c**) The responsiveness curves of the $CH_3NH_3PbI_3/Au/Mg_{0.2}Zn_{0.8}O$ heterojunction SPPDs at 0 V. (**d**) D^* values of the $CH_3NH_3PbI_3/Au$ PDs and $CH_3NH_3PbI_3/Au/Mg_{0.2}Zn0_{.8}O$ heterojunction SPPDs at 0 V.

Figure 5 shows a schematic of the carrier transport mechanism of the $CH_3NH_3PbI_3/Au/Mg_{0.2}Zn_{0.8}O$ heterojunction SPPDs. Such a mechanism can be explained by the band diagram. As can be seen from the figure, the band gaps of $CH_3NH_3PbI_3$ and $Mg_{0.2}Zn_{0.8}O$ are 1.45 and 3.89 eV, respectively. When the $Mg_{0.2}Zn_{0.8}O$ film forms a heterojunction with the $CH_3NH_3PbI_3$ film, the electrons in the former diffuse into the latter. Furthermore, the pores in the $CH_3NH_3PbI_3$ film diffuse into the $Mg_{0.2}Zn_{0.8}O$ film. When they are in equilibrium, a self-built electric field is formed at the heterojunction interface, as shown in Figure 5b. Under illumination, $Mg_{0.2}Zn_{0.8}O$ absorbs UV light to produce photogenerated electrons and holes, while the $CH_3NH_3PbI_3$ film absorbs UV, Vis, and NIR light to produce electrons and holes. Under the action of this self-established electric field, the photogenerated electrons and holes, which are diffused from the dissipative region of the heterojunction, rapidly separate; the electrons drift to the $Mg_{0.2}Zn_{0.8}O$ film, and the holes drift to the $CH_3NH_3PbI_3$ film; thus, a stable external current is generated. Therefore, self-driven $CH_3NH_3PbI_3/Au/Mg_{0.2}Zn_{0.8}O$ devices have a low dark current, a high light current, and high responsiveness. Due to the high recombination probability of the photogenerated electrons and holes in $CH_3NH_3PbI_3$, the photocurrent enhancement of the $CH_3NH_3PbI_3/Au$ PDs is not very significant. However, when the n-$Mg_{0.2}Zn_{0.8}O$ layer and the p-$CH_3NH_3PbI_3$ perovskite material are introduced, a p-n junction with a well-matched band structure is formed, from p-$CH_3NH_3PbI_3$ with a high Fermi level to n-$Mg_{0.2}Zn_{0.8}O$ with a low Fermi level. The Fermi level of p-$CH_3NH_3PbI_3$ in the p region gradually increases, while that of $Mg_{0.2}Zn_{0.8}O$ in the n region gradually decreases, thus increasing the Fermi level [37]. Therefore, when the recombination probability of photogenerated electron-hole pairs in the $CH_3NH_3PbI_3$ layer decreases, the photocurrent of the

$CH_3NH_3PbI_3/Au/Mg_{0.2}Zn_{0.8}O$ heterojunction SPPDs increases significantly [38–42]. Thus, lowering the contact barrier on the surface results in a high photocurrent in the device. Therefore, the heterostructure PD can not only reduce the recombination probability of electrons and holes but also increase the depletion layer width, as shown in Figure 5c; the dark current decreases, so that the $CH_3NH_3PbI_3/Au/Mg_{0.2}Zn_{0.8}O$ heterojunction SPPDs have a high optical gain.

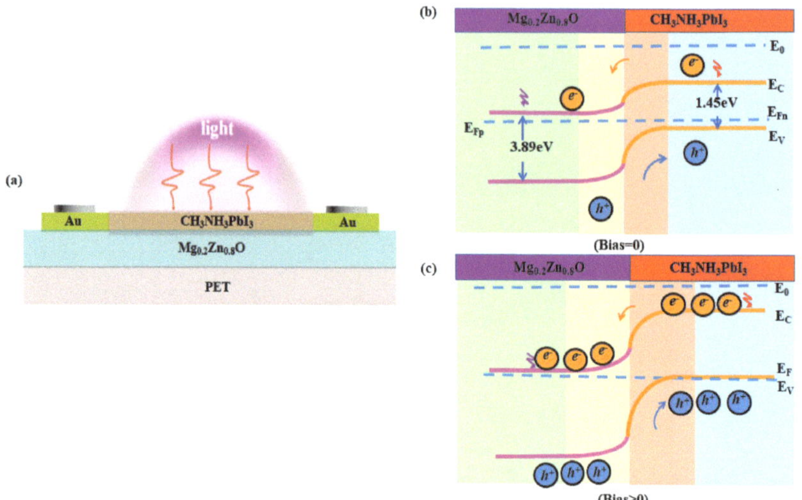

Figure 5. Diagram of the carrier transport mechanism. (**a**) Diagram of the carrier transport mechanism of the $CH_3NH_3PbI_3/Au/Mg_{0.2}Zn0.8O$ SPPDs at 0 V. (**b,c**) Diagram of the carrier transport mechanism of the $CH_3NH_3PbI_3/Au/Mg_{0.2}Zn_{0.8}O$ heterojunction SPPDs under 0 bias and reverse bias.

$D*$ is a measurement of the detector's sensitivity. Assuming that shot noise from the dark current is the major contributor to the total noise, it can be written as [43,44]:

$$D* = \frac{R}{\left(\frac{2qI_d}{A}\right)^{\frac{1}{2}}} \tag{2}$$

where R is the responsivity, A is the area of the detectors, q is the unit charge, and I_d is the dark current.

Due to the suppressed dark current and the enhanced responsivity of the $CH_3NH_3PbI_3/Au/Mg_{0.2}Zn_{0.8}O$ heterojunction SPPDs, as shown in Figure 4d, the best value of $D*$ is as high as 4.7×10^{12} Jones at 780 nm, while it is only 3.0×10^{10} Jones in $CH_3NH_3PbI_3/Au$ devices. Notably, the $CH_3NH_3PbI_3/Au/Mg_{0.2}Zn_{0.8}O$ heterojunction PDs exhibit a pronounced response in the vis-light range as well as in the UV- and NIR-light ranges. It is found that the detectivity of the prepared perovskite SPPDs is greatly improved in the range of 300–800 nm, and a detectivity exceeding 2.34×10^{12} Jones is achieved in most of the range (320–780 nm). These results are comparable with those of other reports, and a detailed comparison is provided in Table 1. The table shows that the detection rate of the device prepared in this paper is higher than that of other devices, which have a detection rate of 4.7×10^{12} Jones and obtain ultra-low magnitude-dark currents compared to other devices, which have a dark current of 1.4×10^{-1} pA. The significant increase in detectivity indicates that the modification on both sides of the active layer results in a reduced trap density as well as improved carrier transport and extraction.

Table 1. The performance parameters of perovskite PDs in this and previously reported work.

Device Structure	R (A/W)	I_{dark}	EQE	D^*	R_0 (mA/W)	Ref
$CH_3NH_3PbI_3/Au/Mg_{0.2}Zn_{0.8}O$	0.58	1.4×10^{-1} pA	120	4.7×10^{12}	1.1	This work
$Ag/NP_3/MAPbI_3/Al$	0.25	0.3 uA	-	1.53×10^{11}	-	[5]
$Al/Si/SiO_2/MAPbI_3/Pt$	-	50 pA	-	8.8×10^{10}	-	[6]
$ZnO/CsPbBr_3$	0.01	-	-	-	-	[16]
$Au/MoO_3/MAPbI_3/ZnO/FTO$	0.05	1 nA	-	4.5×10^{11}	-	[19]

4. Conclusions

In summary, a $CH_3NH_3PbI_3$ film was prepared and combined with an $Mg_{0.2}Zn_{0.8}O$ film fabricated via vacuum magnetron sputtering to realize $CH_3NH_3PbI_3/Au/Mg_{0.2}Zn_{0.8}O$ heterojunction SPPDs. The results show that due to the addition of the $Mg_{0.2}Zn_{0.8}O$ layer, the recombination probability of the photogenerated electron-hole pairs is reduced, and the optical gain is enhanced. Compared with the $CH_3NH_3PbI_3/Au$ PDs, the photoresponsivity is increased by nearly 53.17 times across the entire spectral range, and the EQE of the $CH_3NH_3PbI_3/Au/Mg_{0.2}Zn_{0.8}O$ heterojunction SPPDs is increased from 12.45% to 120%. Remarkably, in the self-supply voltage-detection mode, the SPPDs achieve a high responsivity of up to 1.1 mA/W. The dark current of the $CH_3NH_3PbI_3/Au/Mg_{0.2}Zn_{0.8}O$ heterojunction SPPDs is less than 1.4×10^{-1} pA at 0 V. The best value of the detectivity is as high as 4.7×10^{12} Jones.

Author Contributions: Writing—original draft, M.W.; Writing—review & editing, M.Z.; Writing—review & Funding acquisition, D.J. All authors have read and agreed to the published version of the manuscript.

Funding: This research was fund by [National Natural Science Foundation of China] grant number [62274016], [Scientific and Technological Development Project of Jilin Province] grant number [20210203216SF], and [Scientific and Technological Development Project of Jilin Province] grant number [20210509065RQ].

Institutional Review Board Statement: Not applicable.

Data Availability Statement: The data that support the findings of this study are available from the corresponding author upon reasonable request.

Conflicts of Interest: The authors declare no conflict of interest.

References

1. Khan, M.F.; Rehman, S.; Akhtar, I.; Aftab, S.; Ajma, H.M.S.; Khan, K.; Deok, K.K.; Eom, J. High mobility $ReSe_2$ field effect transistors: Schottky-barrier-height dependent photoresponsivity and broadband light detection with Co decoration. *2D Mater.* **2020**, *7*, 015010. [CrossRef]
2. Wei, Y.C.; Chen, C.; Tan, C.; He, L.; Ren, Z.Z.; Zhang, C.Y.; Peng, S.L.; Zhang, C.Y.; Han, J.Y.; Zhou, H.X.; et al. High-Performance Visible to Near-Infrared Broadband Bi_2O_2Se Nanoribbon Photodetectors. *Adv. Opt. Mater.* **2022**, *10*, 2201396. [CrossRef]
3. Chen, J.S.; Li, Y.; Zhang, Z.Q.; Lu, H. Structure design and properties investigation of Bi_2O_2Se/graphene van der Waals heterojunction from first-principles study. *Surf. Interfaces* **2022**, *33*, 102289. [CrossRef]
4. Kong, L.; Sun, H.T.; Nie, Y.H.; Yan, Y.; Wang, R.Z.; Ding, Q.; Zhang, S.; Yu, H.; He, J.; Luan, G.Y. Luminescent Properties and Charge Compensator Effects of $SrMo_{0.5}W_{0.5}O_4$: Eu^{3+} for White Light LEDs. *Molecules* **2023**, *28*, 2681. [CrossRef]
5. Liu, B.W.; Peng, Y.; Jin, Z.M.; Wu, X.; Gu, H.Y.; Wei, D.S.; Zhu, Y.M.; Zhuang, S.L. Terahertz ultrasensitive biosensor based on wide-area and intense light-matter interaction supported by QBIC. *Chem. Eng. J.* **2023**, *462*, 142347. [CrossRef]
6. Cai, L.Y.; LU, Y.G.; Zhu, H.H. Performance enhancement of on-chip optical switch and memory using $Ge_2Sb_2Te_5$ slot-assisted microring resonator. *Opt. Laser Eng.* **2023**, *102*, 10746. [CrossRef]
7. Leung, S.F.; Ho, K.T.; Kung, P.K.; Hsiao, V.K.; Alshareef, H.N.; Wang, Z.L.; He, J.H. A self-powered and flexible organometallic halide perovskite photodetector with very high detectivity. *Adv. Mater.* **2018**, *30*, 1704611. [CrossRef] [PubMed]
8. Li, Y.; Li, Y.; Zhou, X. Flexible Optoelectronic Devices Based on Hybrid Perovskites. *Mater. Devices* **2022**, *2*, 379–431.
9. Bao, C.; Zhu, W.; Yang, J.; Li, F.; Gu, S.; Wang, Y.; Yu, T.; Zhu, J.; Zhou, Y.; Zou, Z. Highly flexible self-powered organolead trihalide perovskite photodetectors with gold nanowire networks as transparent electrodes. *ACS Appl. Mater. Interfaces* **2016**, *8*, 23868–23875. [CrossRef]

10. Sun, H.; Tian, W.; Cao, F.; Xiong, J.; Li, L. Ultrahigh-performance self-powered flexible double-twisted fibrous broadband perovskite photodetector. *Adv. Mater.* **2018**, *30*, 1706986. [CrossRef]

11. Zhu, Y.; Song, Z.; Zhou, H.; Wu, D.; Lu, R.; Wang, R.; Wang, H. Self-powered, broadband perovskite photodetector based on ZnO microspheres as scaffold layer. *Appl. Surf. Sci.* **2018**, *448*, 23–29. [CrossRef]

12. Pang, T.; Jia, R.; Wang, Y.; Sun, K.; Hu, Z.; Zhu, Y.; Luan, S.; Zhang, Y. Self-powered behavior based on the light-induced self-poling effect in perovskite-based transport layer-free photodetectors. *J. Mater. Chem. C* **2019**, *7*, 609–616. [CrossRef]

13. Chen, Y.; Chen, T.; Dai, L. Layer-by-layer growth of $CH_3NH_3PbI_{3-x}Cl_x$ for highly efficient planar heterojunction perovskite solar cells. *Adv. Mater.* **2015**, *27*, 1053–1059. [CrossRef] [PubMed]

14. Tong, S.; Wu, H.; Zhang, C.; Li, S.; Wang, C.; Shen, J.; Xiao, S.; He, J.; Yang, J.; Sun, J. Large-area and high-performance $CH_3NH_3PbI_3$ perovskite photodetectors fabricated via doctor blading in ambient condition. *Org. Electron.* **2017**, *49*, 347–354. [CrossRef]

15. Hu, X.; Zhang, X.; Liang, L.; Bao, J.; Li, S.; Yang, W.; Xie, Y. High-performance flexible broadband photodetector based on organolead halide perovskite. *Adv. Funct. Mater.* **2014**, *24*, 7373–7380. [CrossRef]

16. Watanabe, Y. Epitaxial all-perovskite ferroelectric field effect transistor with a memory retention. *Appl. Phys. Lett.* **1995**, *66*, 1770–1772. [CrossRef]

17. Xia, H.-R.; Li, J.; Sun, W.-T.; Peng, L.-M. Organohalide lead perovskite based photodetectors with much enhanced performance. *Chem. Commun.* **2014**, *50*, 13695–13697. [CrossRef]

18. Zhang, Y.; Du, J.; Wu, X.; Zhang, G.; Chu, Y.; Liu, D.; Zhao, Y.; Liang, Z.; Huang, J. Ultrasensitive photodetectors based on island-structured $CH_3NH_3PbI_3$ thin films. *ACS Appl. Mater. Interfaces* **2015**, *7*, 21634–21638. [CrossRef] [PubMed]

19. Hu, Q.; Wu, H.; Sun, J.; Yan, D.; Gao, Y.; Yang, J. Large-area perovskite nanowire arrays fabricated by large-scale roll-to-roll micro-gravure printing and doctor blading. *Nanoscale* **2016**, *8*, 5350–5357. [CrossRef]

20. Park, N.; Kang, H.; Park, J.; Lee, Y.; Yun, Y.; Lee, J.-H.; Lee, S.-G.; Lee, Y.H.; Suh, D. Ferroelectric single-crystal gated graphene/hexagonal-BN/ferroelectric field-effect transistor. *ACS Nano* **2015**, *9*, 10729–10736. [CrossRef]

21. Zhuo, S.; Zhang, J.; Shi, Y.; Huang, Y.; Zhang, B. Self-template-directed synthesis of porous perovskite nanowires at room temperature for high-performance visible-light photodetectors. *Angew. Chem.* **2015**, *127*, 5785–5788. [CrossRef]

22. Liu, D.; Kelly, T.L. Perovskite solar cells with a planar heterojunction structure prepared using room-temperature solution processing techniques. *Nat. Photonics* **2014**, *8*, 133–138. [CrossRef]

23. Kim, J.; Kim, G.; Kim, T.K.; Kwon, S.; Back, H.; Lee, J.; Lee, S.H.; Kang, H.; Lee, K. Efficient planar-heterojunction perovskite solar cells achieved via interfacial modification of a solgel ZnO electron collection layer. *J. Mater. Chem. A* **2014**, *2*, 17291–17296. [CrossRef]

24. Dong, X.; Hu, H.; Lin, B.; Ding, J.; Yuan, N. The effect of ALD-ZnO layers on the formation of $CH_3NH_3PbI_3$ with different perovskite precursors and sintering temperatures. *Chem. Commun.* **2014**, *50*, 14405–14408. [CrossRef]

25. Bai, F.; Qi, J.; Li, F.; Fang, Y.; Han, W.; Wu, H.; Zhang, Y. A high-performance self-powered photodetector based on monolayer MoS_2/Perovskite heterostructures. *Adv. Mater. Interfaces* **2018**, *5*, 1701275. [CrossRef]

26. Harrison, S.E. Conductivity and Hall effect of ZnO at low temperatures. *Phys. Rev.* **1954**, *93*, 52. [CrossRef]

27. Hutson, A.R. Hall effect studies of doped zinc oxide single crystals. *Phys. Rev.* **1957**, *108*, 222. [CrossRef]

28. Thomas, D. Interstitial zinc in zinc oxide. *J. Phys. Chem. Solids* **1957**, *3*, 229–237. [CrossRef]

29. Ghosh, J.; Natu, G.; Giri, P. Plasmonic hole-transport-layer enabled self-powered hybrid perovskite photodetector using a modified perovskite deposition method in ambient air. *Org. Electron.* **2019**, *71*, 175–184. [CrossRef]

30. Li, Y.; Zhang, Y.; Li, T.; Li, M.; Chen, Z.; Li, Q.; Zhao, H.; Sheng, Q.; Shi, W.; Yao, J. Ultrabroadband, ultraviolet to terahertz, and high sensitivity $CH_3NH_3PbI_3$ perovskite photodetectors. *Nano Lett.* **2020**, *20*, 5646–5654. [CrossRef]

31. Su, H.; Meng, L.; Liu, Y.; Zhang, Y.; Hu, M.; Yang, Z.; Liu, S.F. Effective electron extraction from active layer for enhanced photodetection of photoconductive type detector with structure of $Au/CH_3NH_3PbI_3/Au$. *Org. Electron.* **2019**, *74*, 197–203. [CrossRef]

32. Gao, L.; Zeng, K.; Guo, J.; Ge, C.; Du, J.; Zhao, Y.; Chen, C.; Deng, H.; He, Y.; Song, H. Passivated single-crystalline $CH_3NH_3PbI_3$ nanowire photodetector with high detectivity and polarization sensitivity. *Nano Lett.* **2016**, *16*, 7446–7454. [CrossRef]

33. Chen, Q.; Zhou, H.; Hong, Z.; Luo, S.; Duan, H.-S.; Wang, H.-H.; Liu, Y.; Li, G.; Yang, Y. Planar heterojunction perovskite solar cells via vapor-assisted solution process. *J. Am. Chem. Soc.* **2014**, *136*, 622–625. [CrossRef]

34. Duan, Y.; Zhang, S.; Cong, M.; Jiang, D.; Liang, Q.; Zhao, X. Performance modulation of a MgZnO/ZnO heterojunction flexible UV photodetector by the piezophototronic effect. *J. Mater. Chem. C* **2020**, *8*, 12917–12926. [CrossRef]

35. Willander, M.; Yang, L.L.; Wadeasa, A.; Ali, S.U.; Asif, M.H.; Zhao, Q.X.; Nur, O. Zinc oxide nanowires: Controlled low temperature growth and some electrochemical and optical nano-devices. *J. Mater. Chem.* **2009**, *19*, 1006–1018. [CrossRef]

36. Youngblood, N.; Chen, C.; Koester, S.J.; Li, M. Waveguide-integrated black phosphorus photodetector with high responsivity and low dark current. *Nat. Photon.* **2015**, *9*, 247–252. [CrossRef]

37. Aharon, S.; Gamliel, S.; El Cohen, B.; Etgar, L. Depletion region effect of highly efficient hole conductor free $CH_3NH_3PbI_3$ perovskite solar cells. *Phys. Chem. Chem. Phys.* **2014**, *16*, 10512–10518. [CrossRef]

38. Jošt, M.; Kegelmann, L.; Korte, L.; Albrecht, S. Monolithic perovskite tandem solar cells: A review of the present status and advanced characterization methods toward 30% efficiency. *Adv. Energy Mater.* **2020**, *10*, 1904102. [CrossRef]

39. Liu, P.; Han, N.; Wang, W.; Ran, R.; Zhou, W.; Shao, Z. High-quality ruddlesden–popper perovskite film formation for high-performance perovskite solar cells. *Adv. Mater.* **2021**, *33*, 2002582. [CrossRef]
40. Dai, X.; Xu, K.; Wei, F. Recent progress in perovskite solar cells: The perovskite layer. *Beilstein J. Nanotech.* **2020**, *11*, 51–60. [CrossRef]
41. Hu, Y.; Schlipf, J.; Wussler, M.; Petrus, M.L.; Jaegermann, W.; Bein, T.; Muüller-Buschbaum, P.; Docampo, P. Hybrid perovskite/perovskite heterojunction solar cells. *ACS Nano* **2016**, *10*, 5999–6007. [CrossRef]
42. Zhang, T.; Long, M.; Qin, M.; Lu, X.; Chen, S.; Xie, F.; Gong, L.; Chen, J.; Chu, M.; Miao, Q. Stable and efficient 3D-2D perovskite-perovskite planar heterojunction solar cell without organic hole transport layer. *Joule* **2018**, *2*, 2706–2721. [CrossRef]
43. Ma, C.; Shi, Y.; Hu, W.; Chiu, M.H.; Liu, Z.; Bera, A.; Li, F.; Wang, H.; Li, L.J.; Wu, T. Heterostructured $WS_2/CH_3NH_3PbI_3$ photoconductors with suppressed dark current and enhanced photodetectivity. *Adv. Mater.* **2016**, *28*, 3683–3689. [CrossRef]
44. Yu, J.; Chen, X.; Wang, Y.; Zhou, H.; Xue, M.; Xu, Y.; Li, Z.; Ye, C.; Zhang, J.; Van Aken, P.A. A high-performance self-powered broadband photodetector based on a $CH_3NH_3PbI_3$ perovskite/ZnO nanorod array heterostructure. *J. Mater. Chem. C* **2016**, *4*, 7302–7308. [CrossRef]

Disclaimer/Publisher's Note: The statements, opinions and data contained in all publications are solely those of the individual author(s) and contributor(s) and not of MDPI and/or the editor(s). MDPI and/or the editor(s) disclaim responsibility for any injury to people or property resulting from any ideas, methods, instructions or products referred to in the content.

Article

Efficient Environmentally Friendly Flexible CZTSSe/ZnO Solar Cells by Optimizing ZnO Buffer Layers

Quanzhen Sun [1,2,†], Jianlong Tang [1,†,‡], Caixia Zhang [1,2,3,*], Yaling Li [1,2], Weihao Xie [1,2], Hui Deng [1,2], Qiao Zheng [1,2], Jionghua Wu [1,2] and Shuying Cheng [1,2,3,*]

[1]   College of Physics and Information Engineering, Institute of Micro-Nano Devices and Solar Cells, Fuzhou University, Fuzhou 350108, China

[2]   Fujian Science & Technology Innovation Laboratory for Optoelectronic Information of China, Fuzhou 350108, China

[3]   Jiangsu Collaborative Innovation Center of Photovoltaic Science and Engineering, Changzhou 213164, China

*   Correspondence: zhangcx@fzu.edu.cn (C.Z.); sycheng@fzu.edu.cn (S.C.)

†   These authors contributed equally to this work.

‡   Current address: Nanping Management Branch of Fujian Expressway Group Co., Ltd., Nanping 354300, China.

Abstract: Flexible CZTSSe solar cells have attracted much attention due to their earth-abundant elements, high stability, and wide application prospects. However, the environmental problems caused by the high toxicity of the Cd in the buffer layers restrict the development of flexible CZTSSe solar cells. Herein, we develop a Cd-free flexible CZTSSe/ZnO solar cell. The influences of the ZnO films on device performances are investigated. The light absorption capacity of flexible CZTSSe solar cells is enhanced due to the removal of the CdS layer. The optimal thickness of the ZnO buffer layers and the appropriate annealing temperature of the CZTSSe/ZnO are 100 nm and 200 °C. Ultimately, the optimum flexible CZTSSe/ZnO device achieves an efficiency of 5.0%, which is the highest efficiency for flexible CZTSSe/ZnO solar cells. The systematic characterizations indicate that the flexible CZTSSe/ZnO solar cells based on the optimal conditions achieved quality heterojunction, low defect density and better charge transfer capability. This work provides a new strategy for the development of the environmentally friendly and low-cost flexible CZTSSe solar cells.

Keywords: Cd-free flexible CZTSSe/ZnO solar cells; ZnO buffer layer; CZTSSe/ZnO heterojunction; buffer layer optimization

Citation: Sun, Q.; Tang, J.; Zhang, C.; Li, Y.; Xie, W.; Deng, H.; Zheng, Q.; Wu, J.; Cheng, S. Efficient Environmentally Friendly Flexible CZTSSe/ZnO Solar Cells by Optimizing ZnO Buffer Layers. *Materials* **2023**, *16*, 2869. https://doi.org/10.3390/ma16072869

Academic Editors: Cristobal Voz and Ioana Pintilie

Received: 21 February 2023
Revised: 20 March 2023
Accepted: 29 March 2023
Published: 4 April 2023

Copyright: © 2023 by the authors. Licensee MDPI, Basel, Switzerland. This article is an open access article distributed under the terms and conditions of the Creative Commons Attribution (CC BY) license (https://creativecommons.org/licenses/by/4.0/).

1. Introduction

$Cu_2ZnSn(S,Se)_4$ (CZTSSe) is one of the most promising absorber materials for thin-film solar cells due to its earth-abundant, non-toxic constituent elements and tunable bandgap [1–6]. The maximum power conversion efficiency (*PCE*) of CZTSSe solar cells can reach 13.0% [7] and 11.2% [8] on rigid substrates and flexible substrates, respectively. Flexible CZTSSe solar cells have great development potential in the field of photovoltaic building integration and indoor photovoltaic applications due to their light weight and flexibility [9–14]. However, the toxic Cd element in the CdS buffer layer of the solar cell causes a series of environmental problems, limiting the application of CZTSSe solar cells. In addition, the narrow energy bandgap (E_g) of CdS (2.4 eV) leads to loss of photons at short wavelengths (less than 520 nm), which accounts for 24% of the entire solar spectrum [15,16]. Incomplete absorption in the visible spectrum of the CdS buffer layer is thought to be responsible for the reduced quantum efficiency of CZTSSe-based solar cells [17]. Therefore, it is necessary to replace the CdS buffer layer in CZTSSe solar cells with a Cd-free and wider E_g buffer layer.

The ZnO semiconductor has been widely applied in a variety of solar cells, such as perovskite solar cells [18], organic solar cells [19], Sb_2Se_3 solar cells [20], CIGS solar cells [21], CZTSSe solar cells, etc., due to advantages which include excellent N-type characteristics, wide E_g (3.3 eV), non-toxicity, and abundant reserves [22]. At present, the buffer layers most commonly used in Cd-free CZTSSe solar cells with over 10% PCE are $Zn_{1-x}Sn_xO$ (ZTO) films obtained by doping Sn elements into ZnO films [23]. Therefore, the development of quality ZnO films is key to fabricating efficient Cd-free CZTSSe devices. Compared with other Cd-free buffer layers, such as $Zn_{1-x}Mg_xO$, In_2O_3, etc. [24,25], the use of an ZnO buffer layer with a simple preparation process and low raw material cost is expected to assist in the development of low-cost environmentally friendly CZTSSe solar cells. Htay et al. used ultrasonic spray pyrolysis (USP) to deposit ZnO films on Cu_2ZnSnS_4 films to fabricate Cd-free solar cells based on CZTS/ZnO heterojunctions [26]. Katagiri et al. fabricated a Cd-free CZTS device with a ZnO buffer layer using atmospheric pressure chemical vapor deposition (A-CVD) [27]. In these reports, the poor density and complex preparation process of the ZnO buffer layers were not conducive to the preparation of low-cost and efficient CZTSSe/ZnO devices. Meanwhile, high-quality interfaces are very important for the improvement of device properties [28,29]. Therefore, improvements in ZnO films and the quality of CZTSSe/ZnO interfaces are beneficial to the development of low-cost and efficient CZTSSe/ZnO solar cells. However, to date, the PCEs of CZTSSe/ZnO solar cells, which are the basic component of CZTSSe/ZTO solar cells, are only 1.3% [30]. Consequently, an investigation of efficient CZTSSe/ZnO solar cells would lay the foundation for the development of low-cost and efficient Cd-free CZTSSe devices. In addition, research on Cd-free buffer layers has mainly been focused on rigid CZTSSe solar cells with soda-lime glass (SLG) as substrates, which can only be applied in limited planar scenarios, such as roof and ground situation. The application of CZTSSe solar cells could be extended to curved scenarios such as solar cars and buildings through the application Cd-free buffer layers to flexible CZTSSe solar cells.

In this work, ZnO buffer layers are used to prepare flexible CZTSSe solar cells, where the most flexible CZTSSe/ZnO device is able to achieve a PCE of 5.0%. The effects of the ZnO film thickness and the CZTSSe/ZnO heterojunction annealing temperature on the photovoltaic properties of Cd-free flexible CZTSSe solar cells are studied in detail. The results show that a flexible CZTSSe/ZnO solar cell with high-quality heterojunction and low defect density can be obtained when using the optimal ZnO buffer layer thickness (100 nm) and the appropriate annealing temperature (200 °C).

2. Materials and Methods

2.1. Fabrication of ZnO Buffer Layers

The ZnO thin film was prepared using the sputtering method, where the RF power source was applied to the ZnO target (99.99%, ZhongNuo Advanced Material (Beijing) Technology Co., Ltd., Beijing, China). The base pressure was below 9.0×10^{-4} Pa. The operating gas was argon. The working gas pressure was 2 Pa. The distance between the target and the sample was 12.5 cm. The deposition rate of the ZnO film was 0.05 nm/s. The thickness of the ZnO film (50, 100 and 150 nm) was controlled by the deposition time, and the flexible CZTSSe/ZnO solar cells were correspondingly named ZnO-50 nm, ZnO-100 nm and ZnO-150 nm, respectively. The annealing temperature of the CZTSSe/ZnO heterojunction was further investigated after optimization of the ZnO film thickness. The ZnO film was annealed on a hot stage at a temperature of 150–250 °C for 30 min after deposition.

2.2. Fabrication of Solar Cells

Flexible CZTSSe/ZnO solar cells based on Mo foil/CZTSSe/ZnO/ITO/Ag were prepared. Ethylenediamine, ethanedithiol, and elemental powders (Cu, Zn, Sn, S, and Se) were used to prepare the precursor solution by heating and stirring at 60–70 °C. The ethylenediamine, ethanedithiol, Sn powder and Se powder were purchased from Alfa Aesar. The Cu powder was purchased from Macklin. The Zn and S powders were purchased from Aladdin. Subsequently, in an Ar-filled glovebox, CZTSSe precursor films were prepared on clean Mo foils using the spin-coating method. Then, the precursor film was placed in a graphite box for the selenization process. During selenization, nitrogen was introduced into the selenization furnace to ensure that the process took place under normal pressure. The selenization conditions were raised from room temperature to 550 °C within 1 min, held at 550 °C for 900 s, and then naturally cooled to obtain the CZTSSe film (2 μm). A detailed description of the device fabrication process can be found in our previous work [31]. The ZnO and ITO (200 nm) films were deposited by RF magnetron sputtering. The Ag electrode (500 nm) was deposited by thermal evaporation. Finally, 9 cells with an active area of 0.21 cm^2 were realized from each sample.

2.3. Characterizations and Measurements

The crystallization of the thin films was achieved by a multifunctional X-ray polycrystalline diffractometer (DY1602). The SEM images were characterized by a focused ion beam scanning electron microscope (Helios G4 CX). The transmittances of the films were measured using a spectrophotometer (Agilent CARY 5000 Scan, Santa Clara, CA, USA). The external quantum efficiency (EQE) spectra were measured using a QTEST HIFINITY 5 IPCE instrument. Under a standard AM1.5 illumination (100 mW/cm^2) from a solar simulator (SUN 2000), a Keithley 2400 source meter was used to measure the current density–voltage (J-V) curves. The space charge limited current (SCLC) characterizations were conducted using a semiconductor characterization system (Fs-Pro, Hong Kong). The photoluminescence (PL) quenching spectra were measured using a Fluorescence Spectrophotometer (F-7000). Capacitance–voltage (C-V) tests were performed using a Keithley 4200 semiconductor parametric instrument. Electrochemical impedance spectra (EIS) were obtained using an electrochemical workstation (SP-200) test setup. The transient photo-voltage (TPV) and transient photocurrent (TPC) signals were recorded using an LED light source system and a digital oscilloscope (Agilent, 1 GHz), and then calculated by fitting.

3. Results and Discussion

In order to solve the problem related to the high toxicity of Cd in conventional flexible CZTSSe solar cells, we designed an environmentally friendly flexible CZTSSe solar cell with a device structure consisting of Mo foil/CZTSSe/ZnO/ITO/Ag (Figure 1a). Figure 1b shows a cross-section SEM image of the flexible CZTSSe device. There is a three-layer structure (CZTSSe/ZnO/ITO) on the Mo foil. The light transmittance of the buffer layers and the light absorption of CZTSSe absorber are compared between the device with CZTSSe/ZnO and the device without CZTSSe/CdS/ZnO. The ZnO and CdS/ZnO layers are grown on soda-lime glass substrates, denoted as ZnO and CdS + ZnO, respectively. Figure 1c shows the transmittance of the ZnO and CdS + ZnO layers. It can be seen that the transmittance of the buffer layer is significantly improved in the range of 400–600 nm after the removal of the CdS layer. This is because the energy band gap (E$_g$) of ZnO (3.3 eV) is wider than that of CdS (2.4 eV), which is conducive to the absorption of more photons and increasing the short-circuit current density (J$_{sc}$) of the device. Next, the EQE curves of the flexible devices with ZnO and CdS+ZnO buffer layers are shown in Figure 1d. It can be seen that the EQE values of the Cd-free CZTSSe solar cell are obviously higher than those of the device with the CdS buffer layer in the range of 400–600 nm, which is consistent with the results obtained for buffer layer transmittance. This indicates that the flexible CZTSSe solar cell with ZnO buffer layer exhibits better light absorption capacity. The integrated J$_{sc}$ of the flexible device with the ZnO buffer layer is greater than that of the flexible device with the CdS + ZnO buffer layer, which is because more photons enter the CZTSSe absorber. The electrical properties of the CdS/ZnO and ZnO buffer layers are analyzed using the space charge limited current (SCLC) characteristic, where the I-V data are acquired from electron-only devices with the structure of Au/buffer layer/Au. The current and voltage of the buffer layers are in accordance with Ohm's law (I = k*V, k is constant) at low voltage. As shown in Figure 1e, the conductivity (σ) values of ZnO and CdS/ZnO buffer layers are 1.56×10^{-8} and 2.55×10^{-7} S·cm^{-1}, respectively. The lower σ of the ZnO buffer layer is not conducive to the fill factor (FF) of the device.

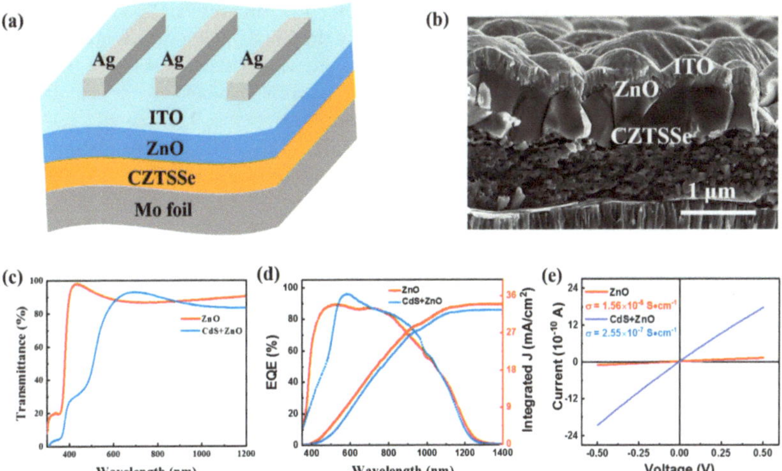

Figure 1. (**a**) The schematic diagram of device structure and (**b**) the cross-section SEM image for the flexible CZTSSe/ZnO solar cell. (**c**) Transmittances of the ZnO and CdS/ZnO buffer layers. (**d**) EQE curves of devices based on ZnO and CdS/ZnO buffer layers. (**e**) SCLC of ZnO and CdS/ZnO buffer layers.

3.1. Optimization of ZnO Buffer Layer Thickness

In conventional flexible CZTSSe solar cells, the CdS and ZnO layers together constitute the N-type region of the efficient solar cells [32]. The removal of CdS film affects the device performance. In addition, after the buffer layer thickness is reduced to around 10 nm, the PCE of the device is only 0.6% (Figure S1, Supplementary Materials). Therefore, the thickness of the ZnO layer is increased to compensate for the buffer layer effect. ZnO films with different thicknesses (50 nm, 100 nm, and 150 nm) were deposited on the CZTSSe absorbers, where the CZTSSe/ZnO heterojunctions were annealed at low temperature (150 °C). The corresponding CZTSSe solar cells are referred to as ZnO-50 nm, ZnO-100 nm, and ZnO-150 nm, respectively. The J-V curves of the devices were determined in order to obtain photovoltaic parameters. The statistical histogram and Gaussian distribution curves of the PCEs of the three devices are shown in Figure 2a. When the thickness of ZnO films is increased from 50 nm to 150 nm, the average PCE of the devices first increases from about 2% to 4% and then decreases to about 3%. The flexible CZTSSe/ZnO solar cells with 100 nm ZnO films exhibit the best performance. The J-V curves of champion devices with different ZnO film thickness are shown in Figure S2a. Both the open-circuit voltage (V_{oc}) and J_{sc} of the ZnO-100 nm device are higher than those of the ZnO-50 nm and ZnO-150 nm devices. The best device—the ZnO-100 nm device—achieves a PCE of 4.4%, with a V_{oc} of 287 mV, a J_{sc} of 34.9 mA/cm^2, and an FF of 44.8%. The charge extraction efficiency of the devices with optimized ZnO layer are studied on the basis of the PL quenching spectra of CdS/ZnO and ZnO buffer layers (different thicknesses) based on CZTSSe absorbers. The CZTSSe/ZnO-100 nm device shows better PL quenching (Figure 2b), indicating that the charge extraction efficiency between CZTSSe absorber and ZnO-100 nm buffer layer is higher [33]. The PL spectra of ZnO buffer layers with different thicknesses are shown in Figure S2b. The PL intensity of the ZnO-100 nm buffer layer is the lowest, reflecting that there are fewer bulk defects in the ZnO-100 nm buffer layer. In order to further study the effect of ZnO thickness on the CZTSSe/ZnO heterojunction, the C-V characteristics were determined in order to obtain information on the built-in electric field and the width of the heterojunction depletion region. Figure 2c shows the C^{-2}-V curves of the CZTSSe/ZnO heterojunction solar cells, where the built-in electric field (V_{bi}) can be obtained on the basis of the intersection of the linear region extension of the curve with the x axis. The V_{bi} values of the ZnO-50 nm, ZnO-100 nm and ZnO-150 nm devices are 0.30 V, 0.32 V and 0.27 V, respectively. The V_{bi} of the ZnO-100 nm device is the greatest, which is consistent with the high V_{oc} of the device. Figure 2d shows curves of the width of the depletion region (W_d) versus carrier concentration (N_D), which can be calculated using the C-V data [32,34]. The W_d of the ZnO-100 nm device reaches 0.31 μm, which is wider than those of the ZnO-50 nm and ZnO-150 nm devices (0.23 and 0.27 μm). Wider values of W_d are beneficial to the collection and transmission of photon-generated carriers, improving the J_{sc} of the device.

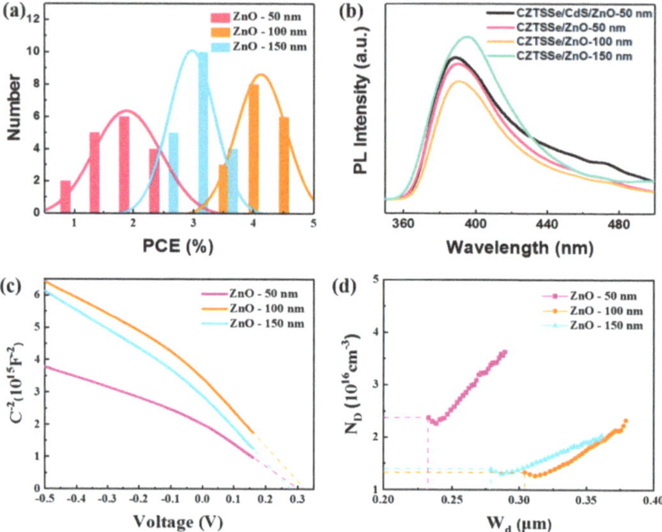

Figure 2. (**a**) The statistical PCE histogram of the flexible CZTSSe solar cells with ZnO layers with different thicknesses. (**b**) PL quenching spectra of the different buffer layers based on CZTSSe absorbers. (**c,d**) The C-V characteristics of the devices: (**c**) C^{-2}-V and (**d**) N_D-W_d curves.

3.2. Optimization of Annealing Temperature of CZTSSe/ZnO Heterojunction

High-quality N-type ZnO buffer layers play an important role in the development of efficient flexible CZTSSe solar cells. The improvement of the quality of ZnO films is an effective way of enhancing the performance of flexible devices. The annealing temperature is one of the important factors affecting the growth of ZnO films [35,36]. Temperatures that are too high can destroy the structure of CZTSSe absorbers, thus deteriorating the device performance. Therefore, the annealing temperature (150 °C, 200 °C and 250 °C) of the CZTSSe/ZnO was adjusted to achieve high-quality ZnO film and a flexible device with improved performance.

AFM and SEM characterizations were performed to characterize the surface roughness and crystal morphology of the CZTSSe/ZnO layers following annealing at different temperatures. The AFM images of three ZnO films deposited on CZTSSe absorbers are shown in Figure 3a–c. The root mean square (RMS) values of the surface roughness of the CZTSSe/ZnO layers produced with three different annealing temperatures are around 60 nm, showing no obvious difference. Figure 3d–f show top-view SEM images of the ZnO films prepared on CZTSSe with annealing at different temperatures. The larger grains (about 1.5 μm diameter) are the CZTSSe absorber. The fine grains (about 10 nm) on the surface are the ZnO film. According to Figure 3d–f, the annealing temperature (150–250 °C) has little effect on the morphology of the CZTSSe absorber.

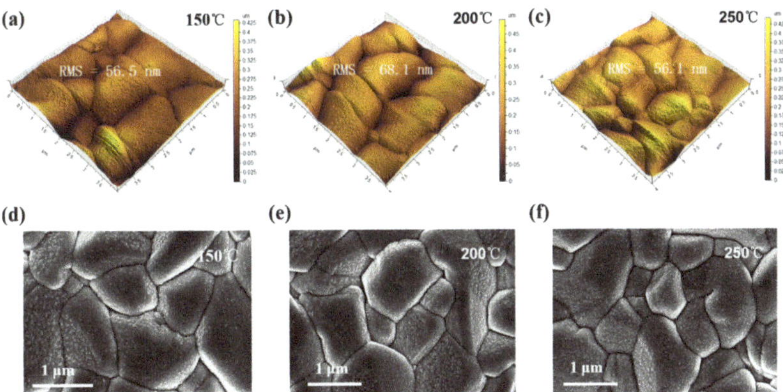

Figure 3. Top-view images of the ZnO films on the CZTSSe absorbers prepared at annealing temperatures of 150, 200 and 250 °C: (**a–c**) AFM images; (**d–f**) SEM images.

The effects of annealing temperature on the crystal structures of the CZTSSe and ZnO were analyzed by XRD characterization. Conventional XRD can only detect the phase of the CZTSSe absorber due to the ZnO film being too thin. Figure 4a shows the XRD patterns of the CZTSSe absorbers after annealing at different temperatures. The (112), (220) and (312) XRD peaks belong to the kesterite-type tetragonal phase. The characteristic peaks at 14.8° (marked ♣) and 30.1° (marked ♠) may be $Sn(S,Se)_2$ and $Cu(S,Se)$, respectively [9,37–39], which are secondary phases. There are no significant differences among the three samples in terms of peak position, peak intensity or half width, indicating that annealing temperature in the range 150–250 °C has not effect on the crystalline phase of the CZTSSe absorber. Subsequently, the crystal structures of the ZnO films on the CZTSSe absorber were analyzed by grazing incidence X-ray diffraction (GIXRD), as shown in Figure 4b. The two characteristic peaks perfectly match those of the ZnO crystal, among which the characteristic peak of 2θ at 34.14° is the strongest, corresponding to the (100) crystal plane of ZnO, and the characteristic peak of 2θ near 62.58° corresponds to the (103) crystal plane of ZnO [40,41]. Detailed data regarding these parameters are summarized in Table S1. With increasing annealing temperature, the full width at half maximum (FWHM) of the characteristic peaks of XRD decreases gradually, indicating that the grain size of the ZnO films is gradually increasing. However, the characteristic peak intensities show different trends. The characteristic peak intensity is higher for the ZnO film annealed at a temperature of 200 °C compared to the films based on annealing temperatures of 150 °C and 250 °C. Therefore, the ZnO film shows good crystallinity at an annealing temperature of 200 °C.

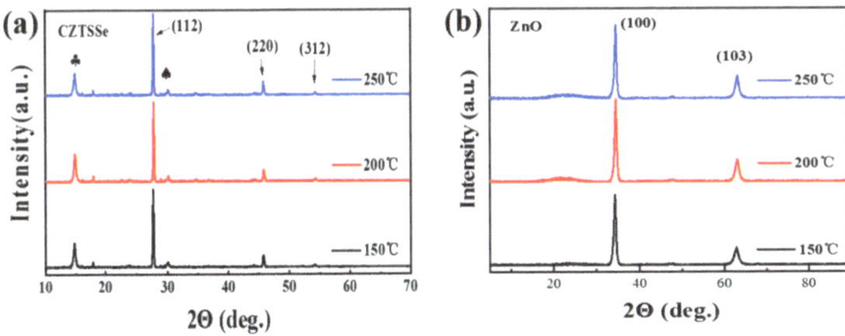

Figure 4. XRD patterns of ZnO/CZTSSe heterojunctions annealed at different temperatures: (**a**) conventional XRD and (**b**) GIXRD patterns. The peaks at 14.8° and 30.1° are marked ♣ and ♠, respectively.

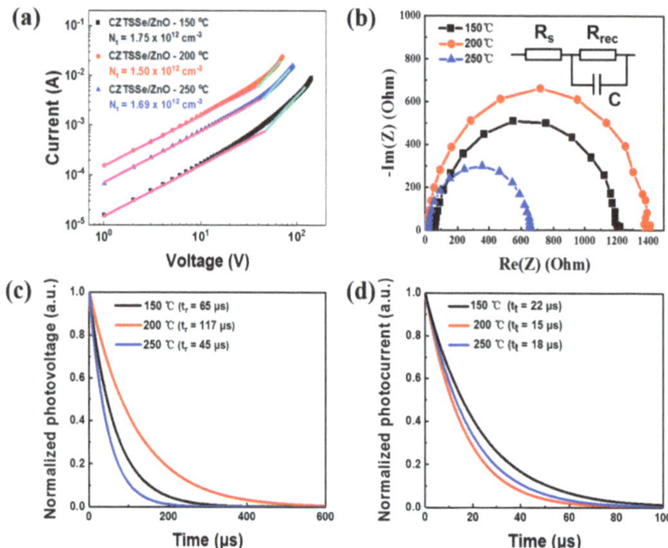

Figure 6. (a) The SCLC of CZTSSe/ZnO films produced with different annealing temperatures. The interfacial characterization of the three flexible CZTSSe/ZnO solar cells: (b) EIS plots, (c) TPV decay curves, and (d) TPC decay curves.

4. Conclusions

In this work, we demonstrated a Cd-free flexible CZTSSe/ZnO solar cell that satisfies the aims of environmental protection and low cost. The effect of the ZnO films on the performance of the flexible device was investigated. The results show that the 100nm thick ZnO film offers the best buffer layer effect. The most appropriate annealing temperature for the ZnO for attaining high performance in the flexible devices was explored. The AFM, SEM and XRD characterizations showed that the annealing treatment (150 °C–250 °C) did not destroy the structure of the CZTSSe absorbers. Meanwhile, the ZnO buffer layer exhibited better crystallinity when using an annealing temperature of 200 °C. The optimal flexible device was able to achieve a PCE of 5%, which is the highest PCE achieved by flexible CZTSSe/ZnO solar cells. The J-V, EIS, TPV and TPC characterizations showed that the optimal flexible CZTSSe/ZnO solar cell was able to realize a high-quality heterojunction, low defect density, and improved charge transfer capability. The investigation of the ZnO buffer layer in the flexible CZTSSe/ZnO solar cells could provide new concepts in the development of low-cost, environmentally friendly and efficient CZTSSe solar cells.

Supplementary Materials: The following supporting information can be downloaded at: https://www.mdpi.com/article/10.3390/ma16072869/s1, Figure S1: J-V curves of the flexible CZTSSe solar cells with 10 nm thick ZnO layers; Figure S2: (a) J-V curves of the flexible CZTSSe solar cells with different thick ZnO layers. (b) PL spectra of ZnO layers with different thickness; Table S1: GIXRD results of ZnO films annealed at different temperatures.

Author Contributions: Q.S. and J.T. contributed equally to this work. Conceptualization, C.Z. and S.C.; validation, Q.S., J.T. and Y.L.; resources, S.C.; data curation, J.T., W.X., H.D. and Q.Z.; writing—original draft preparation, Q.S. and J.T.; writing—review and editing, C.Z., S.C., and J.W.; visualization, Q.S., J.T., Y.L., W.X., H.D., Q.Z. and J.W.; supervision, C.Z. and S.C. All authors have read and agreed to the published version of the manuscript.

Funding: This research was supported by the National Natural Science Foundation of China (grant number: 62074037), the Science and Technology Department of Fujian Province (grant number: 2020I0006), the Fujian Science & Technology Innovation Laboratory for Optoelectronic Information of China (grant number: 2021ZZ124).

Institutional Review Board Statement: Not applicable.

Informed Consent Statement: Not applicable.

Data Availability Statement: Data are contained within the article.

Acknowledgments: The authors thank the Testing Center of Fuzhou University for the test service.

Conflicts of Interest: The authors declare no conflict of interest.

References

1. Sahu, M.; Reddy, V.R.M.; Kim, B.; Patro, B.; Park, C.; Kim, W.K.; Sharma, P. Fabrication of Cu_2ZnSnS_4 Light Absorber Using a Cost-Effective Mechanochemical Method for Photovoltaic Applications. *Materials* **2022**, *15*, 1708. [CrossRef] [PubMed]
2. Jeong, W.-L.; Park, S.-H.; Jho, Y.-D.; Joo, S.-K.; Lee, D.-S. The Role of the Graphene Oxide (GO) and Reduced Graphene Oxide (RGO) Intermediate Layer in CZTSSe Thin-Film Solar Cells. *Materials* **2022**, *15*, 3419. [CrossRef]
3. Xie, W.; Yan, Q.; Sun, Q.; Li, Y.; Zhang, C.; Deng, H.; Cheng, S. A Progress Review on Challenges and Strategies of Flexible $Cu_2ZnSn(S, Se)_4$ Solar Cells. *Sol. RRL* **2023**, *7*, 2201036. [CrossRef]
4. Lin, B.; Sun, Q.; Zhang, C.; Deng, H.; Li, Y.; Xie, W.; Li, Y.; Zheng, Q.; Wu, J.; Cheng, S. Efficient Cd-Free Flexible CZTSSe Solar Cells with Quality Interfaces by Using the $Zn_{1-x}Sn_xO$ Buffer Layer. *ACS Appl. Energy Mater.* **2023**, *6*, 1037–1045. [CrossRef]
5. Rondiya, S.; Jadhav, Y.; Nasane, M.; Jadkar, S.; Dzade, N.Y. Interface Structure and Band Alignment of CZTS/CdS Heterojunction: An Experimental and First-Principles DFT Investigation. *Materials* **2019**, *12*, 4040. [CrossRef]
6. Guo, Q.; Ford, G.M.; Yang, W.-C.; Walker, B.C.; Stach, E.A.; Hillhouse, H.W.; Agrawal, R. Fabrication of 7.2% Efficient CZTSSe Solar Cells Using CZTS Nanocrystals. *J. Am. Chem. Soc.* **2010**, *132*, 17384–17386. [CrossRef]
7. Gong, Y.; Zhu, Q.; Li, B.; Wang, S.; Duan, B.; Lou, L.; Xiang, C.; Jedlicka, E.; Giridharagopal, R.; Zhou, Y.; et al. Elemental de-mixing-induced epitaxial kesterite/CdS interface enabling 13%-efficiency kesterite solar cells. *Nat. Energy* **2022**, *7*, 966–977. [CrossRef]
8. Yang, K.J.; Kim, S.; Kim, S.Y.; Son, D.H.; Lee, J.; Kim, Y.I.; Sung, S.J.; Kim, D.H.; Enkhbat, T.; Kim, J.; et al. Sodium Effects on the Diffusion, Phase, and Defect Characteristics of Kesterite Solar Cells and Flexible $Cu_2ZnSn(S,Se)_4$ with Greater than 11% Efficiency. *Adv. Funct. Mater.* **2021**, *31*, 2102238. [CrossRef]
9. Xu, H.; Ge, S.; Yang, W.; Khan, S.N.; Huang, Y.; Mai, Y.; Gu, E.; Lin, X.; Yang, G. 9.63% efficient flexible $Cu_2ZnSn(S,Se)_4$ solar cells fabricated via scalable doctor-blading under ambient conditions. *J. Mater. Chem. A* **2021**, *9*, 25062–25072. [CrossRef]
10. Sun, Q.-Z.; Jia, H.-J.; Cheng, S.-Y.; Deng, H.; Yan, Q.; Duan, B.-W.; Zhang, C.-X.; Zheng, Q.; Yang, Z.-Y.; Luo, Y.-H.; et al. A 9% efficiency of flexible Mo-foil-based $Cu_2ZnSn(S,Se)_4$ solar cells by improving CdS buffer layer and heterojunction interface. *Chin. Phys. B* **2020**, *29*, 128801. [CrossRef]
11. Yang, K.-J.; Kim, S.; Kim, S.-Y.; Ahn, K.; Son, D.-H.; Kim, S.-H.; Lee, S.-J.; Kim, Y.-I.; Park, S.-N.; Sung, S.-J.; et al. Flexible $Cu_2ZnSn(S,Se)_4$ solar cells with over 10% efficiency and methods of enlarging the cell area. *Nat. Commun.* **2019**, *10*, 2959. [CrossRef]
12. Xie, W.; Sun, Q.; Yan, Q.; Wu, J.; Zhang, C.; Zheng, Q.; Lai, Y.; Deng, H.; Cheng, S. 10.24% Efficiency of Flexible $Cu_2ZnSn(S,Se)_4$ Solar Cells by Pre-Evaporation Selenization Technique. *Small* **2022**, *18*, 2201347. [CrossRef]
13. Yan, Q.; Sun, Q.; Deng, H.; Xie, W.; Zhang, C.; Wu, J.; Zheng, Q.; Cheng, S. Enhancing carrier transport in flexible CZTSSe solar cells via doping Li strategy. *J. Energy Chem.* **2022**, *75*, 8–15. [CrossRef]
14. Zhang, C.; Yang, Z.; Deng, H.; Yan, Q.; Xie, W.; Sun, Q.; Sheng, X.; Cheng, S. Numerical Study to Improve the Back Interface Contact of CZTSSe Solar Cells Using Oxygen-Doped $Mo(Se_{1-x}, O_x)_2$. *J. Phys. Chem. C* **2021**, *125*, 16746–16752. [CrossRef]
15. Nguyen, M.; Ernits, K.; Tai, K.F.; Ng, C.F.; Pramana, S.S.; Sasangka, W.A.; Batabyal, S.K.; Holopainen, T.; Meissner, D.; Neisser, A.; et al. ZnS buffer layer for $Cu_2ZnSn(SSe)_4$ monograin layer solar cell. *Sol. Energy* **2015**, *111*, 344–349. [CrossRef]
16. Larsen, J.K.; Larsson, F.; Törndahl, T.; Saini, N.; Riekehr, L.; Ren, Y.; Biswal, A.; Hauschild, D.; Weinhardt, L.; Heske, C.; et al. Cadmium Free Cu_2ZnSnS_4 Solar Cells with 9.7% Efficiency. *Adv. Energy Mater.* **2019**, *9*, 1900439. [CrossRef]
17. Naghavi, N.; Abou-Ras, D.; Allsop, N.; Barreau, N.; Bücheler, S.; Ennaoui, A.; Fischer, C.H.; Guillen, C.; Hariskos, D.; Herrero, J.; et al. Buffer layers and transparent conducting oxides for chalcopyrite $Cu(In,Ga)(S,Se)_2$ based thin film photovoltaics: Present status and current developments. *Prog. Photovolt. Res. Appl.* **2010**, *18*, 411–433. [CrossRef]
18. Mahmood, K.; Hameed, M.; Rehman, F.; Khalid, A.; Imran, M.; Mehran, M.T. A multifunctional blade-coated ZnO seed layer for high-efficiency perovskite solar cells. *Appl. Phys. A* **2019**, *125*, 83. [CrossRef]
19. Lee, D.; Kang, T.; Choi, Y.-Y.; Oh, S.-G. Preparation of electron buffer layer with crystalline ZnO nanoparticles in inverted organic photovoltaic cells. *J. Phys. Chem. Solids* **2017**, *105*, 66–71. [CrossRef]
20. Wang, L.; Li, D.-B.; Li, K.; Chen, C.; Deng, H.-X.; Gao, L.; Zhao, Y.; Jiang, F.; Li, L.; Huang, F.; et al. Stable 6%-efficient Sb_2Se_3 solar cells with a ZnO buffer layer. *Nat. Energy* **2017**, *2*, 17046. [CrossRef]
21. Arepalli, V.K.; Lee, W.-J.; Chung, Y.-D.; Kim, J. Growth and device properties of ALD deposited ZnO films for CIGS solar cells. *Mater. Sci. Semicond. Process* **2021**, *121*, 105406. [CrossRef]
22. Lin, B.; Sun, Q.; Zhang, C.; Deng, H.; Xie, W.; Tang, J.; Zheng, Q.; Wu, J.; Zhou, H.; Cheng, S. 9.3% Efficient Flexible $Cu_2ZnSn(S,Se)_4$ Solar Cells with High-Quality Interfaces via Ultrathin CdS and $Zn_{0.8}Sn_{0.2}O$ Buffer Layers. *Energy Technol.* **2022**, *10*, 2200571. [CrossRef]

23. Lee, J.; Enkhbat, T.; Han, G.; Sharif, M.H.; Enkhbayar, E.; Yoo, H.; Kim, J.H.; Kim, S.; Kim, J. Over 11 % efficient eco-friendly kesterite solar cell: Effects of S-enriched surface of $Cu_2ZnSn(S,Se)_4$ absorber and band gap controlled (Zn,Sn)O buffer. *Nano Energy* **2020**, *78*, 105206. [CrossRef]

24. Barkhouse, D.A.R.; Haight, R.; Sakai, N.; Hiroi, H.; Sugimoto, H.; Mitzi, D.B. Cd-free buffer layer materials on $Cu_2ZnSn(S_xSe_{1-x})_4$: Band alignments with ZnO, ZnS, and In_2S_3. *Appl. Phys. Lett.* **2012**, *100*, 193904. [CrossRef]

25. Hedibi, A.; Gueddim, A.; Bentria, B. Numerical Modeling and Optimization of ZnO:Al/iZnO/ZnMgO/CZTS Photovoltaic Solar Cell. *Trans. Electr. Electron. Mater.* **2021**, *22*, 666–672. [CrossRef]

26. Htay, M.T.; Hashimoto, Y.; Momose, N.; Sasaki, K.; Ishiguchi, H.; Igarashi, S.; Sakurai, K.; Ito, K. A Cadmium-Free Cu_2ZnSnS_4/ZnO Hetrojunction Solar Cell Prepared by Practicable Processes. *Jpn. J. Appl. Phys.* **2011**, *50*, 032301. [CrossRef]

27. Hironori, K.; Kazuo, J.; Masami, T.; Hideaki, A.; Koichiro, O. The influence of the composition ratio on CZTS-based thin film solar cells. *Mater. Res. Soc. Symp. Proc.* **2009**, *1165*, 1–12.

28. Alam, M.W.; Ansari, M.Z.; Aamir, M.; Waheed-Ur-Rehman, M.; Parveen, N.; Ansari, S.A. Preparation and Characterization of Cu and Al Doped ZnO Thin Films for Solar Cell Applications. *Crystals* **2022**, *12*, 128. [CrossRef]

29. Alam, M.W.; Wang, Z.; Naka, S.; Okada, H. Performance Enhancement of Top-Contact Pentacene-Based Organic Thin-Film Transistors with Bilayer WO_3/Au Electrodes. *Jpn. J. Appl. Phys.* **2013**, *52*, 03BB08. [CrossRef]

30. Li, X.; Su, Z.; Venkataraj, S.; Batabyal, S.K.; Wong, L.H. 8.6% Efficiency CZTSSe solar cell with atomic layer deposited Zn-Sn-O buffer layer. *Sol. Energy Mater. Sol. Cells* **2016**, *157*, 101–107. [CrossRef]

31. Yan, Q.; Cheng, S.; Li, H.; Yu, X.; Fu, J.; Tian, Q.; Jia, H.; Wu, S. High flexible $Cu_2ZnSn(S,Se)_4$ solar cells by green solution-process. *Sol. Energy* **2019**, *177*, 508–516. [CrossRef]

32. Sun, Q.; Deng, H.; Yan, Q.; Lin, B.; Xie, W.; Tang, J.; Zhang, C.; Zheng, Q.; Wu, J.; Yu, J.; et al. Efficiency Improvement of Flexible $Cu_2ZnSn(S,Se)_4$ Solar Cells by Window Layer Interface Engineering. *ACS Appl. Energy Mater.* **2021**, *4*, 14467–14475. [CrossRef]

33. Kattoor, V.; Awasthi, K.; Jokar, E.; Diau, E.W.-G.; Ohta, N. Integral Method Analysis of Electroabsorption Spectra and Electrophotoluminescence Study of $(C_4H_9NH_3)_2PbI_4$ Organic–Inorganic Quantum Well. *J. Phys. Chem. C* **2018**, *122*, 26623–26634. [CrossRef]

34. Yu, X.; Cheng, S.; Yan, Q.; Fu, J.; Jia, H.; Sun, Q.; Yang, Z.; Wu, S. Efficient flexible Mo foil-based $Cu_2ZnSn(S, Se)_4$ solar cells from In-doping technique. *Sol. Energy Mater. Sol. Cells* **2020**, *209*, 110434. [CrossRef]

35. Hirahara, N.; Onwona-Agyeman, B.; Nakao, M. Preparation of Al-doped ZnO thin films as transparent conductive substrate in dye-sensitized solar cell. *Thin Solid Film* **2012**, *520*, 2123–2127. [CrossRef]

36. Iqbal, J.; Jilani, A.; Ziaul Hassan, P.M.; Rafique, S.; Jafer, R.; Alghamdi, A.A. ALD grown nanostructured ZnO thin films: Effect of substrate temperature on thickness and energy band gap. *J. King Saud Univ.-Sci.* **2016**, *28*, 347–354. [CrossRef]

37. Guo, J.; Zhou, W.-H.; Pei, Y.-L.; Tian, Q.-W.; Kou, D.-X.; Zhou, Z.-J.; Meng, Y.-N.; Wu, S.-X. High efficiency CZTSSe thin film solar cells from pure element solution: A study of additional Sn complement. *Sol. Energy Mater. Sol. Cells* **2016**, *155*, 209–215. [CrossRef]

38. Zhao, X.; Kou, D.; Zhou, W.; Zhou, Z.; Meng, Y.; Meng, Q.; Zheng, Z.; Wu, S. Nanoscale electrical property enhancement through antimony incorporation to pave the way for the development of low-temperature processed $Cu_2ZnSn(S,Se)_4$ solar cells. *J. Mater. Chem. A* **2019**, *7*, 3135–3142. [CrossRef]

39. Deng, Y.; Zhou, Z.; Zhang, X.; Cao, L.; Zhou, W.; Kou, D.; Qi, Y.; Yuan, S.; Zheng, Z.; Wu, S. Adjusting the Sn_{Zn} defects in $Cu_2ZnSn(S,Se)_4$ absorber layer via Ge^{4+} implanting for efficient kesterite solar cells. *J. Energy Chem.* **2021**, *61*, 1–7. [CrossRef]

40. Al-Ariki, S.; Yahya, N.A.A.; Al-A'nsi, S.A.A.; Jumali, M.H.H.; Jannah, A.N.; Abd-Shukor, R. Synthesis and comparative study on the structural and optical properties of ZnO doped with Ni and Ag nanopowders fabricated by sol gel technique. *Sci. Rep.* **2021**, *11*, 11948. [CrossRef]

41. Vaiano, V.; Matarangolo, M.; Murcia, J.J.; Rojas, H.; Navío, J.A.; Hidalgo, M.C. Enhanced photocatalytic removal of phenol from aqueous solutions using ZnO modified with Ag. *Appl. Catal. B Environ.* **2018**, *225*, 197–206. [CrossRef]

42. Pockett, A.; Lee, H.K.H.; Coles, B.L.; Tsoi, W.C.; Carnie, M.J. A combined transient photovoltage and impedance spectroscopy approach for a comprehensive study of interlayer degradation in non-fullerene acceptor organic solar cells. *Nanoscale* **2019**, *11*, 10872–10883. [CrossRef]

43. Bisquert, J.; Janssen, M. From Frequency Domain to Time Transient Methods for Halide Perovskite Solar Cells: The Connections of IMPS, IMVS, TPC, and TPV. *J. Phys. Chem. Lett.* **2021**, *12*, 7964–7971. [CrossRef]

44. Chen, Y.; Wu, S.; Li, X.; Liu, M.; Chen, Z.; Zhang, P.; Li, S. Efficient and stable low-cost perovskite solar cells enabled by using surface passivated carbon as the counter electrode. *J. Mater. Chem. C* **2022**, *10*, 1270–1275. [CrossRef]

Disclaimer/Publisher's Note: The statements, opinions and data contained in all publications are solely those of the individual author(s) and contributor(s) and not of MDPI and/or the editor(s). MDPI and/or the editor(s) disclaim responsibility for any injury to people or property resulting from any ideas, methods, instructions or products referred to in the content.

materials MDPI

Article

Lithium-Ion Glass Gating of HgTe Nanocrystal Film with Designed Light-Matter Coupling

Stefano Pierini [1,†], Claire Abadie [1,2,†], Tung Huu Dang [1], Adrien Khalili [1], Huichen Zhang [1], Mariarosa Cavallo [1], Yoann Prado [1], Bruno Gallas [1], Sandrine Ithurria [3], Sébastien Sauvage [4], Jean Francois Dayen [5,6], Grégory Vincent [2] and Emmanuel Lhuillier [1,*]

[1] Sorbonne Université, CNRS, Institut des NanoSciences de Paris, INSP, 75005 Paris, France
[2] ONERA-The French Aerospace Lab, 6 Chemin de la Vauve aux Granges, 91123 Palaiseau, France
[3] Laboratoire de Physique et d'Etude des Matériaux, ESPCI-Paris, PSL Research University, Sorbonne Université, CNRS, 10 Rue Vauquelin, 75005 Paris, France
[4] CNRS, Centre de Nanosciences et de Nanotechnologies, Université Paris-Saclay, 91120 Palaiseau, France
[5] IPCMS-CNRS, Université de Strasbourg, 23 Rue du Loess, 67034 Strasbourg, France
[6] Institut Universitaire de France, 1 Rue Descartes, CEDEX 05, 75231 Paris, France
* Correspondence: el@insp.upmc.fr
† These authors contributed equally to this work.

Abstract: Nanocrystals' (NCs) band gap can be easily tuned over the infrared range, making them appealing for the design of cost-effective sensors. Though their growth has reached a high level of maturity, their doping remains a poorly controlled parameter, raising the need for post-synthesis tuning strategies. As a result, phototransistor device geometry offers an interesting alternative to photoconductors, allowing carrier density control. Phototransistors based on NCs that target integrated infrared sensing have to (i) be compatible with low-temperature operation, (ii) avoid liquid handling, and (iii) enable large carrier density tuning. These constraints drive the search for innovative gate technologies beyond traditional dielectric or conventional liquid and ion gel electrolytes. Here, we explore lithium-ion glass gating and apply it to channels made of HgTe narrow band gap NCs. We demonstrate that this all-solid gate strategy is compatible with large capacitance up to 2 μF·cm^{-2} and can be operated over a broad range of temperatures (130–300 K). Finally, we tackle an issue often faced by NC-based phototransistors:their low absorption; from a metallic grating structure, we combined two resonances and achieved high responsivity (10 A·W^{-1} or an external quantum efficiency of 500%) over a broadband spectral range.

Keywords: field effect transistor; solid electrolyte; HgTe; nanocrystals; nanophotonics; light resonator; infrared detection

check for **updates**

Citation: Pierini, S.; Abadie, C.; Dang, T.H.; Khalili, A.; Zhang, H.; Cavallo, M.; Prado, Y.; Gallas, B.; Ithurria, S.; Sauvage, S.; et al. Lithium-Ion Glass Gating of HgTe Nanocrystal Film with Designed Light-Matter Coupling. *Materials* **2023**, *16*, 2335. https://doi.org/10.3390/ma16062335

Academic Editor: Joan-Josep Suñol

Received: 31 January 2023
Revised: 1 March 2023
Accepted: 8 March 2023
Published: 14 March 2023

Copyright: © 2023 by the authors. Licensee MDPI, Basel, Switzerland. This article is an open access article distributed under the terms and conditions of the Creative Commons Attribution (CC BY) license (https://creativecommons.org/licenses/by/4.0/).

1. Introduction

As colloidal growth of nanoparticles has gained maturity, nanocrystals (NCs) have become viable building blocks for optoelectronics. Using narrow band gap materials, it is possible to address the infrared (IR) part of the electromagnetic spectrum [1,2]. At such wavelengths, NCs offer an interesting alternative to epitaxially grown semiconductors to design cost-effective devices [3]. This has led to the demonstration of efficient light-emitting devices [4–7] as well as effective light sensors [8–10], including focal plane arrays used to image near- and mid-IR light [11,12].

Though a high degree of control is achieved for the growth of this material, its doping remains [13] a challenge, and the material may not behave as an intrinsic semiconductor. This affects the final device performance. Thus, it is of utmost interest to couple NC films with a gating method to enable an agile tunability of the carrier density. As a result, the phototransistor appears to be a remarkable evolution of the photoconductive geometry, since the gate bias can be used as a knob to control the magnitude of the dark current.

Phototransistors are built around a field-effect transistor [14] (FET), in which, in addition to drain/source electrodes and a channel based on a NC film, the gate appears as a central building block. The gate is used as a capacitor and, under applied bias, it generates charges, tuning the carrier density in the FET channel.

Various gate technologies have been coupled to NC films [15]. The most conventional method continues to rely on dielectrics such as SiO_2 [9,16,17]. Since this material is commercially available and technologically mature, it presents the clear benefit of being easy to implement. Moreover, the fast gate bias sweep and the ability to operate at low temperatures are other key advantages of this technology. The low dielectric constant of silica, on the other hand, requires a large bias operation, possibly close to the material breakdown. High-k materials such as alumina, ZrO_x [18,19] or HfO_2 are the natural alternatives [20–23]. They present the advantage of an all-solid strategy similar to silica but require an additional fabrication step with the deposition of an insulating layer. Despite a limited number of examples [24,25], ferroelectric gating coupled with a NC film offers a strategy to achieve even higher capacitance FETs. The main idea behind ferroelectric gating is to take advantage of the divergence of a ferroelectric material's dielectric constant close to its Curie temperature. However, this limitation tends to reduce the operating temperature ranges. When higher gate capacitances are required, gating methods based on ion displacement become the most effective. Contrary to the previous methods, electrolytes allow for bulk gating. This ability to gate a thick film is critical for the design of phototransistors, which require a film thickness suitable to the light absorption depth [26,27]. Electrolytes are already commonly used to control the carrier density of a NC population [28–30]. The usual method is to introduce a liquid or ion gel electrolyte, which may bring additional constraints related to liquid handling and moisture sensitivity; this is why a solid electrolyte appears easier to implement. The use of ionic glasses has been a step in this direction; LaF_3 glass presents, for example, mobile fluoride vacancies which can be used to induce gating [31–33].

When applied to infrared phototransistors, the gate has to fulfill some specific constraints, such as being:

1. Compatible with operations below room temperature;
2. Solid state-based to be conveniently integrated into a device;
3. Associated with a large capacitance to tune the relatively large thermally activated carrier density observed in narrow band gap NCs.

In this perspective, new materials compatible with this series of constraints must be identified. For a long time, the battery field was using solid electrolytes, and their potential for field-effect transistors has been investigated for 2D materials [34–37] and superconductor-based [38] FETs, but they remain unused for NC-based FETs. Here, we explore the potential of this strategy to design a FET based on HgTe NCs [39], presenting an absorption in the extended short-wave infrared. We show in particular that this lithium-ion glass substrate tends to promote hole injection over electron injection compared to what is observed with a silica gate. In the last part of the paper, we propose an update of the design to engineer light-matter coupling in this system and achieve high absorption from a thin NC film. This strategy allows a large responsivity ($R > 10$ $A·W^{-1}$), three orders of magnitude larger than the value obtained without a light management strategy.

2. Materials and Methods

2.1. Nanocrystal Growth

Chemicals: Mercury chloride ($HgCl_2$; Sigma-Aldrich, St. Louis, MO, USA, 99%), **Mercury compounds are highly toxic. Handle them with special care.** Tellurium powder (Te; Sigma-Aldrich, 99.99%), trioctylphosphine (TOP; Alfa Aesar, Haverhill, MA, USA, 90%), oleylamine (OLA; Thermo Fisher Scientific, Illkirch, France, 80–90%), dodecanethiol (DDT; Sigma-Aldrich, 98%), ethanol absolute anhydrous (VWR, Radnor, PA, USA), methanol (VWR, >98%), isopropanol (IPA; VWR), hexane (VWR, 99%), 2-mercaptoethanol (MPOH; Merck, Rahway, NJ, USA, >99%), N,N dimethylformamide (DMF; VWR), toluene (VWR, 99.8%), acetone (VWR), methylisobutylketone (MIBK; VWR, >98.5%)

All chemicals were used without further purification except oleylamine, that was centrifuged for 5 min at 6000 rpm before use.

1 M TOP:Te precursor: The procedure was taken from ref [40].

HgTe NCs synthesis with band-edge at 4000 cm^{-1}: The procedure was inspired by ref [41].

HgTe ink: The procedure was inspired by ref [42]. Before ink deposition, the substrate was placed in an oxygen plasma cleaner for 2 min to promote adhesion. The ink was spin-coated on the substrate at 4000 rpm (with 2000 rpm·s^{-1} acceleration) for 1 min. This gave a homogeneous film between 100 nm and 250 nm, as can be seen in the height profile obtained from the profilometer, shown in Figure S1 (Supplementary Materials). The speed of spin coating and ink concentration were used to tune the film thickness.

2.2. Material Characterization

Infrared spectroscopy: We conducted measurements in attenuated total reflection (ATR) mode using a Fischer Nicolet iS50. The spectra were acquired with a 4 cm^{-1} resolution and averaged 32 times. The photocurrent spectra were obtained while biasing the sample with a Femto DLPCA 200 current amplifier; the signal was then magnified by the same amplifier and the output signal was fed into the spectrometer. For polarized measurements, a polarizer aligned with the electrode digits (or perpendicular to them) was added on the optical path. The background was then acquired in the presence of the polarizer in both polarizations.

Transmission electron microscopy: NC solution was drop-casted onto a TEM grid and dried on a paper. The grid was degassed overnight to suppress solvent and volatile species.

2.3. Electrical Characterization

Electrode fabrication: The solid electrolyte was purchased from MTI (ref EQ-CGCS-LD). It is a ceramic based on Li_2O-Al_2O_3-SiO_2-P_2O_5-TiO_2. The substrate was cleaned with acetone and rinsed with ethanol and isopropanol, before being dried by a nitrogen jet gas. First, LOR 3A, a sacrificial release layer, was spin-coated and baked at 160 °C for 180 s. Then, AZ 1505 resist was spin-coated and annealed at 105 °C for 90 s. Laser lithography (Heidelberg μPG 101) was used to conduct the lithography procedure. The resist was developed for 20 s in AZ 726 developer, and finally rinsed in water. Then, 3 nm of titanium and 47 nm of gold were evaporated using an e-beam evaporator (Plassys MEB 550s). For lift-off, the substrate was dipped in a PG remover bath for 1 h, and finally rinsed using DI water. The substrate was then dried and electrically tested to check that there was no electrical short. The electrodes were interdigitated electrodes, as schematized in Figure S2, with 25 pairs of digits (i.e., 49 spacings). Each digit was 700 μm long and spaced from its neighbor by 10 μm.

Electrical measurements: The drain and source contacts of the sample were cleaned using a cotton swab moistened with acetone. The sample was glued using a silver paste on a substrate of silicon covered with a gold layer and mounted on the cold finger of a cryostat. The gold layer was used as a gate. The transfer curves and the IV curves were measured using a Keithley 2634B SourceMeter; this device controls both the gate-source and the drain-source voltages, measuring the relative currents.

Photocurrent measurements: The time response measurements and the responsivity measurements were performed using the scope mode of a MLFI Lock-In Amplifier (Zurich Instruments, Zürich, Switzerland). In this case, both the drain-source and the gate-source voltages were controlled by the Lock-In Amplifier. Figure S4 schematized the setup used for the responsivity measurements. A wavefunction generator was used to trigger the laser emission and the oscilloscope; the current was converted into voltage by the Femto DLPCA-200 amplifier and fed into the oscilloscope (Tektronix MDO 3102). From the time trace we can extract both the photocurrent amplitude and the time response.

Impedance Measurements: The MLFI Lock-In Amplifier (Zurich Instruments, Zürich, Switzerland) was used to perform impedance measurements in the range from 0.1 Hz to

500 kHz. We used an AC voltage of 100 mV. No DC voltage was applied, exploring the temperature dependence of the impedance. We first deposited two gold layers on both sides of a Li-based substrate. The surface of the deposited gold layer was determined a posteriori using a camera and digital processing of the image. The area of the device was 8.9 mm^2. The device was glued on a substrate of silicon coated with a gold layer using silver paste. The sample was then connected inside the cryostat, and the top and the back sides were connected to the Lock-In Amplifier.

Noise measurement: Current from the device (at 0.7 V bias, kept in the dark) was amplified by a Femto DLPCA-200, then fed into a SRS SR780 spectrum analyzer. The sample was mounted on the cold finger of a closed cycle cryostat.

Detectivity determination: The specific detectivity (in Jones) of the sample was determined using the formula: $D^* = \frac{R\sqrt{A}}{S_I}$, where R (in A·W^{-1}) is the responsivity, S_I is the noise (A/$\sqrt{Hz}$) and A is the area of the device (cm^2).

2.4. Optical Resonator

Determination of the optical index: We used spectrally resolved ellipsometry to determine the complex optical index of the different materials involved in the device and later conduct electromagnetic simulations.

Spectroscopic ellipsometry: The procedure was similar to the one used in ref [43]. The spectroscopic ellipsometry measured the change in the polarization state between the incident and the reflected light. This was done by measuring the angles ψ and Δ. The following relation applies:

$$\rho = \frac{r_p}{r_s} = \left|\frac{r_p}{r_s}\right| e^{i(\delta_p - \delta_s)} = \tan \psi e^{i\Delta} \tag{1}$$

where r_p and r_s are the reflection coefficients of p and s polarized light, respectively, and where δ_p and δ_p are the phase shifts in reflection in p and s polarizations, respectively. The measurements were performed on a V-VASE ellipsometer (J.A. Woollam) in the 500–2000 nm range with steps of 10 nm and with angles of incidence of $50°$, $60°$ and $70°$. For the Li-based substrate, the real part is shown in Figure S6 and was later used as a free parameter to fit experimental data; we found that $n = 1.8$ gives a reasonable fit. To account for the diffusive character of the substrate, we used k = 1 for the substrate.

Complementarily, note that for the gold complex optical index, we used ref [44]. Figure S7 provides the imaginary part of the complex optical index of the HgTe NC film. The real part of the optical index was taken as constant and equal to 2.2 according to ref [43].

Electromagnetic simulation: Simulations were conducted using COMSOL 5.6, a software using the Finite Element Method. The array of resonators was modelled using the RF module in 2D geometry. Floquet periodic boundary conditions were used to describe the periodicity in one unit cell. On both sides (top and bottom), we defined perfectly matched layers (PML) to absorb all outgoing waves and prevent nonphysical reflections. The absorption came from emw. Qe, a value corresponding in COMSOL to the power density dissipated in W·m^{-3}. In air, there is no diffracted order for wavelengths above the electrode period. For shorter wavelengths, the energy propagating in the diffracted orders is absorbed by PML. On top of the resonator inside air, a port condition was used to define the incident wave, either in TE or TM polarization. An automatic mesh was used in these simulations with a predefined "extremely fine" mesh, which means that the maximum element size was 140 nm, except for the PML, where a mapped mesh was used with a distribution of 12 elements.

To enhance the device absorption, we kept the same interdigitated device geometry and just updated some geometrical factors, as shown in Figures S8 and S12. We tuned the period, the size of the digit and the film thickness so that light absorption resonances were generated. Along the TE polarization, we designed a guided mode resonance. This mode is dispersive and split in two modes with angles as revealed by the dispersion map along the TE polarization (see Figure S9a). In TM mode, we observed one main resonance that we designed to be red shifted compared to the TE mode. This mode is nondispersive (see TM dispersion map in Figure S9b), and while it is not affected by the digit size (i.e., no shift of the resonance peak for various digit sizes, as shown in Figure S11), it can strongly be affected by the film thickness (i.e., a strong shift of the resonance peak away from the exciton peak, as shown in Figure S10), which suggests a vertical Fabry-Pérot resonance.

Fabrication of the device with resonator The fabrication of the optimized photoresponsive device was accomplished in two steps of lithography. First, the main contact pads were fabricated using optical lithography, and then the interdigitated pattern was fabricated using e-beam patterning.

Electrodes for electrical contacts: A lithium-ion glass substrate was initially rinsed with acetone and isopropanol to remove dust and organic contaminants. An additional O_2 plasma cleaning step was performed for 5 min. The TI Prime adhesion promoter was spin-coated onto the substrate and baked on a hot plate at 120 °C for 2 min. Subsequently, the substrate was spin-coated with AZ5214E photoresist and baked at 110 °C for 1 min 30 s. The first UV exposure through a mask was 1 s, as the substrate was very strongly scattering light, and this was followed by the image reversal bake step at 125 °C for 2 min. A flood exposure for 40 s without the mask was then performed. The resist film was developed in AZ726 MIF with a development time of 30 s, followed by DI water rinse, N_2 gun and finally an oxygen plasma over 5 min. The next step was to deposit 5 nm of Cr and 80 nm of Au on the substrate with a thermal evaporator. Afterwards, the substrate stayed for 2 h in acetone for lift-off. The obtained pattern is shown Figure S12a.

Fabrication of gold nano-stripes: The procedure was similar to the one used in ref [45]. We used a Raith eLine e-beam lithography system for the fabrication of the nano-stripes. First, a 400 nm thick PMMA layer was spin-coated onto the substrate patterned with macroscopic electrodes. The resist was then baked at 180 °C for 2 min. We deposited 10 nm of Al with a thermal evaporator in order to reduce the charging effect. The operating bias of the electron beam was set to 20 kV, and the aperture was set to 10 μm. The dose was set to 170 $\mu C \cdot cm^{-2}$. The resist was first dipped for a duration of 15 s in a KOH solution in order to remove the aluminum layer, and then it was developed in a solution of MIBK: isopropanol (1:3 in volume) for 45 s and rinsed with isopropanol. An oxygen plasma over 5 min was done to remove traces of resist. 5 nm Cr and 80 nm Au were then deposited with a thermal evaporator. The substrate stayed in hot acetone (45 °C) for 2 h after letting it overnight for lift-off. The obtained pattern is shown in Figure S12b–d.

For resonator measurements, a thicker film compared to transistor measurements was needed (>200 nm). Thicker films can be obtained mostly by updating the spin coating condition, which is conducted at reduced speed. The ink is spin-coated on the substrate at 1500 rpm (at 500 $rpm \cdot s^{-1}$ acceleration) for a duration of 5 min (after initial DMF deposition on the substrate with the same speed in order to help adhesion), followed by a fast spin coating step to dry the edges (3000 rpm for 2 min, with 500 $rpm.s^{-1}$). This gives a homogeneous deposition around 225 nm; see Figure S1 for an image of the electrode after functionalization with the NC film.

3. Results and Discussion

3.1. Material of Interest with Short Wave Infrared Band Gap

To design a short wave-infrared (SWIR) absorbing phototransistor, we proceeded in two informative steps. In the first step, we relied on a doped silica substrate. We used HgTe NCs as the material for the FET channel, since HgTe combines both a tunable absorption edge in the short- and mid-wave infrared together with stable photoconductive properties. The nanocrystals were obtained using the Keuleyan's procedure [41]. HgTe NCs present a tripodic shape according to electronic microscopy (Figure 1b) and a band edge at 2 μm with a cut-off wavelength around 2.5 μm (see the absorption spectrum in Figure 1a). In addition to the electronic interband transition attributed to HgTe, the absorption spectrum depicts a doublet that is attributed to the C-H bond resonance at around 2900 cm^{-1}. Several steps of cleaning were conducted in order to reduce the relative magnitude of the C-H resonance close to the one of the NC exciton. When a similar magnitude was obtained, thin conductive and photoconductive films of such NCs were obtained by preparing an ink [16,42]. In this case, the native insulating long ligands (dodecanethiol) were stripped off and replaced by a mixture of a short thiol (mercaptoethanol) and ions (HgCl$_2$) dissolved in dimethylformamide (DMF). The obtained ink can be spin-coated to form a thin film. The film thickness can be tuned with the deposition conditions in the 100 to 300 nm range for the device under investigation in this study.

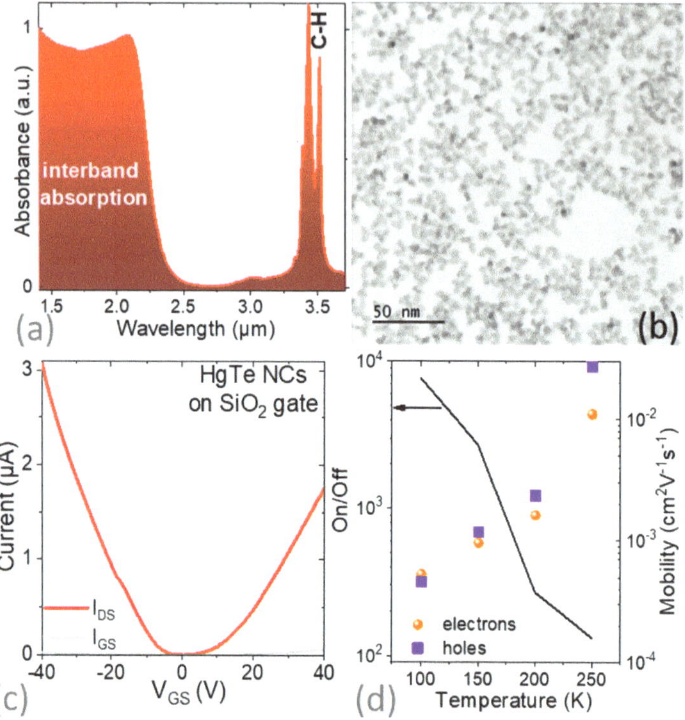

Figure 1. (**a**) Absorption spectrum of the HgTe NCs measured in attenuated total reflection mode. The doublet at around 3.4 µm is related to the resonance of the C-H bond. (**b**) Transmission electron microscopy image of the HgTe NCs. (**c**) Transfer curve (drain I$_{DS}$ and gate current I$_{GS}$ as a function of applied gate bias under the constant drain-source bias of 100 mV) measured at 150 K for a field-effect transistor whose channel is made of a HgTe NC film while the gate is SiO$_2$ (300 nm SiO$_2$ on top of highly doped Si). (**d**) On/off ratio (solid line) and carrier mobility (scatter) as a function of the temperature for a FET whose channel is made of HgTe NCs and the gate of 300 nm SiO$_2$.

Once deposited on a doped Si/SiO$_2$ substrate on which gold interdigitated electrodes have been fabricated, the material presented an ambipolar character. We observed both holes (under negative bias) and electrons (under positive gate bias), as shown in the transfer curve (see Figure 1c). The current modulation was around two orders of magnitude at 250 K and increased to almost four decades at 100 K (see Figure 1d). This increase in current modulation was mostly due to a reduction in the off current caused by the reduced carrier density activation when the temperature drops.

The carrier mobility can be determined from the transfer curve with:

$$\mu = \frac{L}{W C_\Sigma V_{DS}} \frac{\partial I_{DS}}{\partial V_{GS}} \qquad (2)$$

where L is the electrode spacing (10 µm), W is the electrode length (49 × 2.5 mm), C_Σ is the surface capacitance as measured in Figure 2b and V_{DS} is the drain source bias. We noticed a clear thermal activation of the mobility (Figure 1d), which was consistent with the hopping conduction occurring in a disordered array of polydisperse NCs. Electron and hole mobilities were similar in magnitude and dropped by a factor of 50 as the temperature was reduced from 250 to 100 K.

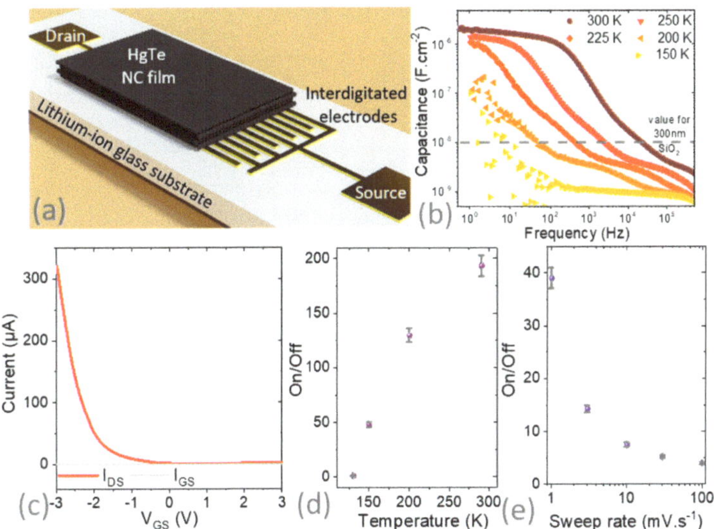

Figure 2. (**a**) Schematic of the FET whose channel is made of HgTe NC film and whose gate is made of a lithium-ion glass substrate. The substrate is used as a gate and the gate electrode is taken on the back side of the substrate. (**b**) Capacitance of the lithium-ion glass substrate as a function of the signal frequency for various temperatures. (**c**) Transfer curve for the FET depicted in part a at room temperature (V_{DS} = 500 mV). On/off ratio (i.e., ratio of the maximum current over the minimum current in the transfer curve) for the same FET as a function of temperature (**d**) and gate bias sweep rate at 200 K (**e**). Error bars in part d and e are determined by several repetitions of the transfer curve.

3.2. Gating Using Solid Electrolyte

In the second step, we designed a FET using a lithium-ion glass substrate. The latter is commercially available from MTI and is a ceramic of oxide materials (Li$_2$O-Al$_2$O$_3$-SiO$_2$-P$_2$O$_5$-TiO$_2$) in which the small Li$^+$ cations can be displaced under the application of a gate bias. On such substrates, we fabricated the same set of interdigitated electrodes as the one used for silica gating (see a schematic of the device in Figures 2a and S2). The benefit of such gates is best revealed by impedance measurements, shown in Figure 2b. At high frequency, the capacitance is low. In this regime, ions cannot move, and only the dielectric behavior of the substrate is observed. As a result of the substrate being thick (several hundred µm), the resulting capacitance is even weaker than in thin silica films. At low frequency, on the other hand, ions can move to form an ionic interfacial layer, and the obtained capacitance becomes high (up to 2 µF·cm^{-2}). The transition frequency between these two regimes depends on the temperature and reflects the thermally activated transport of Li$^+$ ions within the substrate. In practice, ion displacement limited the gate sweep rate range, and the strongest modulations were obtained with low bias sweep rates

(1 mV·s^{-1} typically, see Figure 2e). Note that the capacitance value obtained is similar to the one obtained with a liquid or ion gel electrolyte, while the device can be air-operated without handling any liquid.

As bias was applied over the lithium-ion glass substrate, current modulation was observed in the HgTe NC film (see Figures 2c, S3 and S4). Current modulation was observed for temperatures down to 130 K (see Figure 2d); below this value, the ions froze. This enables the operation of the FET from room temperature down to this value, which is a broader range than the typical one of liquid electrolyte (acetonitrile often used as solvent freezes at −45 °C) or ionic glass [31–33] (operating range 180–300 K). However, despite the higher capacitance than with silica, the on/off ratio is not improved due to higher leakage inherent to the gating method based on ion displacement. Table 1 summarizes some figures of merit relative to various gate technologies for a HgTe NC-based field effect transistor.

Table 1. Typical ranges of use for solid state gating of HgTe nanocrystal-based field effect transistors.

Gate Technology	Dielectric	Ionic Glass	Ion Gel Electrolyte	Solid State Li Electrolyte
Temperature range	4 K–300 K	180 K–260 K	300 K	150–300 K
Sweep-rate range	Fast (several V·s^{-1})	Intermediate (0.1 V·s^{-1})	Slow (1 mV·s^{-1})	Better below 10 mV·s^{-1}
Subthreshold slope	3400 mV/decade	1200 mV/decade	152 mV/decade	1200 mV/decade
Gate voltage range	<60 V (dielectric breakdown)	Up to 10 V at 200 K	<3 V (electrochemical stability of the electrolyte)	Tested up to 7 V

A striking feature is the observation of p-type only conduction (Figure 2c), whereas the material was ambipolar when coupled to silica gate. This means that the lithium-ion glass substrate behaves as a carrier selective charge injector. This behavior likely results from an asymmetry in the mobility of Li$^+$ cations and their vacancies in the lithium-ion glass substrate. We have estimated the hole mobility at 250 K to be 6 × 10^{-3} cm^2·V^{-1}·s^{-1}. This is typically three times lower than the value measured for SiO$_2$ gates; this reflects a carrier density dependence of the mobility.

As the purpose of this device is its use as a phototransistor, we tested its potential under illumination at 1.55 µm. As expected for such NC films, the material was photoresponsive and we observed a responsivity around R = 5 mA·W^{-1} (floating gate), which can be tuned by applying a gate bias, as shown in Figure 3a. Though higher responsivity (up to R = 11 mA·W^{-1}) can be obtained under negative gate bias, the strongest photocurrent modulation is obtained under positive gate bias since this operating condition makes the material more intrinsic. As often observed in NC-based phototransistors, the change in the photocurrent magnitude comes with a change in the response dynamics [27] (see Figure 3b). Faster responses are obtained when majority carriers are removed (i.e., positive gate bias here). These changes in dynamics reflect the change in trap filling when the gate bias is applied [46]. With Li gating, the response time spans from 500 µs to 5 ms depending on the applied gating bias. The former value is itself a decade longer than the value obtained with the silica gate, which is also associated with a much weaker photoresponse. In other words, the FET configuration enables tuning of the device photoresponse while leaving the gain bandwidth product almost unaffected.

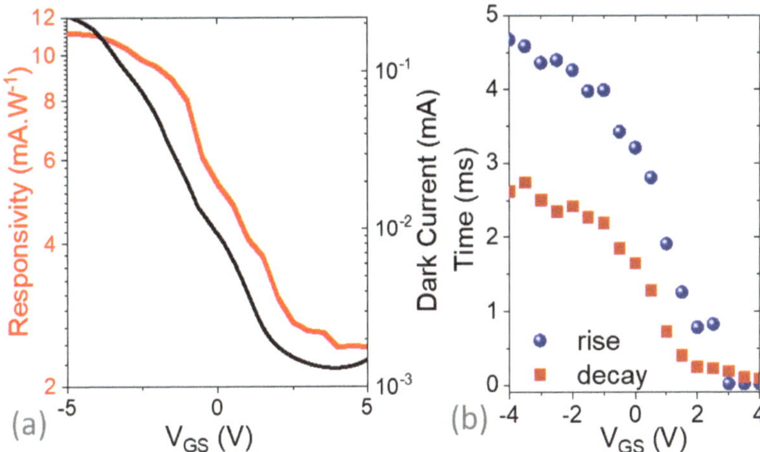

Figure 3. (**a**) Dark current (black line) and responsivity (red line) for the device depicted in Figure 2a as a function of the applied gate bias. (**b**) Rise and decay time (defined as the time for the signal to change from 10% to 90% of its final value) as a function of applied gate bias (see Figure S5 for a time trace). Illumination was ensured by a 1.55 µm laser diode at 110 mW.cm^{-2}; spectra were averaged 500 times The drain-source bias was set to 500 mV. Measurements were conducted at 250 K.

3.3. Introduction of a Photonic Structure to Enhance the Absorption

As with most NC-based light sensors in this configuration, the device faces a limitation. Due to hopping conduction, the mobility and the diffusion length are limited. As a result, the effective light absorption only occurs over a distance much shorter than the light absorption depth (>µm for HgTe NC in the SWIR [47]). Electromagnetic simulations revealed that the film absorbs around 10% of the incident light at the NC band edge. To overcome this limitation, we updated the design of the device to couple the NCs to a photonics structure [48], while maintaining the same interdigitated geometry (Figure S6). The initial device relied on interdigitated electrodes with 10 µm wide digits and 10 µm spacing between digits. We then tuned the geometrical factors (film thickness, period and size of the digits) in order to maximize the absorption in magnitude and linewidth. Compared to the previous work by Gréboval et al. [25], also dealing with a phototransistor coupled to a light resonator, the change in substrate from a single crystal of SrTiO$_3$ to a Li-based ceramic which strongly scatters light (Figures S4 and S5) raised the need for an additional design.

We first matched a resonance in the TM polarization (magnetic field parallel to the electrode ribbons) with the band-edge energy. This first resonance, at 2.2 µm, is a vertical Fabry-Perot resonance highlighted by its lack of dispersion (see Figure S7b). The energy of this resonance is driven by the film thickness (Figure S8b) but is very weakly affected by the digit width (Figure S9). The obtained mode is spatially located on top of the gold pad (Figure 4c) as the cavity is formed between the gold and the air interface. By doing so, we minimized the loss in the metal, which is estimated to be around 5.5% of the total absorption. Details about the electromagnetic design can be found in the supporting information.

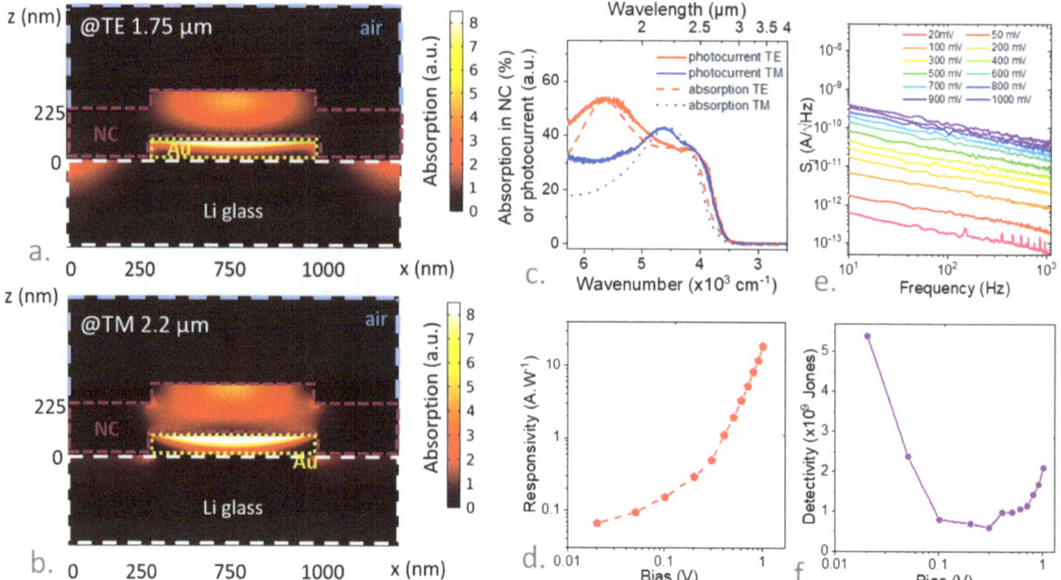

Figure 4. (**a**) (resp (**b**)) Simulated absorption map at the resonance of the TE @1.75 μm (resp TM at 2.2 μm) polarization. (**c**) Simulated absorption (dashed lines) and experimental photocurrent (solid lines) spectra within the NC film (i.e., excluding losses in contact) for TE and TM polarization. (**d**) Responsivity measured at 180 K for a signal at 1 kHz as a function of applied bias for the device with light resonators under broadband radiation resulting from a blackbody at 980 °C. (**e**) Noise current spectral density as a function of signal frequency measured under various applied biases. (**f**) Specific detectivity measured at 180 K (for a signal at 1 kHz) as a function of applied bias for the device with light resonators.

Furthermore, by tuning the period of the grating we can induce a second resonance in the TE polarization (electric field parallel to the electrodes ribbons). We chose the period of the grating and digit size to slightly blueshift this resonance with respect to the band edge (around 1.75 μm). By doing so, we broadened the global response of the device. This mode is a guided mode resonance [49–51] (GMR), whose energy is driven by the period of the grating. This mode is dispersive, as revealed by the dispersion map (see Figure S7a), and spatially located at the top of the gold electrode as revealed by the simulated absorption map (Figure 4a). The obtained absorption within the NCs (i.e., leading to the generation of the photocurrent) is expected to reach 40% for non-polarized light (see Figure 4c).

Since the period of the grating was set at 1.5 μm for a 750 nm wide digit, we used e-beam lithography for its fabrication (see Figure S10). The obtained photoresponse spectrum was now strongly polarized (see Figure 4c), with a TE mode clearly blue shifted with respect to the TM mode. Compared to the simulation, the photocurrent presented a good match both in energy and linewidth as long as we accounted for the scattering of the Li substrate. It is worth noting that the photocurrent spectrum was also significantly red shifted from the absorption shown in Figure 1a. This shift was the combination of two effects: (i) first, we conducted a ligand exchange procedure that tends to increase inter-NC coupling and lead to a redshift. In addition (ii), the photocurrent was measured at low temperature (180 K), while the HgTe NC band gap redshifts upon cooling [47,52].

The responsivity of the device was now much higher and reached 10 A·W^{-1} (Figures 4d and S14) under moderate electric field: this value is three orders of magnitude larger than the one reported for the same film deposited on conventional interdigitated electrodes without resonances. As has already been pointed out by Chu et al. [51], the enhanced responsivity splits into two contributions: absorption enhancement is responsible here for a factor of 4 in the performance increase; the remaining factor of 250 is due to photogain [53] (i.e., recirculation of one of the carriers while the other one stays trapped). Indeed, the external quantum efficiency of the device is around 500% ($R = 10\ A \cdot W^{-1}$ for 2.5 μm cut-off wavelength and under broadband blackbody radiation).

In a nanocrystal array, charge transport occurs through hopping. Under illumination, electron-hole pairs get generated and one carrier can easily get trapped while the second one is transported to the electrode. To ensure sample neutrality, the electrode will reinject this carrier, which can recirculate several times until the trapped charge gets recombined [54]. The gain magnitude is the ratio of the trapped carrier lifetime to the transit time. The latter is directly connected to the device size. Thus, a smaller device favors gain [55].

The factor of 250 for the gain value is further subdivided into a factor of 10 that arises from an increase in the electric field (i.e., a 10 times narrowing of the electrode spacing) and a factor of 25 that we ascribe to a more efficient hopping transport as the device size approaches the diffusion length [51]. Noise in this device is limited by a $1/f$ contribution [56,57] (see the inset of Figure 4e), which currently limits the detectivity around $D^* \approx 5 \times 10^9$ Jones and corresponds to a noise equivalent power (NEP) at around 1 pW·Hz$^{-1/2}$ for a signal at 1 kHz. This is, nevertheless, a factor of 5 ($D^* = 10^9$ jones) above the value obtained for the same material (same band gap and same surface chemistry) without a resonator compared to that achieved for silica [51].

4. Conclusions

To summarize, we have demonstrated that a lithium-ion glass substrate can be used as an all-solid back gate for a NC-based FET. This gate leads to a capacitance as high as 2 μF.cm^{-2}, similar to a liquid electrolyte while considerably easing the device's manipulation and its future integration. Moreover, the gating remains effective over a broad range of temperatures from room temperature down to 130 K, matching the targeted operating temperature range for short- and mid-wave infrared sensors. Interestingly, the gating favors hole injection compared to silica gating. In a second step, we demonstrated a light management strategy for a phototransistor based on this lithium-ion glass. We demonstrated the combination of two light resonators, one based on a guided-mode resonance and the second one relying on a Fabry-Pérot mode. This combination of two resonances enables broad band enhancement in light absorption and responsivity as high as 10 A.W^{-1}; this value is a thousand times larger than the one obtained for interdigitated electrodes without any light management strategy. This enhancement can be split over a factor of 4 for absorption and a factor of 250 for photogating.

Supplementary Materials: The following supporting information can be downloaded at: https://www.mdpi.com/article/10.3390/ma16062335/s1, Figure S1: thickness determination of HgTe film using profilometer. Figure S2: schematic of the field effect transistor. Figure S3: Transfer curve of the Li-gated FET at various temperatures. Figure S4: schematic of the responsivity setup. Figure S5: Photoresponse under illumination. Figure S6: refractive index of the Li-based ceramic. Figure S7: extinction coefficient spectra of the HgTe NC film. Figure S8: Schematic of device with resonator. Figure S9: Dispersion map in TE and TM polarizations. Figure S10: Absorption spectra for various film thicknesses. Figure S11: Absorption spectra for various gold digit sizes. Figure S12: Microscopy images of the resonator. Figure S13: Microscopy images of the resonator functionalized by NC film. Figure S14: IV curve of the resonator device. Figure S15: Responsivity of the resonator device as a function of frequency and power.

Author Contributions: J.F.D. fabricated the electrodes without resonators, A.K., H.Z. and M.C. built the setups for optoelectronic characterization, S.P. conducted the FET characterization, Y.P. and S.I. grew the nanoparticles, C.A. and B.G. conducted ellipsometry measurements, C.A., T.H.D. and G.V. designed the light resonators and fabricated them, E.L. and S.S. designed and funded the project and wrote the paper with input of all authors. All authors have read and agreed to the published version of the manuscript.

Funding: The project is supported by ERC grant blackQD (grant n° 756225) Ne2Dem (grant n° 853049) and AQDtive (grant n° 101086358). We acknowledge the use of clean-room facilities from the "Centrale de Proximité Paris-Centre" and from the STnano platform. This work has been supported by the Region Ile-de-France in the framework of DIM Nano-K (grant dopQD). This work was supported by French state funds managed by the ANR within the Investissements d'Avenir programme under reference ANR-11-IDEX-0004-02 and Labex NIE (ANR-11-LABX-0058, and ANR-10-IDEX-0002-02), and more specifically within the framework of the Cluster of Excellence MATISSE and also by the grant IPER-Nano2 (ANR-18CE30-0023-01), Copin (ANR-19-CE24-0022), Frontal (ANR-19-CE09-0017), Graskop (ANR-19-CE09-0026), NITQuantum (ANR-20-ASTR-0008-01), Bright (ANR-21-CE24-0012-02), MixDferro (ANR-21-CE09-0029) and Quicktera (ANR-22-CE09-0018).

Data Availability Statement: The data used to support the findings of this study are included within the article and its supplementary information.

Conflicts of Interest: The authors declare no conflict of interest.

References

1. Wang, Y.; Gu, Y.; Cui, A.; Li, Q.; He, T.; Zhang, K.; Wang, Z.; Li, Z.; Zhang, Z.; Wu, P.; et al. Fast Uncooled Mid-Wavelength Infrared Photodetectors with Heterostructures of van Der Waals on Epitaxial HgCdTe. *Adv. Mater.* **2022**, *34*, 2107772. [CrossRef] [PubMed]

2. Lu, H.; Carroll, G.M.; Neale, N.R.; Beard, M.C. Infrared Quantum Dots: Progress, Challenges, and Opportunities. *ACS Nano* **2019**, *13*, 939–953. [CrossRef] [PubMed]

3. Rogalski, A. Progress in Quantum Dot Infrared Photodetectors. In *Quantum Dot Photodetectors*; Tong, X., Wu, J., Wang, Z.M., Eds.; Lecture Notes in Nanoscale Science and Technology; Springer International Publishing: Cham, Switzerland, 2021; pp. 1–74. ISBN 978-3-030-74270-6.

4. Pradhan, S.; Di Stasio, F.; Bi, Y.; Gupta, S.; Christodoulou, S.; Stavrinadis, A.; Konstantatos, G. High-Efficiency Colloidal Quantum Dot Infrared Light-Emitting Diodes via Engineering at the Supra-Nanocrystalline Level. *Nat. Nanotechnol.* **2019**, *14*, 72–79. [CrossRef] [PubMed]

5. Qu, J.; Weis, M.; Izquierdo, E.; Mizrahi, S.G.; Chu, A.; Dabard, C.; Gréboval, C.; Bossavit, E.; Prado, Y.; Péronne, E.; et al. Electroluminescence from Nanocrystals above 2 µm. *Nat. Photon.* **2022**, *16*, 38–44. [CrossRef]

6. Sayevich, V.; Robinson, Z.L.; Kim, Y.; Kozlov, O.V.; Jung, H.; Nakotte, T.; Park, Y.-S.; Klimov, V.I. Highly Versatile Near-Infrared Emitters Based on an Atomically Defined HgS Interlayer Embedded into a CdSe/CdS Quantum Dot. *Nat. Nanotechnol.* **2021**, *16*, 673–679. [CrossRef]

7. Geiregat, P.; Houtepen, A.J.; Sagar, L.K.; Infante, I.; Zapata, F.; Grigel, V.; Allan, G.; Delerue, C.; Van Thourhout, D.; Hens, Z. Continuous-Wave Infrared Optical Gain and Amplified Spontaneous Emission at Ultralow Threshold by Colloidal HgTe Quantum Dots. *Nat. Mater.* **2018**, *17*, 35–42. [CrossRef]

8. Cryer, M.E.; Browning, L.A.; Plank, N.O.V.; Halpert, J.E. Large Photogain in Multicolor Nanocrystal Photodetector Arrays Enabling Room-Temperature Detection of Targets Above 100 °C. *ACS Photonics* **2020**, *7*, 3078–3085. [CrossRef]

9. Chen, M.; Lu, H.; Abdelazim, N.M.; Zhu, Y.; Wang, Z.; Ren, W.; Kershaw, S.V.; Rogach, A.L.; Zhao, N. Mercury Telluride Quantum Dot Based Phototransistor Enabling High-Sensitivity Room-Temperature Photodetection at 2000 nm. *ACS Nano* **2017**, *11*, 5614–5622. [CrossRef]

10. Dong, Y.; Chen, M.; Yiu, W.K.; Zhu, Q.; Zhou, G.; Kershaw, S.V.; Ke, N.; Wong, C.P.; Rogach, A.L.; Zhao, N. Solution Processed Hybrid Polymer: HgTe Quantum Dot Phototransistor with High Sensitivity and Fast Infrared Response up to 2400 Nm at Room Temperature. *Adv. Sci.* **2020**, *7*, 2000068. [CrossRef]

11. Bossavit, E.; Qu, J.; Abadie, C.; Dabard, C.; Dang, T.; Izquierdo, E.; Khalili, A.; Gréboval, C.; Chu, A.; Pierini, S.; et al. Optimized Infrared LED and Its Use in an All-HgTe Nanocrystal-Based Active Imaging Setup. *Adv. Opt. Mater.* **2022**, *10*, 2101755. [CrossRef]

12. Pejović, V.; Georgitzikis, E.; Lee, J.; Lieberman, I.; Cheyns, D.; Heremans, P.; Malinowski, P.E. Infrared Colloidal Quantum Dot Image Sensors. *IEEE Trans. Electron. Devices* **2021**, *69*, 2840–2850. [CrossRef]

13. Erwin, S.C.; Zu, L.; Haftel, M.I.; Efros, A.L.; Kennedy, T.A.; Norris, D.J. Doping Semiconductor Nanocrystals. *Nature* **2005**, *436*, 91–94. [CrossRef] [PubMed]

14. Talapin, D.V.; Murray, C.B. PbSe Nanocrystal Solids for N- and p-Channel Thin Film Field-Effect Transistors. *Science* **2005**, *310*, 86–89. [CrossRef]

15. Shulga, A.G.; Kahmann, S.; Dirin, D.N.; Graf, A.; Zaumseil, J.; Kovalenko, M.V.; Loi, M.A. Electroluminescence Generation in PbS Quantum Dot Light-Emitting Field-Effect Transistors with Solid-State Gating. *ACS Nano* **2018**, *12*, 12805–12813. [CrossRef] [PubMed]

16. Lan, X.; Chen, M.; Hudson, M.H.; Kamysbayev, V.; Wang, Y.; Guyot-Sionnest, P.; Talapin, D.V. Quantum Dot Solids Showing State-Resolved Band-like Transport. *Nat. Mater.* **2020**, *19*, 323–329. [CrossRef]

17. Lhuillier, E.; Keuleyan, S.; Zolotavin, P.; Guyot-Sionnest, P. Mid-Infrared HgTe/As₂S₃ Field Effect Transistors and Photodetectors. *Adv. Mater.* **2013**, *25*, 137–141. [CrossRef]

18. Chung, D.S.; Lee, J.-S.; Huang, J.; Nag, A.; Ithurria, S.; Talapin, D.V. Low Voltage, Hysteresis Free, and High Mobility Transistors from All-Inorganic Colloidal Nanocrystals. *Nano Lett.* **2012**, *12*, 1813–1820. [CrossRef]

19. Lim, B.T.; Cho, J.; Cheon, K.H.; Sim, K.M.; Shin, K.; Chung, D.S. Zirconium Oxide Dielectric Layer Grown by a Surface Sol–Gel Method for Low-Voltage, Hysteresis-Free, and High-Mobility Polymer Field Effect Transistors. *Org. Electron.* **2016**, *28*, 1–5. [CrossRef]

20. Kim, H.; Cho, K.; Kim, D.W.; Lee, H.R.; Kim, S. Bottom- and Top-Gate Field-Effect Thin-Film Transistors with p Channels of Sintered HgTe Nanocrystals. *Appl. Phys. Lett.* **2006**, *89*, 173107. [CrossRef]

21. Kim, D.-W.; Jang, J.; Kim, H.; Cho, K.; Kim, S. Electrical Characteristics of HgTe Nanocrystal-Based Thin Film Transistors Fabricated on Flexible Plastic Substrates. *Thin Solid Film.* **2008**, *516*, 7715–7719. [CrossRef]

22. Gréboval, C.; Dabard, C.; Konstantinov, N.; Cavallo, M.; Chee, S.-S.; Chu, A.; Dang, T.H.; Khalili, A.; Izquierdo, E.; Prado, Y.; et al. Split-Gate Photodiode Based on Graphene/HgTe Heterostructures with a Few Nanosecond Photoresponse. *ACS Appl. Electron. Mater.* **2021**, *3*, 4681–4688. [CrossRef]

23. Chee, S.-S.; Gréboval, C.; Magalhaes, D.V.; Ramade, J.; Chu, A.; Qu, J.; Rastogi, P.; Khalili, A.; Dang, T.H.; Dabard, C.; et al. Correlating Structure and Detection Properties in HgTe Nanocrystal Films. *Nano Lett.* **2021**, *21*, 4145–4151. [CrossRef]

24. Shulga, A.G.; Piveteau, L.; Bisri, S.Z.; Kovalenko, M.V.; Loi, M.A. Double Gate PbS Quantum Dot Field-Effect Transistors for Tuneable Electrical Characteristics. *Adv. Electron. Mater.* **2016**, *2*, 1500467. [CrossRef]

25. Gréboval, C.; Chu, A.; Magalhaes, D.V.; Ramade, J.; Qu, J.; Rastogi, P.; Khalili, A.; Chee, S.-S.; Aubin, H.; Vincent, G.; et al. Ferroelectric Gating of Narrow Band-Gap Nanocrystal Arrays with Enhanced Light–Matter Coupling. *ACS Photonics* **2021**, *8*, 259–268. [CrossRef]

26. Liu, H.; Keuleyan, S.; Guyot-Sionnest, P. N- and p-Type HgTe Quantum Dot Films. *J. Phys. Chem. C* **2012**, *116*, 1344–1349. [CrossRef]

27. Livache, C.; Izquierdo, E.; Martinez, B.; Dufour, M.; Pierucci, D.; Keuleyan, S.; Cruguel, H.; Becerra, L.; Fave, J.L.; Aubin, H.; et al. Charge Dynamics and Optoelectronic Properties in HgTe Colloidal Quantum Wells. *Nano Lett.* **2017**, *17*, 4067–4074. [CrossRef]

28. Houtepen, A.J.; Vanmaekelbergh, D. Orbital Occupation in Electron-Charged CdSe Quantum-Dot Solids. *J. Phys. Chem. B* **2005**, *109*, 19634–19642. [CrossRef]

29. Roest, A.L.; Houtepen, A.J.; Kelly, J.J.; Vanmaekelbergh, D. Electron-Conducting Quantum-Dot Solids with Ionic Charge Compensation. *Faraday Discuss.* **2004**, *125*, 55–62. [CrossRef]

30. Mulder, J.T.; du Fossé, I.; Alimoradi Jazi, M.; Manna, L.; Houtepen, A.J. Electrochemical P-Doping of $CsPbBr_3$ Perovskite Nanocrystals. *ACS Energy Lett.* **2021**, *6*, 2519–2525. [CrossRef]

31. Noumbé, U.N.; Gréboval, C.; Livache, C.; Chu, A.; Majjad, H.; Parra López, L.E.; Mouafo, L.D.N.; Doudin, B.; Berciaud, S.; Chaste, J.; et al. Reconfigurable 2D/0D p–n Graphene/HgTe Nanocrystal Heterostructure for Infrared Detection. *ACS Nano* **2020**, *14*, 4567–4576. [CrossRef]

32. Gréboval, C.; Noumbe, U.; Goubet, N.; Livache, C.; Ramade, J.; Qu, J.; Chu, A.; Martinez, B.; Prado, Y.; Ithurria, S.; et al. Field-Effect Transistor and Photo-Transistor of Narrow-Band-Gap Nanocrystal Arrays Using Ionic Glasses. *Nano Lett.* **2019**, *19*, 3981–3986. [CrossRef] [PubMed]

33. Gréboval, C.; Noumbé, U.N.; Chu, A.; Prado, Y.; Khalili, A.; Dabard, C.; Dang, T.H.; Colis, S.; Chaste, J.; Ouerghi, A.; et al. Gate Tunable Vertical Geometry Phototransistor Based on Infrared HgTe Nanocrystals. *Appl. Phys. Lett.* **2020**, *117*, 251104. [CrossRef]

34. Wang, H.; Liu, Q.; Feng, X.; Zhang, Z.; Wang, K.; Liu, Z.; Dai, J.-F. An Ambipolar Transistor Based on a Monolayer WS_2 Using Lithium Ions Injection. *Mater. Res. Express* **2020**, *7*, 076302. [CrossRef]

35. Philippi, M.; Gutiérrez-Lezama, I.; Ubrig, N.; Morpurgo, A.F. Lithium-Ion Conducting Glass Ceramics for Electrostatic Gating. *Appl. Phys. Lett.* **2018**, *113*, 033502. [CrossRef]

36. Zhao, J.; Wang, M.; Li, H.; Zhang, X.; You, L.; Qiao, S.; Gao, B.; Xie, X.; Jiang, M. Lithium-Ion-Based Solid Electrolyte Tuning of the Carrier Density in Graphene. *Sci. Rep.* **2016**, *6*, 34816. [CrossRef]

37. Alam, M.H.; Xu, Z.; Chowdhury, S.; Jiang, Z.; Taneja, D.; Banerjee, S.K.; Lai, K.; Braga, M.H.; Akinwande, D. Lithium-Ion Electrolytic Substrates for Sub-1V High-Performance Transition Metal Dichalcogenide Transistors and Amplifiers. *Nat. Commun.* **2020**, *11*, 3203. [CrossRef]

38. Zhu, C.S.; Cui, J.H.; Lei, B.; Wang, N.Z.; Shang, C.; Meng, F.B.; Ma, L.K.; Luo, X.G.; Wu, T.; Sun, Z.; et al. Tuning Electronic Properties of $FeSe_{0.5}Te_{0.5}$ Thin Flakes Using a Solid Ion Conductor Field-Effect Transistor. *Phys. Rev. B* **2017**, *95*, 174513. [CrossRef]

39. Gréboval, C.; Chu, A.; Goubet, N.; Livache, C.; Ithurria, S.; Lhuillier, E. Mercury Chalcogenide Quantum Dots: Material Perspective for Device Integration. *Chem. Rev.* **2021**, *121*, 3627–3700. [CrossRef]

40. Prado, Y.; Qu, J.; Gréboval, C.; Dabard, C.; Rastogi, P.; Chu, A.; Khalili, A.; Xu, X.Z.; Delerue, C.; Ithurria, S.; et al. Seeded Growth of HgTe Nanocrystals for Shape Control and Their Use in Narrow Infrared Electroluminescence. *Chem. Mater.* **2021**, *33*, 2054–2061. [CrossRef]

41. Keuleyan, S.; Lhuillier, E.; Guyot-Sionnest, P. Synthesis of Colloidal HgTe Quantum Dots for Narrow Mid-IR Emission and Detection. *J. Am. Chem. Soc.* **2011**, *133*, 16422–16424. [CrossRef]

42. Martinez, B.; Ramade, J.; Livache, C.; Goubet, N.; Chu, A.; Gréboval, C.; Qu, J.; Watkins, W.L.; Becerra, L.; Dandeu, E.; et al. HgTe Nanocrystal Inks for Extended Short-Wave Infrared Detection. *Adv. Opt. Mater.* **2019**, *7*, 1900348. [CrossRef]

43. Rastogi, P.; Chu, A.; Dang, T.H.; Prado, Y.; Gréboval, C.; Qu, J.; Dabard, C.; Khalili, A.; Dandeu, E.; Fix, B.; et al. Complex Optical Index of HgTe Nanocrystal Infrared Thin Films and Its Use for Short Wave Infrared Photodiode Design. *Adv. Opt. Mater.* **2021**, *9*, 2002066. [CrossRef]

44. Olmon, R.L.; Slovick, B.; Johnson, T.W.; Shelton, D.; Oh, S.-H.; Boreman, G.D.; Raschke, M.B. Optical Dielectric Function of Gold. *Phys. Rev. B* **2012**, *86*, 235147. [CrossRef]

45. Dang, T.H.; Khalili, A.; Abadie, C.; Gréboval, C.; Cavallo, M.; Zhang, H.; Bossavit, E.; Utterback, J.K.; Dandeu, E.; Prado, Y.; et al. Nanocrystal-Based Active Photonics Device through Spatial Design of Light-Matter Coupling. *ACS Photonics* **2022**, *9*, 2528–2535. [CrossRef]

46. Lhuillier, E.; Robin, A.; Ithurria, S.; Aubin, H.; Dubertret, B. Electrolyte-Gated Colloidal Nanoplatelets-Based Phototransistor and Its Use for Bicolor Detection. *Nano Lett.* **2014**, *14*, 2715–2719. [CrossRef]

47. Lhuillier, E.; Keuleyan, S.; Guyot-Sionnest, P. Optical Properties of HgTe Colloidal Quantum Dots. *Nanotechnology* **2012**, *23*, 175705. [CrossRef]

48. Chen, M.; Lu, L.; Yu, H.; Li, C.; Zhao, N. Integration of Colloidal Quantum Dots with Photonic Structures for Optoelectronic and Optical Devices. *Adv. Sci.* **2021**, *8*, 2101560. [CrossRef]

49. Maës, C.; Vincent, G.; Flores, F.G.-P.; Cerutti, L.; Haïdar, R.; Taliercio, T. Infrared Spectral Filter Based on All-Semiconductor Guided-Mode Resonance. *Opt. Lett.* **2019**, *44*, 3090–3093. [CrossRef]

50. Sharon, A.; Rosenblatt, D.; Friesem, A.A. Resonant Grating–Waveguide Structures for Visible and near-Infrared Radiation. *J. Opt. Soc. Am. A* **1997**, *14*, 2985–2993. [CrossRef]

51. Chu, A.; Gréboval, C.; Goubet, N.; Martinez, B.; Livache, C.; Qu, J.; Rastogi, P.; Bresciani, F.A.; Prado, Y.; Suffit, S.; et al. Near Unity Absorption in Nanocrystal Based Short Wave Infrared Photodetectors Using Guided Mode Resonators. *ACS Photonics* **2019**, *6*, 2553–2561. [CrossRef]

52. Moghaddam, N.; Gréboval, C.; Qu, J.; Chu, A.; Rastogi, P.; Livache, C.; Khalili, A.; Xu, X.Z.; Baptiste, B.; Klotz, S.; et al. The Strong Confinement Regime in HgTe Two-Dimensional Nanoplatelets. *J. Phys. Chem. C* **2020**, *124*, 23460–23468. [CrossRef]

53. Guo, N.; Xiao, L.; Gong, F.; Luo, M.; Wang, F.; Jia, Y.; Chang, H.; Liu, J.; Li, Q.; Wu, Y.; et al. Light-Driven WSe$_2$-ZnO Junction Field-Effect Transistors for High-Performance Photodetection. *Adv. Sci.* **2020**, *7*, 1901637. [CrossRef] [PubMed]

54. Konstantatos, G.; Sargent, E.H. PbS Colloidal Quantum Dot Photoconductive Photodetectors: Transport, Traps, and Gain. *Appl. Phys. Lett.* **2007**, *91*, 173505. [CrossRef]

55. Chu, A.; Gréboval, C.; Prado, Y.; Majjad, H.; Delerue, C.; Dayen, J.-F.; Vincent, G.; Lhuillier, E. Infrared Photoconduction at the Diffusion Length Limit in HgTe Nanocrystal Arrays. *Nat. Commun.* **2021**, *12*, 1794. [CrossRef]

56. Liu, H.; Lhuillier, E.; Guyot-Sionnest, P. 1/f Noise in Semiconductor and Metal Nanocrystal Solids. *J. Appl. Phys.* **2014**, *115*, 154309. [CrossRef]

57. Lai, Y.; Li, H.; Kim, D.K.; Diroll, B.T.; Murray, C.B.; Kagan, C.R. Low-Frequency (1/f) Noise in Nanocrystal Field-Effect Transistors. *ACS Nano* **2014**, *8*, 9664–9672. [CrossRef]

Disclaimer/Publisher's Note: The statements, opinions and data contained in all publications are solely those of the individual author(s) and contributor(s) and not of MDPI and/or the editor(s). MDPI and/or the editor(s) disclaim responsibility for any injury to people or property resulting from any ideas, methods, instructions or products referred to in the content.

materials

Review

Lead Chalcogenide Colloidal Quantum Dots for Infrared Photodetectors

Xue Zhao [1,†], Haifei Ma [1,†], Hongxing Cai [2], Zhipeng Wei [2], Ying Bi [3], Xin Tang [1] and Tianling Qin [1,*]

1   School of Optics and Photonics, Beijing Institute of Technology, Beijing 100081, China; 3220210544@bit.edu.cn (X.Z.); 3220220493@bit.edu.cn (H.M.); xintang@bit.edu.cn (X.T.)
2   Physics Department, Changchun University of Science and Technology, Changchun 130022, China; caihx@cust.edu.cn (H.C.); weizp@cust.edu.cn (Z.W.)
3   Beijing Institute of Aerospace Systems Engineering, Beijing 100076, China; centres@foxmail.com
*   Correspondence: 3220215096@bit.edu.cn
†   These authors contribute equally to this work.

Abstract: Infrared detection technology plays an important role in remote sensing, imaging, monitoring, and other fields. So far, most infrared photodetectors are based on InGaAs and HgCdTe materials, which are limited by high fabrication costs, complex production processes, and poor compatibility with silicon-based readout integrated circuits. This hinders the wider application of infrared detection technology. Therefore, reducing the cost of high-performance photodetectors is a research focus. Colloidal quantum dot photodetectors have the advantages of solution processing, low cost, and good compatibility with silicon-based substrates. In this paper, we summarize the recent development of infrared photodetectors based on mainstream lead chalcogenide colloidal quantum dots.

Keywords: colloidal quantum dots; infrared photodetector; lead chalcogenide; mercury chalcogenide

Citation: Zhao, X.; Ma, H.; Cai, H.; Wei, Z.; Bi, Y.; Tang, X.; Qin, T. Lead Chalcogenide Colloidal Quantum Dots for Infrared Photodetectors. *Materials* **2023**, *16*, 5790. https://doi.org/10.3390/ma16175790

Academic Editor: Antonio Polimeni

Received: 12 July 2023
Revised: 1 August 2023
Accepted: 21 August 2023
Published: 24 August 2023

Copyright: © 2023 by the authors. Licensee MDPI, Basel, Switzerland. This article is an open access article distributed under the terms and conditions of the Creative Commons Attribution (CC BY) license (https://creativecommons.org/licenses/by/4.0/).

1. Introduction

Infrared photoelectric technologies use infrared materials to convert infrared radiation energy into electrical signals. Infrared photodetectors are widely used in autonomous driving, machine vision, security surveillance, and other fields [1–5]. In recent years, with the continuous development of infrared detection technology, more and more infrared photodetectors have been developed. Currently, the mainstream infrared photodetectors are mainly based on narrow-gap semiconductors such as InGaAs [6,7], HgCdTe [8–10], and other materials, which have high sensitivity, wide-band detection capabilities, long-time stability, and good reliability [11]. However, they still have problems such as high cost, a complex epitaxial growth process, and poor compatibility with silicon-based readout integrated circuit chips [12–14], limiting their wider application.

The emergence of colloidal quantum dots (CQDs) provides a feasible way for the development of infrared photodetectors. As a new type of photoelectric material, CQDs have the advantages of a wide spectral tuning range, low cost of preparation by thermal injection, and solution processing [15–17], and CQDs have great application potential in many practical applications. In addition, CQDs can be directly integrated with silicon-based readout integrated circuits through solution processing, which reduces the fabrication cost and difficulty [18,19]. As a result, CQD infrared detectors have been a research hotspot in recent years. Among CQD infrared detectors [20], lead-based [21] CQD photodetectors have been demonstrated to have high sensitivity and response speed. Lead-based CQDs include PbTe, PbSe, and PbS CQDs. Their detection bands are mainly in the near-infrared and visible light range. Among them, PbSe and PbS CQDs have already been applied to large-array readout integrated circuits [22]. However, lead-based CQDs have the problems of easy oxidation and interference from humidity in the air [23].

In this paper, we discuss the progress of lead chalcogenide colloidal quantum dots for infrared photodetectors. First, we introduce the CQD synthesis method and illustrate the passivation effect of different ligand exchanges on the surface of CQDs. In addition, the development of the device structure of CQD photodetectors is also introduced in detail, and the advancements of photodetectors combining lead chalcogenide CQDs with organic materials, two-dimensional materials, and other materials are discussed.

2. PbTe CQD-Based Photodetectors

2.1. Synthesis of PbTe CQDs

Lead-based CQDs have received significant attention in recent years due to their unique optical and electronic properties. PbX (Te, Se, S) possess a cubic crystal lattice structure [24–26] and are narrow-band-gap semiconductors with 0.32 eV, 0.28 eV, and 0.41 eV, respectively. As a result, PbX (Te, Se, S) CQDs exhibit size-tunable infrared band gaps from visible to infrared. Benefitting from mature chemical synthesis technology, they are attractive for a wide range of applications, including infrared photodetectors, photovoltaics, and thermoelectric devices. PbTe CQDs, in particular, possess a large exciton Bohr radius (46 nm) [27], a high dielectric constant (1000) [28], and a high multiexciton generation yield.

In 1995, Reynoso et al. first grew PbTe CQDs in doped glass, and the absorption wavelength could be adjusted at 1.1–2.0 μm by changing the heat time and temperature [29]. The application potential of PbTe CQDs in optoelectronic devices was demonstrated. In 2006, Murphy et al. reported a synthesis method for spherical and cubic PbTe CQDs [30], whose size distribution could reach 7%, and the diameter of synthesized CQDs ranged from 2.6 to 18 nm. The first exciton transition was achieved from 1009 to 2054 nm (Figure 1a). The photoluminescence quantum yield of spherical PbTe CQDs was as high as 52 ± 2%. In the same year, Urban et al. synthesized monodispersed PbTe CQDs [31] with obvious advantages in size tunability and solution processability compared to the early CQDs in glass (Figure 1b). In addition, the conductivity of the PbTe CQDs film could be increased by 9–10 orders of magnitude through chemical treatment. Therefore, PbTe CQDs were expected to be applied in photodetectors.

Aiming at the problem of PbTe CQDs' susceptibility to oxidation, in 2009, Lambert et al. proposed a core–shell structure of PbTe/CdTe [32]. The CdTe shell was grown around PbTe using the cation exchange method (Figure 1c). It was observed that PbTe and CdTe had the same lattice orientation, achieving seamless matching. This method effectively resolved the oxidation problem of PbTe CQDs, but it had an anisotropy problem in the exchange process. To improve the method for the synthesis of PbTe CQDs, in 2012, Pan et al. proposed a synthesis method for monodisperse hydrophobic PbTe CQDs [27]. Using oleylamine as the capping ligand and solvent allowed the hydrophobic PbTe CQDs to be easily transformed into CQDs with different ligands. The application of PbTe CQDs in biomedicine was possible. PbTe CQDs could be changed from hydrophobic to hydrophilic through ligand exchange with 4-mercaptopyridine. In addition, the synthesized PbTe CQDs were found to be air-stable.

To enhance the photoluminescence (PL) performance of PbTe CQDs, Protesescu et al. reported the PL properties of core–shell PbTe/CdTe CQDs in 2016 [33]. PbTe/CdTe CQDs were demonstrated to have stable near-infrared emission within the 1–3 μm range. During the cation exchange process, with the addition of excess cadmium oleate, Cd could replace Pb to shrink the PbTe core, resulting in a shift of the PL peak to a shorter wavelength (Figure 1d). Compared with pure PbTe CQDs, the PL stability of PbTe/CdTe CQDs was significantly improved, even in ambient air.

In 2019, Peters et al. studied the fundamental chemistry of PbTe CQDs [34]. The relationship between the band gap and the NC diameter was measured, and the results showed that the energy absorbed by the primary optical absorption had a 1/d relationship with the diameter of the CQDs. And the connection mode of surface ligand oleic acid and PbTe CQDs was mostly chelating bidentate coordination. In 2020, Miranti et al. proposed a core–shell structure of PbTe/PbS CQDs, which established a type II heterojunction that enabled PbTe/PbS CQDs to carry out electron transport exclusively [35] (Figure 1f). The structure exhibited maximum electron mobility of 0.62 cm^2/Vs with an n-channel current modulation ratio of 104, which was significantly higher than that of PbTe CQDs and PbS CQDs. Enhancing n-type transport made core–shell PbTe/PbS promising for applications in the thermoelectric and electron transport layers of photovoltaic devices.

PbTe CQDs are highly sensitive to oxygen. Modifying the surface of PbTe CQDs or using other CQDs to wrap PbTe CQDs could be considered in future development to prevent the interference of external oxygen. And through the combination of other CQDs and PbTe CQDs, the expansion of the detection band and performance of PbTe CQDs was expected.

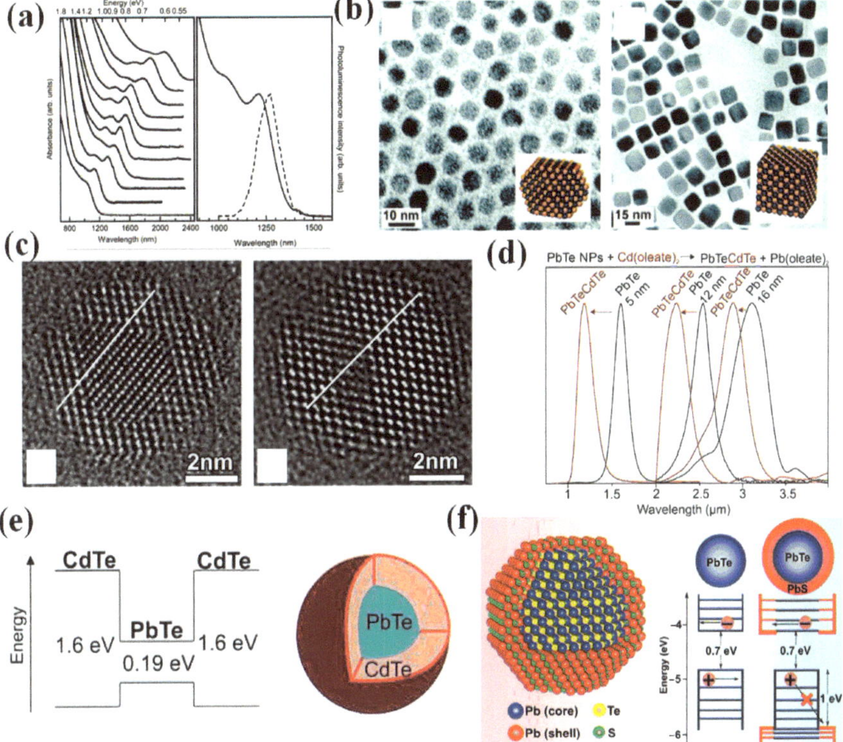

Figure 1. (**a**) The left inset shows the infrared absorption spectra of spherical PbTe CQDs with different sizes, and the right inset shows the absorption (solid line) and photoluminescence (dotted line) spectra of PbTe CQDs with a size of 2.9 nm [30]. Copyright 2006, Journal of the American Chemical Society. (**b**) Transmission electron microscopy (TEM) images and model diagrams of cuboctahedron and cubic PbTe CQDs [31]. Copyright 2006, Journal of the American Chemical Society. (**c**) TEM image of PbTe/CdTe CQDs in the <110> crystal plane [32]. Copyright 2009, Chemistry of Materials. (**d**) PL spectrum of PbTe/CdTe CQDs. (**e**) The energy band diagram of PbTe/CdTe CQDs [33]. Copyright 2016, ChemPhysChem. (**f**) The model and energy band diagram of PbTe/PbS CQDs [35]. Copyright 2020, ACS Nano.

2.2. PbTe CQD Photodetectors

The exceptional electronic and optical properties, along with the controllable synthesis, made PbTe CQDs a promising candidate for a wide range of applications in various devices.

By acquiring a deeper understanding of the coordination chemistry between oleic acid and PbTe CQDs, we can further optimize the synthesis and properties of PbTe CQDs. This endeavor will lay a robust foundation for their development in future electronic and optoelectronic devices.

The weak binding energy of PbTe CQDs, easy oxidation in air, and low photoconductivity limited their application in photodetectors. In 2016, Lin et al. proposed a PbTe CQD photodetector treated with tetrabutylammonium iodine (TBAI) [36] (Figure 2a,b). The TABI ligand exchange effectively passivated the surface of PbTe CQDs, reducing the average spacing between CQDs from 4.2 nm to 2.3 nm. As a result, the TABI-treated PbTe CQDs film transformed from a nonlinear Schottky contact to a symmetrical Ohmic contact. The maximum responsivity of the device could reach 1.9 mA/W. Under 150 Hz laser irradiation, the response time of the device reached 0.39 ms and the recovery time reached 0.49 ms, which effectively improved the response speed of the device (Figure 2c,d). Moreover, the detector with TBAI-treated PbTe CQDs had a stable photoresponse in air due to the protective effect of the upper polymethyl methacrylate (PMMA) layer.

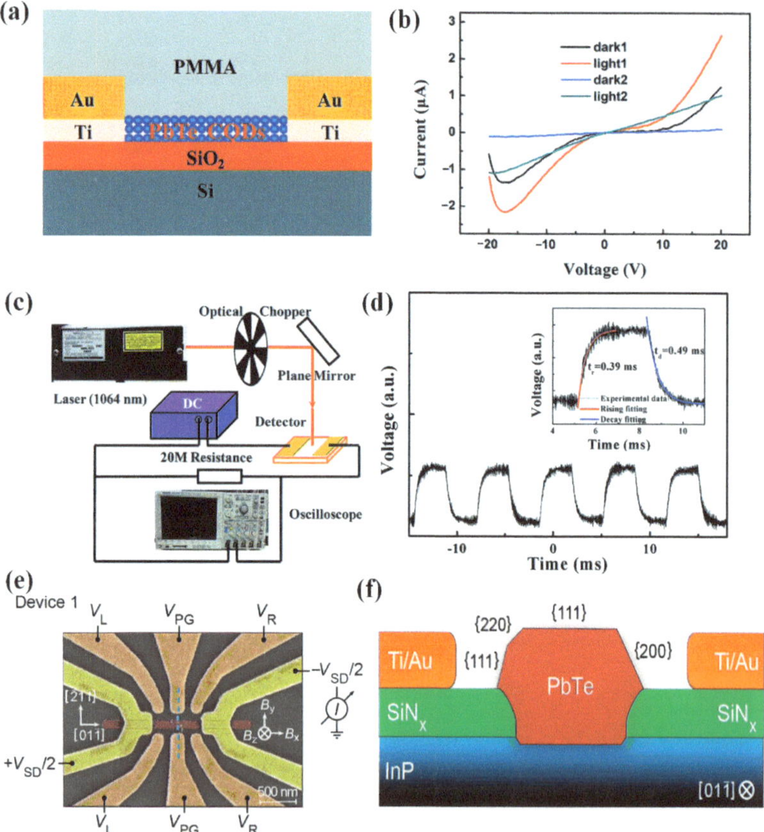

Figure 2. (**a**) Schematic diagram of the structure of the PbTe photodetector. (**b**) I–V curve of the PbTe photodetector. (**c**) Schematic diagram of the experimental setup for testing the photoresponse of the PbTe photodetector. (**d**) Photoresponse curve of the PbTe photodetector under laser irradiation frequency of 150 Hz [36]. Copyright 2016, RSC Advances. (**e**) Scanning electron microscope (SEM) image of PbTe device with crystal and magnetic field. (**f**) Schematic diagram of PbTe device [37]. Copyright 2022, Nano Letters.

In 2022, Kate et al. proposed a device structure for selectively growing PbTe CQDs on InP and characterized the CQD devices in both zero and finite magnetic fields (Figure 2e,f) [37]. The Coulomb charge stability analysis revealed large even–odd spacing at zero magnetic field. Further study of the devices under a limited magnetic field showed that PbTe CQDs exhibit a strong Rashba spin–orbit interaction.

The application of PbTe CQD photodetectors had been limited due to the issue of easy oxidation. To overcome this challenge, the investigation of surface ligand exchange for PbTe CQDs could be considered to protect CQDs from oxidation and enhance their stability. Alongside ligand exchange, exploring other surface passivation strategies was also important. This included the utilization of inorganic passivation materials such as metal oxides and chalcogenides, as well as the development of hybrid passivation methods combining organic and inorganic materials. The study of surface passivation strategies for PbTe CQDs plays a crucial role in improving their stability and enabling their practical application in optoelectronic devices.

3. PbSe CQD-Based Photodetectors

3.1. Synthesis of PbSe CQDs

PbSe CQDs are a material with strong quantum confinement, having an exciton Bohr radius of 46 nm [38]. This characteristic contributes to stronger electron coupling, which leads to improved charge carrier transport within the PbSe CQD film.

In 1997, Lipovskii et al. successfully synthesized size-controllable PbSe CQDs with a size between 2 and 15 nm and a size distribution of about 7% [39]. In 2004, Pietryga et al. studied the mid-infrared PL properties of PbSe CQDs [40]. The experimental results showed that the synthesized PbSe CQDs had excellent fluorescence tunability and narrow size distribution (Figure 3a,b). In 2007, Baek et al. synthesized monodisperse PbSe CQDs using solution chemistry and realized the use of two different ligands of acetic acid (AA) and hexanoic acid (HA) to control the size of CQDs [41] (Figure 3c). The experimental results showed that AA and HA could replace part of the oleate on the surface of PbSe CQDs, resulting in a reduction in steric hindrance and a reduction in the average size of CQDs. To enhance the antioxidant ability of PbSe CQDs, in 2012, Hughes et al. proposed a method of ligand exchange for PbSe CQDs using alkyl selenides [42] (Figure 3d). Alkyl selenides could completely replace the oleic acid ligands on the surface of CQDs. Compared with the oleic acid ligand, the combination between the Pb atom on the surface of the alkyl selenide ligand and the alkyl selenide ligand was more stable, thereby improving the oxidation resistance of the surface of CQDs.

In 2017, Campos et al. proposed a method for the synthesis of PbSe CQDs through N, N, N′-Trisubstituted selenourea precursors [43]. The size of CQDs (1.7–6.6 nm) could be well regulated, with a narrow size distribution of 0.5–2% and a narrow spectral absorption peak (Figure 3e). Experiments had proved that the N, N, N′-Trisubstituted selenourea had a slower and more controllable conversion reaction than the N, N′-Trisubstituted selenourea and the N, N, N′, N′-tetrame-thylselenourea. The proposal of trisubstituted selenourea provided a feasible strategy for other CQDs synthesized from selenium precursors. PbSe CQDs usually required complex and tedious synthesis steps. In 2020, Liu et al. proposed a method for the direct synthesis of PbSe CQDs at room temperature in one step, which simplified the synthesis steps and costs [44] (Figure 3f). The experimental results confirmed that CQDs had excellent surface passivation ability.

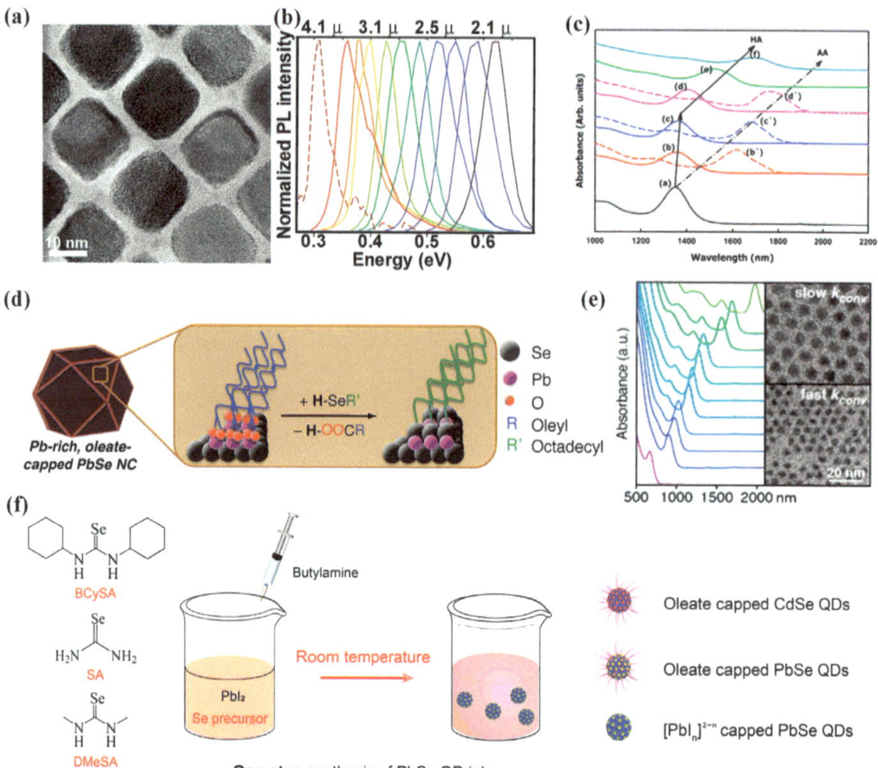

Figure 3. (**a**) TEM image of PbSe CQDs. (**b**) PL spectra of CQDs with different sizes at room temperature (size distribution = 10%) [40]. Copyright 2004, Journal of the American Chemical Society. (**c**) Infrared absorption spectra of PbSe CQDs with different contents of AA and HA. The molar ratios of PbO:Se:oleic acid (OA): HA = 1:3:4.5:X, where X = 0.002, 0.02, 0.1, 0.2, 0.5, 1.0, (a–f). The molar ratio of PbO:Se:OA:AA = 1:3:4.5:X, where X = 0.02, 0.1, 0.2, (b'–d') [41]. Copyright 2007, Journal of Colloid Interface Science. (**d**) Process for surface ligand exchange of PbSe CQDs using alkyl selenides [42]. Copyright 2012, ACS Nano. (**e**) Absorption spectrum and TEM images of PbSe CQDs synthesized using N, N, N'-Trisubstituted selenourea [43]. Copyright 2017, Journal of the American Chemical Society. (**f**) Procedure for the direct synthesis of PbSe CQDs using a one-step approach [44]. Copyright 2020, ACS Energy Letters.

3.2. PbSe CQD Photodetectors

Due to the gradual maturity of PbSe CQD synthesis technology, photodetectors based on PbSe CQDs were studied. In 2010, Sarasqueta et al. first proposed an ITO-PbSe-Al infrared photodetector prepared with PbSe CQDs, and the responsivity could reach 0.67 A/W [45] (Figure 4a). The PbSe surface was passivated by chemical treatment, which effectively reduced the dark current of the device. The performance of the devices under different ligand exchanges was also explored, and the current density of the device treated with benzonedthiol (BDT) was 2 orders of magnitude higher than that of the untreated device. Treatment with BDT showed a 20-fold increase in hole mobility and an 80-fold increase in electron mobility. In 2016, Fu et al. enhanced the stability of synthesized PbSe CQDs by ammonium chloride treatment and prepared Au/PbSe/PMMA/Au photodetectors [46] (Figure 4b). The responsivity of the device reached 64.17 mA/W, and the detectivity reached 5.08×10^{10} Jones at 980 nm. This work on PbSe CQDs demonstrated a highly promising approach for the development of infrared detectors.

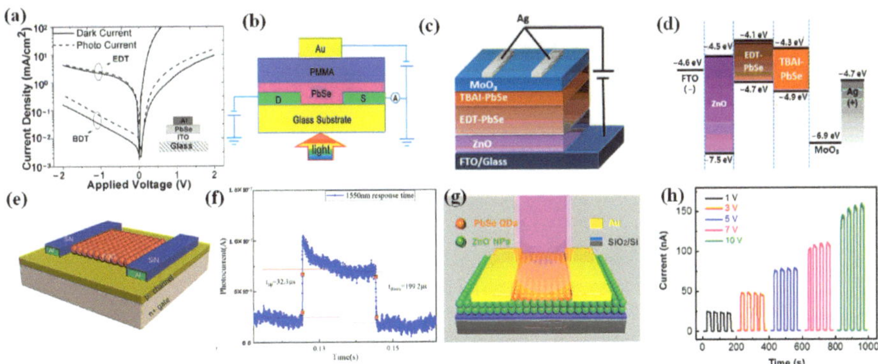

Figure 4. (**a**) I-V curves of the PbSe CQD photodetectors treated with EDT and BDT; the inset is the device schematic of the photodetector [45]. Copyright 2010, Chemistry of Materials. (**b**) Cross-sectional schematic of a FET-based photodetector [46]. Copyright 2016, Nanotechnology. (**c**) Schematic of a device based on bilayer PbSe CQDs. (**d**) Schematic diagram of energy bands based on bilayer PbSe CQDs [47]. Copyright 2019, ACS Applied Materials & Interfaces. (**e**) Schematic diagram of the device of Si: PbSe CQDs. (**f**) Response time curve of Si: PbSe CQDs [48]. Copyright 2021, Journal of materials science. (**g**) Schematic diagram of PbSe CQD photodetector. (**h**) Photoresponse of PbSe CQD photodetector with different voltage biases [49]. Copyright 2021, ACS Applied Materials & Interfaces.

To extend the detection band of PbSe CQD photodetectors, in 2019, Zhu et al. proposed a PbSe CQD photodetector with a detection range from the ultraviolet to the mid-infrared region (350–2500 nm) [47] (Figure 4c,d). The hole-trapping-induced photomultiplication effect was realized by p-type EDT (ethanedithiol)-PbSe CQDs and an n-type TABI-PbSe CQDs double-layer film. Experimental results showed that the photodetector had a quantum efficiency of 450% in the visible region and a detectivity of more than 10^{12} Jones at room temperature. In the infrared region, the quantum efficiency could reach 120% and the detectivity could reach 4×10^{11} Jones. Such photodetectors provided a feasible way for the development of uncooled, broadband photodetectors.

In order to achieve PbSe CQD compatibility with silicon-based substrates, in 2021, Chen et al. proposed a photodetector using PbSe CQDs in Si, with an infrared spectral response range of 450–1550 nm [48] (Figure 4e,f). The device achieved a responsivity of 648.7 A/W and a response time of 32.3 µs at 1550 nm. The detectivity reached 7.48×10^{10} Jones, and the external quantum efficiency reached $6.47 \times 10^{4}\%$. The device had the advantages of being low-cost and easy to integrate with silicon-based readout circuits. In the same year, Peng et al. proposed a near-infrared detector using PbSe CQDs [49] (Figure 4g,h). PbSe CQDs were created using the method of one-step synthesis at room temperature and directly introduced iodide ions, which realized in situ passivation of iodide ions and avoided redundant ligand exchange processes. The device had a responsivity of 970 mA/W and a detectivity of 1.86×10^{11} Jones at 808 nm. Compared with the pure PbSe CQD detector, the photocurrent of the heterojunction detector increased from 7.6×10^{-9} A to 7.4×10^{-8} A, and the dark current decreased from 1.3×10^{-9} A to 7.7×10^{-11} A, effectively reducing the dark current of the device.

Since pure PbSe CQDs were easily oxidized, which led to the degradation of the device performance, the concept of combining PbSe CQDs with other materials had been proposed. A composite material was prepared with materials such as organic polymers and metal oxides, enabling the interaction between different materials to optimize device performance. It provided a novel way to achieve an extended band and improve device stability.

In 2015, Wang et al. proposed a near-infrared photodetector with a field-effect transistor structure combining PbSe CQDs and poly (3-hexylthiophene-2, 5-diyl) (P3HT) [50] (Figure 5a). Responsivity and detectivity reached 500 A/W and 5.05×10^{12} Jones, respectively. Two-dimensional materials became a research hotspot because of their excellent optical and electrical properties. Two-dimensional materials have high carrier performance, so combining with CQDs could make up for the shortcomings of CQDs. In this combination, a quantum dot layer is used as a photosensitive material, and a two-dimensional material can improve the mobility of carriers and other properties. Therefore, devices combining zero-dimensional and two-dimensional materials could effectively improve the performance of detectors. In 2019, Luo et al. reported a device combining PbSe CQDs and Bi_2O_2Se [51] (Figure 5b,c), realizing 2 μm short-wave detection. The responsivity was greater than 103 A/W, and the response time could reach 4 ms. High sensitivity and quick response were realized. In 2022, Peng et al. reported a photodetector integrated with PbSe CQDs and two-dimensional material MoS_2 [52] (Figure 5j). The photo–dark current ratio of the heterogeneous photodetector could reach 10^2, the maximum responsivity could reach 23.5 A/W, and the maximum detectivity could reach 3.17×10^{10} Jones under 635 nm illumination. And the detection band of the detector could be extended to the near-infrared region. Under the illumination of 808 nm, it achieved a responsivity of 19.7 A/W and a detectivity of 2.65×10^{10} Jones. The proposed heterojunction photodetector improved the performance of photodetectors based on zero- and two-dimensional materials.

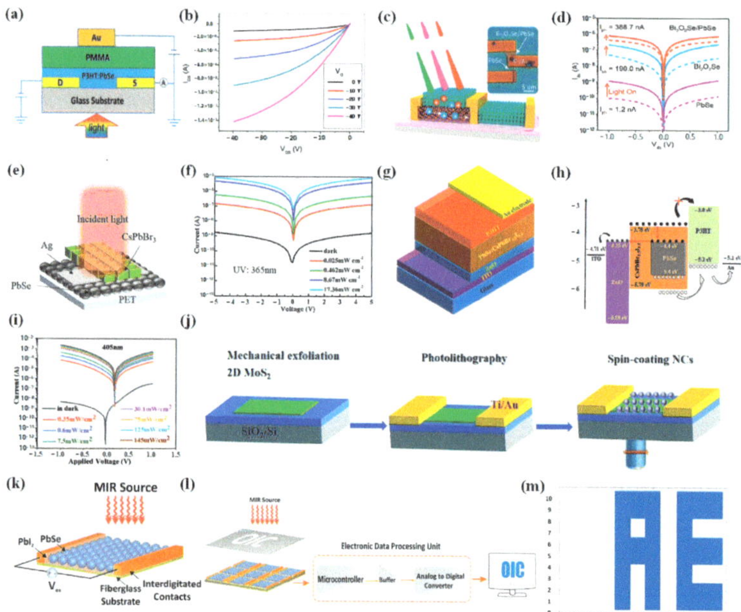

Figure 5. (**a**) Schematic diagram of P3HT: PbSe CQD photodetector. (**b**) Output characteristics of FET-based photodetector Au(Gate)/PMMA/P3HT:PbSe/Au(Source, Drain) in dark [50]. Copyright 2015, IEEE Photonics Technology Letters. (**c**) Schematic illustration of PbSe/Bi_2O_2Se photodetector. (**d**) I−V curves of bare Bi_2O_2Se, PbSe, and PbSe/Bi_2O_2Se hybrid photodetectors under dark and illumination (532 nm, 3.7 mW/cm^2) [51]. Copyright 2019, ACS Nano. (**e**) Schematic diagram of $CsPbBr_3$/PbSe photodetector. (**f**) Typical I–V curves of the $CsPbBr_3$/PbSe heterostructure-based PDs under 365 nm [53]. Copyright 2021, Journal of Materials Science & Technology. (**g**) Schematic diagram of PbSe/$CsPbBr_{1.5}I_{1.5}$. (**h**) Energy band structure diagram of PbSe/$CsPbBr_{1.5}I_{1.5}$. (**i**) I–V curves of photodetector ITO/ZnO/PbSe:$CsPbBr_{1.5}I_{1.5}$/P3HT/Au in dark and under 405 nm [54]. Copyright 2022, Advanced Functional Materials. (**j**) Schematic diagram of PbSe/MoS_2 fabrication process [52]. Copyright 2022, Journal of Materials Chemistry C. (**k**) Schematic diagram of PbSe/PbI_2 photodetector. (**l**) Schematic diagram of the imaging process of the camera. (**m**) "AE" logos obtained by the camera [55]. Copyright 2021, Scientific Reports.

Due to the rise of perovskite materials, in 2021, Hu et al. proposed a flexible broadband photodetector based on a $CsPbBr_3$/PbSe CQD heterostructure, which took advantage of the excellent properties of perovskite nanocrystalline materials and detected a wide range of wavelengths, from ultraviolet to long-wave infrared [53] (Figure 5e,f). The responsivity of the device was 7.17 A/W and the detectivity was 8.97×10^{12} Jones under 365 nm light and 5 V voltage. Response rise and decay times were 0.5 ms and 0.78 ms, respectively. Moreover, the flexible detector could retain 91.2% of its initial performance even after being bent thousands of times. It provided a feasible strategy for the development of flexible devices. In 2022, Sulaman et al. proposed a self-powered broadband photodetector combining PbSe CQDs with $CsPbBr_{1.5}I_{1.5}$ [54] (Figure 5g–i). The responsivity could reach 6.16 A/W, and the detectivity could reach 5.96×10^{13} Jones.

Due to the excellent electrical and optical properties of PbSe CQDs, in 2021, Dortaj et al. suggested a 10×10-pixel high-speed mid-infrared (3–5 µm) camera based on $PbSe/PbI_2$ core–shell CQDs [55] (Figure 5k–m). The procedure involved spin-coating $PbSe/PbI_2$ CQDs onto the substrate of interdigitated electrodes, followed by surface passivation with epoxy resin. Subsequently, the imaging results were obtained on the display through the readout circuit, as shown in Figure 5l,m. Experimental results showed that the response rise time of the detector could reach 100 ns. The camera could achieve 1 million frames per second, so high-resolution images were obtained.

As a new type of photoelectric conversion device, PbSe CQD photodetectors offer several advantages, including fast response, high sensitivity, low cost, and easy preparation. The performance of PbSe CQDs could be further improved by combining them with other materials. To further improve performance, one approach is to employ a new ligand solution for surface modification of PbSe CQDs. Additionally, the integration of new materials or optical devices can also be considered to achieve performance enhancements.

4. PbS CQD-Based Photodetectors

4.1. Synthesis of PbS CQDs

PbS CQDs have a wide tunable band-gap range (0.6–1.6 eV), a high molar broadband absorption coefficient ($\approx 10^6$ M^{-1} cm^{-1}), and a large Bohr exciton radius (~18 nm) [27]. These characteristics make PbS CQDs excellent candidates for low-cost broad-spectrum photodetectors.

In 1990, Nenadovic et al. first prepared 4 nm PbS CQDs in an aqueous solution [56]. In 2011, Lingley et al. proposed a new method of ligand exchange by replacing the oleic acid ligands on the surface of PbS CQDs with nonanoic and dodecanoic acid ligands [57] (Figure 6a). And a high quantum efficiency (55%) could be maintained. In 2013, Zhang et al. synthesized PbS CQDs with good size distribution using H_2S as a sulfur source, and the stability and reproducibility of CQDs prepared by this method were better [58] (Figure 6b). In addition, the long ligand of oleic acid on the surface of PbS CQDs was replaced by a short ligand (butylamine), which could effectively extend the carrier lifetime. In 2017, Lin et al. proposed a liquid-phase ligand exchange method for PbS CQDs, transferring PbS CQDs to the polar solvent DMF for ligand exchange [59] (Figure 6c). The ligand-exchanged PbS CQDs could achieve enhanced mobility and were stable for several months, which laid a solid foundation for the development of photodetectors in the future. In the same year, to produce stable PbS CQDs, Shestha et al. proposed a method of pre-combining thiol ligands with Pb^{2+} before ligand exchange [60], which effectively improved the ligand exchange efficiency of PbS CQDs (Figure 6d). Experimental results showed that 78% of the original oleic acid-terminated CQD photoluminescence quantum efficiency could be maintained after ligand exchange in PbS CQDs by Pb-thiolate. To further study the influence of reaction conditions on PbS CQDs, in 2020, Shuklov et al. proposed a method to synthesize 1.7–2.05 µm PbS CQDs through a mixture of oleic acid and oleylamine [61]. The experimental study showed that the reaction temperature could significantly affect the size of PbS CQDs, while the reaction time had little effect on the PbS CQDs. Good monodisperse PbS CQDs could be obtained when Pb: S = 3: 1. At present, most of the syntheses of PbS CQDs use PbO as the precursor of lead, but it increases the hydroxyl ligands on the surface of PbS CQDs. In 2023, Wang et al. used lead (II) acetylacetonate as the lead precursor to reduce the influence of the hydroxyl ligands on the surface of PbS CQDs [62], and the synthesized PbS CQDs could achieve better binding to iodine ligands during the ligand exchange (Figure 6e). Thus, the surface passivation of PbS CQDs was improved, and the carrier transport ability was enhanced.

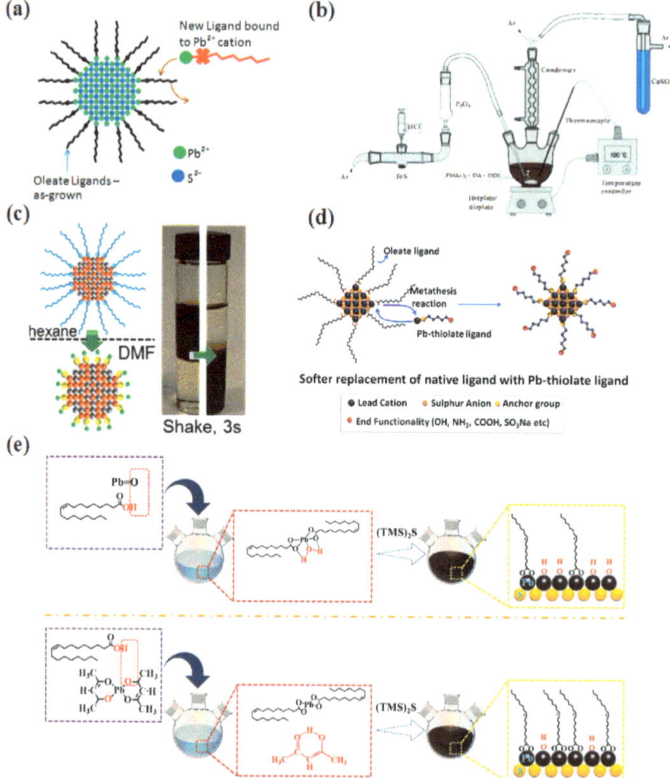

Figure 6. (**a**) Schematic diagram of ligand exchange [57]. Copyright 2011, Nano Letters. (**b**) Diagram of the synthesis setup for PbS CQDs [58]. Copyright 2013, CrystEngComm. (**c**) Schematic diagram of the ligand exchange process. The right inset is the phase transfer photo of PbS CQDs [59]. Copyright 2017, Journal of the American Chemical Society. (**d**) Schematic diagram of the ligand exchange using Pb-thiolate [60]. Copyright 2017, Small. (**e**) Schematic diagram of PbS CQD synthesis process using lead oxide and lead acetylacetonate [62]. Copyright 2023, Advanced Science.

4.2. PbS CQD Photodetectors

PbS CQDs are widely used in photodetectors because of their simple preparation, spectral tunability, and wide wavelength response. In 2006, Konstantatos et al. reported an infrared photodetector based on PbS CQDs [63] (Figure 7a,b). The normalized detectivity of the device at 1.3 μm at room temperature reached 1.8×10^{13} Jones, and the responsivity reached 10^3 A/W. Its performance was comparable to that of epitaxially grown InGaAs detectors. In 2014, Liu et al. first reported a flexible NO_2 gas sensor based on PbS CQDs [64] (Figure 7c,d), which achieved high sensitivity and reproducible performance at room temperature. The experimental results showed that p-type doping was achieved by adding NO_2 on the surface of PbS CQDs. The smaller binding energy was favorable for the rapid desorption of NO_2 from the surface of PbS CQDs, which led to the rapid recovery of the signal. In order to promote the development of photoconductive devices, in 2016, Iacovo et al. proposed a photoconductive device based on PbS CQDs [65] (Figure 7e). The device achieved high responsivity and detectivity at 1.3 μm under 1 V bias, and the maximum values could reach 30 A/W and 2×10^{10} Jones. It was expected to realize applications in silicon integrated circuits.

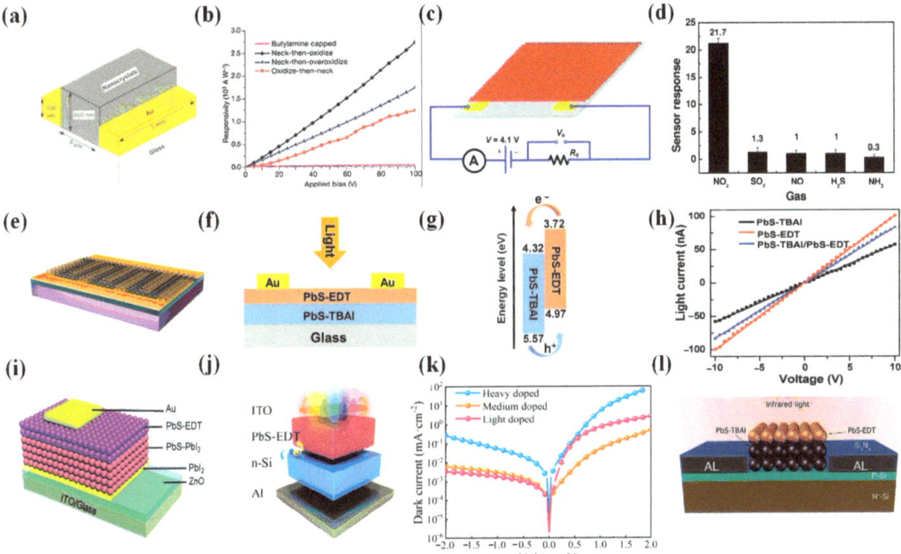

Figure 7. (**a**) Schematic diagram of PbS CQD device. (**b**) Responsivity as a function of applied bias. Necked nanocrystal devices show comparable responsivities, consistent with similar carrier mobilities and trap state lifetimes [63]. Copyright 2006, Nature. (**c**) Schematic illustration of NO_2 gas sensors. (**d**) Response of PbS CQDs sensors to different gases (50 ppm) [64]. Copyright 2014, Advanced Materials. (**e**) Schematic diagram of PbS CQD photodetector on interdigitated contacts [65]. Copyright 2016, Scientific Reports. (**f**) Schematic diagram of PbS-EDT/PbS-TABI photodetector. (**g**) Band diagram of PbS-EDT/PbS-TABI photodetector. (**h**) I–V curves of the three devices under a white light illumination of 0.20 mW/cm² [66]. Copyright 2017, Advance Materials. (**i**) Schematic diagram of PbS-EDT/PbS-PbI$_2$ photodetector [67]. Copyright 2017, RSC Advances. (**j**) Schematic diagram of PbS CQD photodetector. (**k**) Dark current density of Si/PbS PD made using light-doped, medium-doped, and heavy-doped Si wafers [68]. Copyright 2020, ACS Applied Materials & Interfaces. (**l**) Schematic diagram of PbS-EDT/PbS-TABI/Si photodetector [69]. Copyright 2020, Nanotechnology.

Due to the low performance of single-layer CQDs, in 2017, Ren et al. proposed a photodetector combining PbS-TABI and PbS-EDT [66] (Figure 7f–h). The detectivity of the double-layer CQD device could reach 1.71×10^{12} Jones under 580 nm illumination, which was about 3 times higher than that of the single-layer CQD device. The photo–dark current ratio of the bilayer device was 152.35, much larger than that of PbS-EDT (13.41) and PbS-TABI devices (36.9). In the same year, Qiao et al. further improved the double-layer CQD device and proposed the photodetector structure of PbI$_2$/PbS-PbI$_2$/PbS-EDT [67] (Figure 7i). The detector could increase the photocurrent while reducing the dark current and achieved a detectivity of 1.3×10^{13} Jones and a responsivity of 0.43 A/W. Since silicon-based photodetectors were gradually applied to the field of photodetection, in 2020, Xu et al. proposed a photodetector combining n-type silicon with p-type PbS CQDs [68] (Figure 7j,k). The device had a lower band offset and better charge trapping. Under 1540 nm illumination, a detectivity of 1.47×10^{11} Jones and a responsivity of 0.264 A/W were achieved. In the same year, Shi et al. proposed a silicon-compatible PbS CQD photodetector that could be integrated into a chip [69] (Figure 7l). The detectivity reached 3.95×10^{12} Jones, and the external quantum efficiency could reach 4.96×10^5%. This photodetector could realize a broadband response wavelength in the 405–1550 nm range. The mature silicon processing technology and rational band optimization of CQDs further enhanced the capabilities of PbS CQD-based photodetectors.

Detectors based on pure PbS CQDs were still limited in the detection band, so the purpose of extending the band and improving performance was achieved by combining them with other materials. In 2009, Krisztina et al. proposed a photodetector combined with [6,6]-phenyl-C61-butyric acid methyl ester (PCBM) and PbS CQDs (Figure 8a), and its spectral range covered the visible and near-infrared regions [70]. The detectivity obtained at 1200 nm was 2.5×10^{10} Jones. In 2014, He et al. reported a flexible photodetector combined with Ag nanocrystals (NCs) and PbS CQDs (Figure 8b–e), which could effectively reduce dark current and enhance photocurrent [71]. Experiments proved that adding 0.5% to 1% Ag NCs could effectively improve the detection ability of the device. Ag NCs could capture photogenerated electrons, effectively prolonging the lifetime of carriers, thereby improving the photocurrent. The detectivity of the flexible device could reach 1.7×10^{10} Jones. The combination of Ag NCs and CQDs opened up a new strategy for future detector development.

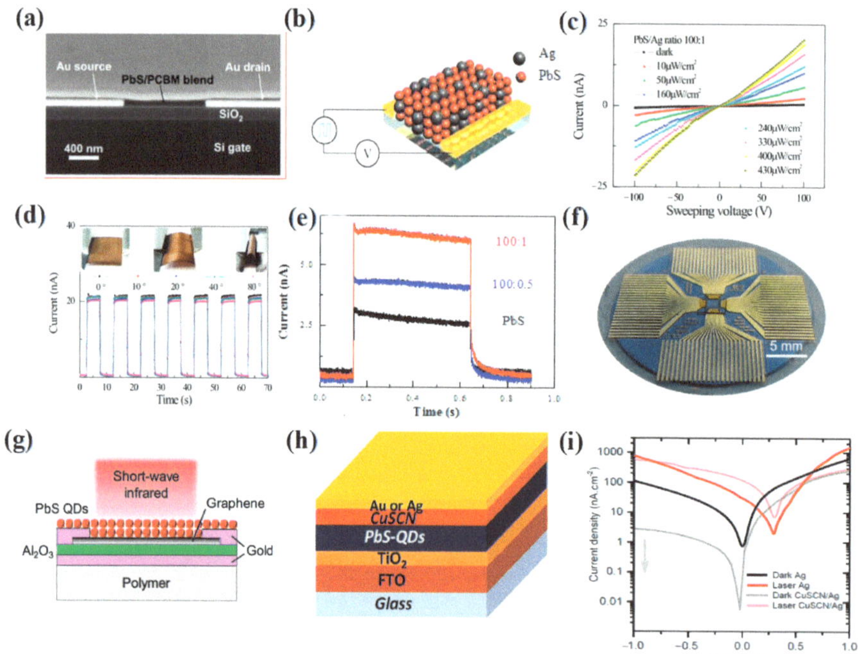

Figure 8. (**a**) SEM of the PbS/PCBM device [70]. Copyright 2009, Advanced Materials. (**b**) Schematic diagram of PbS CQD/Ag NC photodetector. (**c**) I–V characteristics of the 1% Ag NC composite device under different levels of light illumination ranging from 10 to 430 µW/cm². (**d**) Current as a function of time. The inset shows the detector bent at different angles. (**e**) Dynamic current–time (I–T) curves of PbS CQD/Ag NC composite (100:0, 100:0.5, and 100:1) photodetectors under 36 µW/cm² illumination and 40 V voltage [71]. Copyright 2014, ACS Photonics. (**f**) Photograph of PbS/graphene photodetector on a silicon wafer. (**g**) Schematic diagram of PbS/graphene [72]. Copyright 2017, ACS Nano. (**h**) Schematic diagram of PbS/CuSCN photodetector. (**i**) I–V characteristics of the photodiodes fabricated with and without the CuSCN film in the dark and under illumination with a 532 nm laser at 0.5 mW/cm² [73]. Copyright 2020, ACS Photonics.

To realize low-cost and high-performance optoelectronic devices, in 2017, Bessonov et al. proposed a method of coupling PbS CQDs with $CH_3NH_3PbI_3$ (Figure 8f,g), and the detectivity could reach 5×10^{12} Jones [72]. The addition of $CH_3NH_3PbI_3$ enabled PbS CQDs to generate greater gains. To further reduce the dark current of the device, in 2020, Ka et al. reported a strategy combining copper thiocyanate (CuSCN) and PbS CQDs to reduce the dark current of PbS CQD-based photodetectors [73] (Figure 8h,i). The detectivity of the prepared photodiode could reach 10^{11} Jones. The dark current was reduced by 2 orders of magnitude, effectively reducing the dark current of the photoelectric device.

Since PbS CQDs could realize photodetection in the visible to infrared region, and CQDs had the advantages of solution processability and substrate compatibility, PbS CQDs were applied in readout circuits. In 2015, Klem et al. proposed a short-wavelength readout circuit photodiode based on PbS CQDs, which consisted of a linear 320-pixel array [74]. The dark current density at room temperature was 6.8 nA/cm^2, and the specific detectivity could reach 10^{12}–10^{13} Jones, which was comparable to that of the InGaAs photodiode.

Due to the rise of graphene materials, in 2017, Goossens et al. proposed a 388×288 array complementary metal oxide semiconductor (CMOS) imaging system combining PbS CQDs and graphene [75] (Figure 9a,b). The CQD layer and the graphene layer formed a vertical heterojunction; the graphene layer trapped holes, and the CQD layer trapped electrons. The detectivity reached up to 1012 Jones in the range of 300–2000 nm.

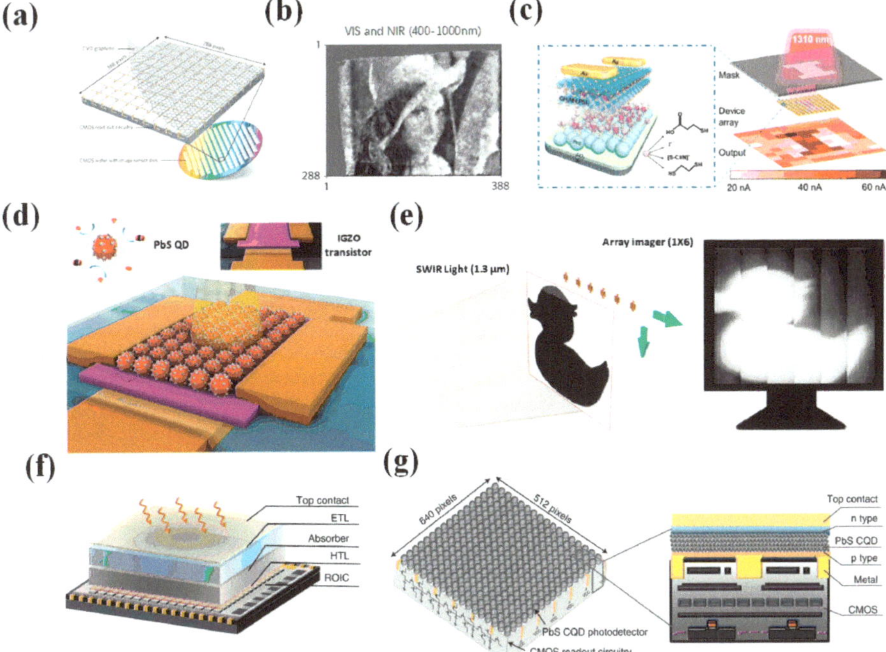

Figure 9. (**a**) Schematic diagram of graphene transfer process on 388×288 array image sensor. (**b**) Visible and near-infrared photograph of a standard image reference "Lena" printed in black and white on paper illuminated with an LED desk lamp [75]. Copyright 2017, Nature Photonics. (**c**) Device structure of $PbS/CH_3NH_3PbI_3$ photodetector. The right inset shows photoelectric imaging of the letter "I" under near-infrared light [76]. Copyright 2019, ACS Applied Materials & Interfaces. (**d**) Schematic diagram of PbS CQD/InGaZnO photodetector. (**e**) Imaging schematic diagram of the PbS CQD/InGaZnO photodetector [77]. Copyright 2020, ACS Photonics. (**f**) Schematic diagram of PbS CQD readout circuit. (**g**) Schematic diagram of PbS CQD imager and cross-section of PbS CQD imager [21]. Copyright 2022, Nature Electronics.

In 2019, Zhang et al. further optimized a device combining PbS CQDs with $CH_3NH_3PbI_3$ and fabricated a 10×10 array photodetector [76] (Figure 9c). It realized a wide-band detection capability in the ultraviolet–visible–near-infrared region, the responsivities could reach 255 A/W and 1.58 A/W at 365 nm and 940 nm, respectively, the detectivities at 365 nm and 940 nm were 4.9×10^{13} Jones and 3.0×10^{11} Jones, and the response time was 42 ms. The appearance of this detector laid the foundation for the development of high-sensitivity broadband photodetectors and imagers. In the same year, Georgitzikis et al. proposed an imager combining a polymer (organic) with PbS CQDs [78]. The photodiode combined with the polymer and PbS CQDs was integrated into a 512×768-pixel focal plane array to realize imaging in visible and infrared environments. The maximum detectivity could reach 2.3×10^{12} Jones, the minimum rise response time was 13 µs, and the minimum fall response time was 30 µs. High-resolution and high-sensitivity infrared imaging was realized. In 2020, Choi et al. proposed a 1×6 linear array photodetector combined with PbS CQDs and InGaZnO [77] (Figure 9d). At 1310 nm, the detector achieved a responsivity of 10^4 A/W and a detectivity of 10^{12} Jones. Imaging of the pattern was finally achieved (Figure 9e). The surface modification of PbS CQDs with TABI increased the stability and oxidation resistance of the devices. However, the devices surface-modified with EDT lost their detection ability within 2 weeks. In 2022, Liu et al. proposed a 640×512-pixel high-resolution imager based on PbS CQDs, with a spectral range of 400–1300 nm [21] and a detectivity of 2.1×10^{12} Jones, and the spatial resolution of the imager could reach 40 line-pairs per millimeter (Figure 9f,g). Experiments showed that the imager realized vein imaging and substance identification, which promoted the development of PbS CQDs. The PbS CQD imager had the advantages of high sensitivity, high resolution, fast response, and multispectral imaging.

Consequently, compared with single-photon detectors, array detectors have broad development prospects. But in future development, further cost reduction and smaller size are required. Meanwhile, during the synthesis process, PbS CQDs can be susceptible to oxidation and volatility, necessitating the implementation of additional protective measures in future developments.

In addition to the aforementioned CQD detectors, lead chalcogenide photodetectors also encompass bulk semiconductor photodetectors. However, lead chalcogenide CQD photodetectors have numerous advantages over lead chalcogenide bulk photodetectors.

(1) The band structure of lead chalcogenide bulk materials is relatively fixed and difficult to control, resulting in a limited spectral tuning range. In contrast, the band structure of lead chalcogenide CQDs can be adjusted by tuning their size, thereby expanding the spectral tuning range and leading to broader potential applications.

(2) Lead chalcogenide bulk semiconductor thin films are usually prepared using the chemical bath deposition method [79,80], which poses challenges in integrating bulk materials with silicon-based readout circuits. In contrast, CQDs are synthesized using a thermal injection method, leading to lower manufacturing costs for CQD detectors. Moreover, CQDs can be directly integrated with silicon-based readout circuits through solution processing, thereby expanding the potential applications of lead chalcogenide CQD photodetectors.

(3) Photodetectors based on lead chalcogenide bulk materials need to undergo high-temperature sensitization at 300–600 °C in a specific atmosphere, such as an oxygen-rich and iodine-rich atmosphere [81]. However, the existing sensitization process lacks repeatability, stability, and uniformity, thereby restricting their application [82,83]. Lead chalcogenide CQD photodetectors can operate at room temperature, reducing the manufacturing cost and difficulty.

(4) Lead chalcogenide CQD photodetectors can be self-assembled in vertical or horizontal directions, forming more complex structures. The feature provides lead chalcogenide CQD photodetectors with a distinct advantage in terms of integration and multi-channel detection.

5. Conclusions

In this review, we reviewed the development of lead chalcogenide CQD photodetectors in recent years. Table 1 summarizes the performance of lead-based CQD photodetectors. The lead-based CQDs include PbTe, PbSe, and PbS CQDs. One can see that lead chalcogenide CQD photodetectors have developed quickly, developing from single-pixel photodetectors to large array imagers. However, compared with traditional bulk semiconductor photodetectors, CQD-based photodetectors still face challenges, such as low detectivity and large dark current. There are several challenges that need to be addressed:

Table 1. The summary of CQD-based photodetectors.

Year	Photoactive Material	Detection Range (nm)	Detectivity (Jones)	Responsivity (A/W)	Rise Decay Time	Refs.
2006	PbS CQDs	1300	1.8×10^{13}	10^3	--	[63]
2009	PbS CQDs/PCBM	1200	2.5×10^{10}	1.6	--	[70]
2010	PbSe CQDs	1400	--	0.67	--	[45]
2014	PbS CQDs/Ag NCs	1100	1.7×10^{10}	0.0038	--	[71]
2015	PbSe CQDs/P3HT	980	5.05×10^{12}	500	--	[50]
2016	PbSe CQDs	980	5.08×10^{10}	0.06417	--	[46]
2016	PbTe CQDs	1064	--	0.0019	0.39 ms 0.49 ms	[36]
2016	PbS CQDs	1300	2×10^{10}	30	160 ms 3 s	[65]
2017	PbS-EDT/PbS-TABI	580	1.71×10^{12}	0.25	3.63 ms 29.56 ms	[66]
2017	PbS-EDT/PbS-PbI$_2$	850	10^{13}	0.43	5.3 μs 4.9 μs	[67]
2017	PbS CQDs/CH$_3$NH$_3$PbI$_3$	520	5×10^{12}	2×10^5	10 ms 0.5s	[72]
2019	PbS CQDs/CH$_3$NH$_3$PbI$_3$	365/940	4.9×10^{13}@365 nm 3.0×10^{11}@940 nm	255@365 nm 1.58@940 nm	42 ms --	[76]
2019	PbSe-TABI/PbSe-EDT	1300/2400	10^{12}@1300 nm 10^{11}@2400 nm	0.05–0.2	140 μs 410 μs	[47]
2019	PbSe CQDs/Bi$_2$O$_2$Se	2000	--	10^3	4 ms	[51]
2020	PbS-TABI/PbS-EDT	1540	1.47×10^{11}	0.264	2.04 μs 5.34 μs	[69]
2020	PbS CQDs/CuSCN	532	10^{11}	--	50 μs 110 μs	[73]
2021	PbSe CQDs	1550	7.48×10^{10}	648.7	32.3 μs 73.2 μs	[48]
2021	PbSe CQDs	808	1.86×10^{11}	0.97	0.34 s 0.67 s	[49]
2021	PbSe CQDs/CsPbBr$_3$	365	8.97×10^{12}	7.17	0.5 ms 0.78 ms	[53]
2022	PbSe CQDs/MoS$_2$	635/808	3.17×10^{10}@635 nm 2.65×10^{10}@808 nm	23.5@635 nm 19.7@808 nm	0.36s@635 nm; 0.38s@808 nm 0.52s@635 nm; 0.86s@808 nm	[52]
2022	PbSe CQDs/CsPbBr$_{1.5}$I$_{1.5}$	532	5.96×10^{13}	6.16	350 ms 375 ms	[54]
2022	PbS CQDs	940	2.1×10^{12}	--	1.15 μs 0.49 μs	[19]

(1)   CQD surface passivation. CQD have large surface to volume ratio. As a result, they are overly sensitive to the environment. Surface modification with organic or inorganic ligands could improve CQD stability and protect their physical properties. For instance, PbTe CQDs, are susceptible to oxidation, making them less suitable for photodetector applications. To address the problem of oxidation, surface modification

techniques utilizing organic or inorganic ligands can be employed to enhance the stability and photoelectric conversion efficiency of PbTe CQDs. This improvement is expected to enhance the overall photoelectric performance and lifespan of photodetectors. In addition, the directional assembly of CQDs and the fine regulation of their optical properties can be achieved through surface modification.

(2) Dark current reduction on CQD-based photodetectors. Compared with InGaAs and HgCdTe based photodetectors, CQD-based photodetectors typically suffer the disadvantage on large dark current. The dark current is usually generated by the surface defects on the CQDs, which can trap and recombine charges. Additionally, thermal excitation in CQD-based devices can lead to dark current generation. To solve this problem, reducing the surface defects and band tail regulation should be the key. In addition, transport property improvement would also be useful such as doping density and mobility modification.

(3) Large array photodetectors. At present, most research focus on single-pixel CQD detectors. However, in real application, it is usually necessary to use array detectors. Large area array photodetectors can be prepared by nanoimprinting technology and micro-nano processing technology. There are many technical challenges need to be solved.

(4) Broad band photodetectors. At present, the main research on lead based CQD photodetector could only cover near-infrared to short-wave infrared. More research is necessary to promote the progress on broad band photodetection. For example, combining PbS CQDs with graphene, perovskite, and other materials can achieve detection in the visible and near-infrared bands. Therefore, combining CQDs with other materials achieves the purpose of broad-spectrum detection.

Author Contributions: Conceptualization, T.Q. and X.Z.; investigation, X.Z., H.M. and H.C.; data curation, X.Z. and Z.W.; writing—original draft preparation, X.Z., H.M. and Y.B.; writing—review and editing, T.Q. and X.T.; supervision, T.Q.; funding acquisition, X.T. All authors have read and agreed to the published version of the manuscript.

Funding: This research was funded by National Natural Science Foundation of China (U22A2081, 62105022).

Institutional Review Board Statement: Not applicable.

Informed Consent Statement: Not applicable.

Data Availability Statement: Not applicable.

Conflicts of Interest: The authors declare no conflict of interest.

References

1. Gmachl, C.; Capasso, F.; Sivco, D.L.; Cho, A.Y. Recent progress in quantum cascade lasers and applications. *Rep. Prog. Phys.* **2001**, *64*, 1533. [CrossRef]

2. Lohse, S.E.; Murphy, C.J. Applications of Colloidal Inorganic Nanoparticles: From Medicine to Energy. *J. Am. Chem. Soc.* **2012**, *134*, 15607–15620. [CrossRef] [PubMed]

3. Lhuillier, E.; Guyot-Sionnest, P. Recent Progresses in Mid Infrared Nanocrystal based Optoelectronics. *IEEE J. Sel. Top. Quantum Electron.* **2017**, *23*, 6000208. [CrossRef]

4. Yadav, P.V.K.; Ajitha, B.; Kumar Reddy, Y.A.; Sreedhar, A. Recent advances in development of nanostructured photodetectors from ultraviolet to infrared region: A review. *Chemosphere* **2021**, *279*, 130473. [CrossRef] [PubMed]

5. Lyu, X. Recent Progress on Infrared Detectors: Materials and Applications. In Proceedings of the 4th International Conference on Electronic Science and Automation Control (ESAC), Bangkok, Thailand, 19–20 November 2022; Volume 27, pp. 191–200.

6. Rogalski, A.; Martyniuk, P.; Kopytko, M. InAs/GaSb type-II superlattice infrared detectors: Future prospect. *Appl. Phys. Rev.* **2017**, *4*, 31304. [CrossRef]

7. DeWames, R.E.; Schuster, J. Performance and limitations of NIR and extended wavelength eSWIR InP/InGaAs image sensors. In Proceedings of the Quantum Sensing and Nano Electronics and Photonics XVII, San Francisco, CA, USA, 2–6 February 2020.

8. Rogalski, A.; Martyniuk, P.; Kopytko, M. Type-II superlattice photodetectors versus HgCdTe photodiodes. *Prog. Quantum Electron.* **2019**, *68*, 100228. [CrossRef]

9.   Bhan, R.K.; Dhar, V. Recent infrared detector technologies, applications, trends and development of HgCdTe based cooled infrared focal plane arrays and their characterization. *Opto-Electron. Rev.* **2019**, *27*, 174–193. [CrossRef]
10. Wang, X.; Wang, M.; Liao, Y.; Zhang, H.; Zhang, B.; Wen, T.; Yi, J.; Qiao, L. Molecular-beam epitaxy-grown HgCdTe infrared detector: Material physics, structure design, and device fabrication. *Sci. China Phys. Mech. Astron.* **2023**, *66*, 237302. [CrossRef]
11. Chen, J.; Chen, J.; Li, X.; He, J.; Yang, L.; Wang, J.; Yu, F.; Zhao, Z.; Shen, C.; Guo, H.; et al. High-performance HgCdTe avalanche photodetector enabled with suppression of band-to-band tunneling effect in mid-wavelength infrared. *npj Quantum Mater.* **2021**, *6*, 103. [CrossRef]
12. Rothman, J. Physics and Limitations of HgCdTe APDs: A Review. *J. Electron. Mater.* **2018**, *47*, 5657–5665. [CrossRef]
13. Ciura, Ł.; Kopytko, M.; Martyniuk, P. Low-frequency noise limitations of InAsSb-, and HgCdTe-based infrared detectors. *Sens. Actuators A Phys.* **2020**, *305*, 111908. [CrossRef]
14. Batty, K.; Steele, I.; Copperwheat, C. Laboratory and On-sky Testing of an InGaAs Detector for Infrared Imaging. *Publ. Astron. Soc. Pac.* **2022**, *134*, 65001. [CrossRef]
15. Livache, C.; Martinez, B.; Goubet, N.; Gréboval, C.; Qu, J.; Chu, A.; Royer, S.; Ithurria, S.; Silly, M.G.; Dubertret, B.; et al. A colloidal quantum dot infrared photodetector and its use for intraband detection. *Nat. Commun.* **2019**, *10*, 2125. [CrossRef] [PubMed]
16. De Arquer, F.P.G.; Talapin, D.V.; Klimov, V.I.; Arakawa, Y.; Bayer, M.; Sargent, E.H. Semiconductor quantum dots: Technological progress and future challenges. *Science* **2021**, *373*, eaaz8541. [CrossRef] [PubMed]
17. Liu, M.; Yazdani, N.; Yarema, M.; Jansen, M.; Wood, V.; Sargent, E.H. Colloidal quantum dot electronics. *Nat. Electron.* **2021**, *4*, 548–558. [CrossRef]
18. Qiu, Y.; Zhou, X.; Tang, X.; Hao, Q.; Chen, M. Micro Spectrometers Based on Materials Nanoarchitectonics. *Materials* **2023**, *16*, 2253.
19. Yan, N.; Qiu, Y.; He, X.; Tang, X.; Hao, Q.; Chen, M. Plasmonic Enhanced Nanocrystal Infrared Photodetectors. *Materials* **2023**, *16*, 3216. [CrossRef]
20. Wang, J.; Chen, J. High-sensitivity silicon: PbS quantum dot heterojunction near-infrared photodetector. *Surf. Interfaces* **2022**, *30*, 101945. [CrossRef]
21. Liu, J.; Liu, P.; Chen, D.; Shi, T.; Qu, X.; Chen, L.; Wu, T.; Ke, J.; Xiong, K.; Li, M.; et al. A near-infrared colloidal quantum dot imager with monolithically integrated readout circuitry. *Nat. Electron.* **2022**, *5*, 443–451. [CrossRef]
22. Zhang, S.; Bi, C.; Qin, T.; Liu, Y.; Cao, J.; Song, J.; Huo, Y.; Chen, M.; Hao, Q.; Tang, X. Wafer-Scale Fabrication of CMOS-Compatible Trapping-Mode Infrared Imagers with Colloidal Quantum Dots. *ACS Photonics* **2023**, *10*, 673–682. [CrossRef]
23. Kim, J.Y.; Voznyy, O.; Zhitomirsky, D.; Sargent, E.H. 25th Anniversary Article: Colloidal Quantum Dot Materials and Devices: A Quarter-Century of Advances. *Adv. Mater.* **2013**, *25*, 4986–5010. [CrossRef] [PubMed]
24. Pan, Y.; Bai, H.; Pan, L.; Li, Y.; Tamargo, M.C.; Sohel, M.; Lombardi, J.R. Size controlled synthesis of monodisperse PbTe quantum dots: Using oleylamine as the capping ligand. *J. Mater. Chem.* **2012**, *22*, 23593–23601. [CrossRef]
25. Yan, L.; Zhang, Y.; Zhang, T.; Feng, Y.; Zhu, K.; Wang, D.; Cui, T.; Yin, J.; Wang, Y.; Zhao, J.; et al. Tunable near-Infrared Luminescence of PbSe Quantum Dots for Multigas Analysis. *Anal. Chem.* **2014**, *86*, 11312–11318. [CrossRef] [PubMed]
26. Zhang, J.; Crisp, R.W.; Gao, J.; Kroupa, D.M.; Beard, M.C.; Luther, J.M. Synthetic Conditions for High-Accuracy Size Control of PbS Quantum Dots. *J. Phys. Chem. Lett.* **2015**, *6*, 1830–1833. [CrossRef]
27. Wise, F.W. Lead Salt Quantum Dots: The Limit of Strong Quantum Confinement. *Acc. Chem. Res.* **2000**, *33*, 773–780. [CrossRef]
28. Paufler, P. Landolt-Börnstein. Numerical data and functional relationships in science and technology. New Series, Editor in Chief: K. H. Hellwege. Group III, Crystal and Solid State Physics, Vol. 7, Crystal Structure Data of Inorganic Compounds, W. Pies, A. Weiss, Part b, Key Elements O, S, Se, Te, b3: Key Elements S, Se, Te, Editors: K. H. Hellwege, A.M. Hellwege, Springer-Verlag Berlin 1982, XXVII, 435 Seiten. Leinen, Preis: DM 740. *Cryst. Res. Technol.* **1983**, *18*, 1318.
29. Reynoso, V.C.S.; de Paula, A.M.; Cuevas, R.F.; Medeiros Neto, J.A.; Cesar, C.L.; Barbosa, L.C. PbTe quantum dot doped glasses with absorption edge in the 1.5 μm wavelength region. *Electron. Lett.* **1995**, *31*, 1013–1015. [CrossRef]
30. Murphy, J.E.; Beard, M.C.; Norman, A.G.; Ahrenkiel, S.P.; Johnson, J.C.; Yu, P.; Mićić, O.I.; Ellingson, R.J.; Nozik, A.J. PbTe Colloidal Nanocrystals: Synthesis, Characterization, and Multiple Exciton Generation. *J. Am. Chem. Soc.* **2006**, *128*, 3241–3247. [CrossRef]
31. Urban, J.J.; Talapin, D.V.; Shevchenko, E.V.; Murray, C.B. Self-Assembly of PbTe Quantum Dots into Nanocrystal Superlattices and Glassy Films. *J. Am. Chem. Soc.* **2006**, *128*, 3248–3255. [CrossRef]
32. Lambert, K.; Geyter, B.D.; Moreels, I.; Hens, Z. PbTe | CdTe Core | Shell Particles by Cation Exchange, a HR-TEM study. *Chem. Mater.* **2009**, *21*, 778–780. [CrossRef]
33. Protesescu, L.; Zünd, T.; Bodnarchuk, M.I.; Kovalenko, M.V. Air-Stable, Near- to Mid-Infrared Emitting Solids of PbTe/CdTe Core-Shell Colloidal quantum dots. *ChemPhysChem* **2016**, *17*, 670–674. [CrossRef] [PubMed]
34. Peters, J.L.; de Wit, J.; Vanmaekelbergh, D. Sizing Curve, Absorption Coefficient, Surface Chemistry, and Aliphatic Chain Structure of PbTe Nanocrystals. *Chem. Mater.* **2019**, *31*, 1672–1680. [CrossRef] [PubMed]
35. Miranti, R.; Shin, D.; Septianto, R.D.; Ibáñez, M.; Kovalenko, M.V.; Matsushita, N.; Iwasa, Y.; Bisri, S.Z. Exclusive Electron Transport in Core@Shell PbTe@PbS Colloidal Semiconductor Nanocrystal Assemblies. *ACS Nano* **2020**, *14*, 3242–3250. [CrossRef] [PubMed]
36. Lin, Z.; Yang, Z.; Wang, P.; Wei, G.; He, A.; Guo, W.; Wang, M. Schottky-Ohmic Converted Contact, Fast-response, Infrared PbTe Photodetector with Stable Photoresponse in Air. *RSC Adv.* **2016**, *6*, 107878–107885. [CrossRef]

37. Ten Kate, S.C.; Ritter, M.F.; Fuhrer, A.; Jung, J.; Schellingerhout, S.G.; Bakkers, E.P.A.M.; Riel, H.; Nichele, F. Small Charging Energies and g-Factor Anisotropy in PbTe Quantum Dots. *Nano Lett.* **2022**, *22*, 7049–7056. [CrossRef]
38. Ma, W.; Swisher, S.L.; Ewers, T.; Engel, J.; Ferry, V.E.; Atwater, H.A.; Alivisatos, A.P. Photovoltaic Performance of Ultra small PbSe Quantum Dots. *ACS Nano* **2011**, *5*, 8140–8147. [CrossRef]
39. Lipovskii, A.; Kolobkova, E.; Petrikov, V.; Kang, I.; Olkhovets, A.; Krauss, T.; Thomas, M.; Silcox, J.; Wise, F.; Shen, Q.; et al. Synthesis and characterization of PbSe quantum dots in phosphate glass. *Appl. Phys. Lett.* **1997**, *71*, 3406–3408. [CrossRef]
40. Pietryga, J.M.; Schaller, R.D.; Werder, D.; Stewart, M.H.; Klimov, V.I.; Hollingsworth, J.A. Pushing the Band Gap Envelope: Mid-Infrared Emitting Colloidal PbSe Quantum Dots. *J. Am. Chem. Soc.* **2004**, *126*, 11752–11753. [CrossRef]
41. Baek, I.C.; Seok, S.I.; Pramanik, N.C.; Jana, S.; Lim, M.A.; Ahn, B.Y.; Lee, C.J.; Jeong, Y.J. Ligand-dependent particle size control of PbSe quantum dots. *J. Colloid Interface Sci.* **2007**, *310*, 163–166. [CrossRef]
42. Hughes, B.K.; Ruddy, D.A.; Blackburn, J.L.; Smith, D.K.; Bergren, M.R.; Nozik, A.J.; Johnson, J.C.; Beard, M.C. Control of PbSe Quantum Dot Surface Chemistry and Photophysics Using an Alkylselenide Ligand. *ACS Nano* **2012**, *6*, 5498–5506. [CrossRef]
43. Campos, M.P.; Hendricks, M.P.; Beecher, A.N.; Walravens, W.; Swain, R.A.; Cleveland, G.T.; Hens, Z.; Sfeir, M.Y.; Owen, J.S. A Library of Selenourea Precursors to PbSe Nanocrystals with Size Distributions near the Homogeneous Limit. *J. Am. Chem. Soc.* **2017**, *139*, 2296–2305. [CrossRef] [PubMed]
44. Liu, Y.; Li, F.; Shi, G.; Liu, Z.; Lin, X.; Shi, Y.; Chen, Y.; Meng, X.; Lv, Y.; Deng, W.; et al. PbSe Quantum Dot Solar Cells Based on Directly Synthesized Semiconductive Inks. *ACS Energy Lett.* **2020**, *5*, 3797–3803. [CrossRef]
45. Sarasqueta, G.; Choudhury, K.R.; So, F. Effect of Solvent Treatment on Solution-Processed Colloidal PbSe Nanocrystal Infrared Photodetectors. *Chem. Mater.* **2010**, *22*, 3496–3501. [CrossRef]
46. Fu, C.; Wang, H.; Song, T.; Zhang, L.; Li, W.; He, B.; Sulaman, M.; Yang, S.; Zou, B. Stability enhancement of PbSe quantum dots via post-synthetic ammonium chloride treatment for a high-performance infrared photodetector. *Nanotechnology* **2016**, *27*, 65201. [CrossRef] [PubMed]
47. Zhu, T.; Zheng, L.; Yao, X.; Liu, L.; Huang, F.; Cao, Y.; Gong, X. Ultrasensitive Solution-Processed Broadband PbSe Photodetectors through Photomultiplication Effect. *ACS Appl. Mater. Interfaces* **2019**, *11*, 9205–9212. [CrossRef]
48. Chen, P.; Wu, Z.; Shi, Y.; Li, C.; Wang, J.; Yang, J.; Dong, X.; Gou, J.; Wang, J.; Jiang, Y. High-performance silicon-based PbSe-CQDs infrared photodetector. *J. Mater. Sci. Mater. Electron.* **2021**, *32*, 9452–9462. [CrossRef]
49. Peng, M.; Liu, Y.; Li, F.; Hong, X.; Liu, Y.; Wen, Z.; Liu, Z.; Ma, W.; Sun, X. Room-Temperature Direct Synthesis of PbSe Quantum Dot Inks for High-Detectivity Near-Infrared Photodetectors. *ACS Appl. Mater. Interfaces* **2021**, *13*, 51198–51204. [CrossRef]
50. Wang, H.; Li, Z.; Fu, C.; Yang, D.; Zhang, L.; Yang, S.; Zou, B. Solution-Processed PbSe Colloidal Quantum Dot-Based Near-Infrared Photodetector. *IEEE Photonics Technol. Lett.* **2015**, *27*, 612–615. [CrossRef]
51. Luo, P.; Zhuge, F.; Wang, F.; Lian, L.; Liu, K.; Zhang, J.; Zhai, T. PbSe Quantum Dots Sensitized High-Mobility Bi_2O_2Se Nanosheets for High-Performance and Broadband Photodetection Beyond 2 μm. *ACS Nano* **2019**, *13*, 9028–9037. [CrossRef]
52. Peng, M.; Tao, Y.; Hong, X.; Liu, Y.; Wen, Z.; Sun, X. One-step synthesized PbSe nanocrystal inks decorated 2D MoS_2 heterostructure for high stability photodetectors with photoresponse extending to near-infrared region. *J. Mater. Chem. C* **2022**, *10*, 2236–2244. [CrossRef]
53. Hu, J.; Yang, S.; Zhang, Z.; Li, H.; Perumal Veeramalai, C.; Sulaman, M.; Saleem, M.I.; Tang, Y.; Jiang, Y.; Tang, L.; et al. Solution-processed, flexible and broadband photodetector based on $CsPbBr_3$/PbSe quantum dot heterostructures. *J. Mater. Sci. Technol.* **2021**, *68*, 216–226. [CrossRef]
54. Sulaman, M.; Yang, S.; Bukhtiar, A.; Tang, P.; Zhang, Z.; Song, Y.; Imran, A.; Jiang, Y.; Cui, Y.; Tang, L.; et al. Hybrid Bulk-Heterojunction of Colloidal Quantum Dots and Mixed-Halide Perovskite Nanocrystals for High-Performance Self-Powered Broadband Photodetectors. *Adv. Funct. Mater.* **2022**, *32*, 2201527. [CrossRef]
55. Dortaj, H.; Dolatyari, M.; Zarghami, A.; Alidoust, F.; Rostami, A.; Matloub, S.; Yadipour, R. High-speed and high-precision $PbSe/PbI_2$ solution process mid-infrared camera. *Sci. Rep.* **2021**, *11*, 1533. [CrossRef] [PubMed]
56. Nenadovic, M.T.; Comor, M.I.; Vasic, V.; Micic, O.I. Transient Bleaching of Small PbS Colloids. Influence of Surface Properties. *J. Phys. Chem. A* **1990**, *16*, 6390–6396.
57. Lingley, Z.; Lu, S.; Madhukar, A. A High Quantum Efficiency Preserving Approach to Ligand Exchange on Lead Sulfide Quantum Dots and Interdot Resonant Energy Transfer. *Nano Lett.* **2011**, *11*, 2887–2891. [CrossRef] [PubMed]
58. Zhang, D.; Song, J.; Zhang, J.; Wang, Y.; Zhang, S.; Miao, X. A facile and rapid synthesis of lead sulfide colloidal quantum dots using in situ generated H2S as the sulfur source. *CrystEngComm* **2013**, *15*, 2532–2536. [CrossRef]
59. Lin, Q.; Yun, H.J.; Liu, W.; Song, H.; Makarov, N.S.; Isaienko, O.; Nakotte, T.; Chen, G.; Luo, H.; Klimov, V.I.; et al. Phase-Transfer Ligand Exchange of Lead Chalcogenide Quantum Dots for Direct Deposition of Thick, Highly Conductive Films. *J. Am. Chem. Soc.* **2017**, *139*, 6644–6653. [CrossRef]
60. Shestha, A.; Yin, Y.; Andersson, G.G.; Spooner, N.A.; Qiao, S.; Dai, S. Versatile PbS Quantum Dot Ligand Exchange Systems in the Presence of Pb-Thiolates. *Small* **2017**, *13*, 1602956. [CrossRef]
61. Shuklov, I.A.; Toknova, V.F.; Demkin, D.V.; Lapushkin, G.I.; Nikolenko, L.M.; Lizunova, A.A.; Brichkin, S.B.; Vasilets, V.N.; Razumov, V.F. A New Approach to the Synthesis of Lead Sulfide Colloidal Quantum Dots in a Mixture of Oleylamine and Oleic Acid. *High Energy Chem.* **2020**, *54*, 183–188. [CrossRef]
62. Wang, C.; Wang, Y.; Jia, Y.; Wang, H.; Li, X.; Liu, S.; Liu, X.; Zhu, H.; Wang, H.; Liu, Y.; et al. Precursor Chemistry Enables the Surface Ligand Control of PbS Quantum Dots for Efficient Photovoltaics. *Adv. Sci.* **2023**, *10*, 2204655. [CrossRef]

63. Konstantatos, G.; Howard, I.; Fischer, A.; Hoogland, S.; Clifford, J.; Klem, E.; Levina, L.; Sargent, E.H. Ultrasensitive solution-cast quantum dot photodetectors. *Nature* **2006**, *442*, 180–183. [CrossRef] [PubMed]
64. Liu, H.; Li, M.; Voznyy, O.; Hu, L.; Fu, Q.; Zhou, D.; Xia, Z.; Sargent, E.H.; Tang, J. Physically Flexible, Rapid-Response Gas Sensor Based on Colloidal Quantum Dot Solids. *Adv. Mater.* **2014**, *26*, 2718–2724. [CrossRef] [PubMed]
65. De Iacovo, A.; Venettacci, C.; Colace, L.; Scopa, L.; Foglia, S. PbS Colloidal Quantum Dot Photodetectors operating in the near infrared. *Sci. Rep.* **2016**, *6*, 37913. [CrossRef] [PubMed]
66. Ren, Z.; Sun, J.; Li, H.; Mao, P.; Wei, Y.; Zhong, X.; Hu, J.; Yang, S.; Wang, J. Bilayer PbS Quantum Dots for High-Performance Photodetectors. *Adv. Mater.* **2017**, *29*, 1702055. [CrossRef]
67. Qiao, K.; Cao, Y.; Yang, X.; Khan, J.; Deng, H.; Zhang, J.; Farooq, U.; Yuan, S.; Song, H. Efficient interface and bulk passivation of PbS quantum dot infrared photodetectors by PbI$_2$ incorporation. *RSC Adv.* **2017**, *7*, 52947–52954. [CrossRef]
68. Xu, K.; Xiao, X.; Zhou, W.; Jiang, X.; Wei, Q.; Chen, H.; Deng, Z.; Huang, J.; Chen, B.; Ning, Z. Inverted Si:PbS Colloidal Quantum Dot Heterojunction-Based Infrared Photodetector. *ACS Appl. Mater. Interfaces* **2020**, *12*, 15414–15421. [CrossRef]
69. Shi, Y.; Wu, Z.; Xiang, Z.; Chen, P.; Li, C.; Zhou, H.; Dong, X.; Gou, J.; Wang, J.; Jiang, Y. Silicon-based PbS-CQDs infrared photodetector with high sensitivity and fast response. *Nanotechnology* **2020**, *31*, 485206. [CrossRef]
70. Szendrei, K.; Cordella, F.; Kovalenko, M.V.; Böberl, M.; Hesser, G.; Yarema, M.; Jarzab, D.; Mikhnenko, O.V.; Gocalinska, A.; Saba, M.; et al. Solution-Processable Near-IR Photodetectors Based on Electron Transfer from PbS Nanocrystals to Fullerene Derivatives. *Adv. Mater.* **2009**, *21*, 683–687. [CrossRef]
71. He, J.; Qiao, K.; Gao, L.; Song, H.; Hu, L.; Jiang, S.; Zhong, J.; Tang, J. Synergetic Effect of Silver Nanocrystals Applied in PbS Colloidal Quantum Dots for High-Performance Infrared Photodetectors. *ACS Photonics* **2014**, *1*, 936–943. [CrossRef]
72. Bessonov, A.A.; Allen, M.; Liu, Y.; Malik, S.; Bottomley, J.; Rushton, A.; Medina-Salazar, I.; Voutilainen, M.; Kallioinen, S.; Colli, A.; et al. Compound Quantum Dot–Perovskite Optical Absorbers on Graphene Enhancing Short-Wave Infrared Photodetection. *ACS Nano* **2017**, *11*, 5547–5557. [CrossRef]
73. Ka, I.; Gerlein, L.F.; Asuo, I.M.; Bouzidi, S.; Gedamu, D.M.; Pignolet, A.; Rosei, F.; Nechache, R.; Cloutier, S.G. Solution-Processed p-Type Copper Thiocyanate (CuSCN) Enhanced Sensitivity of PbS-Quantum-Dots-Based Photodiode. *ACS Photonics* **2020**, *7*, 1628–1635. [CrossRef]
74. Klem, E.J.D.; Gregory, C.; Temple, D.; Lewis, J.S. PbS colloidal quantum dot photodiodes for low-cost SWIR sensing. In Proceedings of the SPIE Defense + Security Symposium, Baltimore, MD, USA, 20–24 April 2015.
75. Goossens, S.; Navickaite, G.; Monasterio, C.; Gupta, S.; Piqueras, J.J.; Pérez, R.; Burwell, G.; Nikitskiy, I.; Lasanta, T.; Galán, T.; et al. Broadband image sensor array based on graphene–CMOS integration. *Nat. Photonics* **2017**, *11*, 366–371. [CrossRef]
76. Zhang, J.; Xu, J.; Chen, T.; Gao, X.; Wang, S. Toward Broadband Imaging: Surface-Engineered PbS Quantum Dot/Perovskite Composite Integrated Ultrasensitive Photodetectors. *ACS Appl. Mater. Interfaces* **2019**, *11*, 44430–44437. [CrossRef] [PubMed]
77. Choi, H.T.; Kang, J.; Ahn, J.; Jin, J.; Kim, J.; Park, S.; Kim, Y.; Kim, H.; Song, J.D.; Hwang, G.W.; et al. Zero-Dimensional PbS Quantum Dot–InGaZnO Film Heterostructure for Short-Wave Infrared Flat-Panel Imager. *ACS Photonics* **2020**, *7*, 1932–1941. [CrossRef]
78. Georgitzikis, E.; Malinowski, P.E.; Li, Y.; Lee, J.; Süss, A.; Frazzica, F.; Maes, J.; Gielen, S.; Verstraeten, F.; Boulenc, P.; et al. Organic- and QD-based image sensors integrated on 0.13 μm CMOS ROIC for high resolution, multispectral infrared imaging. In Proceedings of the 2019 International Image Sensor Workshop (IISW), Snowbird, UT, USA, 23–27 June 2019.
79. Kassim, A.; Ho, S.M.; Abdullah, A.H.; Nagalingam, S. XRD, AFM and UV-Vis optical studies of PbSe thin films produced by chemical bath deposition method. *Sci. Iran.* **2010**, *17*, 139–143.
80. Begum, A.; Hussain, A.; Rahman, A. Effect of deposition temperature on the structural and optical properties of chemically prepared nanocrystalline lead selenide thin films. *Beilstein J. Nanotechnol.* **2012**, *3*, 438–443. [CrossRef]
81. Kumar, P.; Pfeffer, M.; Schweda, E.; Eibl, O.; Qiu, J.; Shi, Z. PbSe mid-IR photoconductive thin films (part I): Phase analysis of the functional layer. *J. Alloys Compd.* **2017**, *724*, 316–326. [CrossRef]
82. Sun, J.; Zhang, Y.; Fan, Y.; Tang, X.; Tan, G. Strategies for boosting thermoelectric performance of PbSe: A review. *Chem. Eng. J.* **2022**, *431*, 133699. [CrossRef]
83. Zhao, X.; Tang, X.; Li, T.; Chen, M. Trap-mode PbSe mid-infrared photodetector with decreased-temperature processing method. *Infrared Phys. Technol.* **2023**, *133*, 104788. [CrossRef]

Disclaimer/Publisher's Note: The statements, opinions and data contained in all publications are solely those of the individual author(s) and contributor(s) and not of MDPI and/or the editor(s). MDPI and/or the editor(s) disclaim responsibility for any injury to people or property resulting from any ideas, methods, instructions or products referred to in the content.

MDPI AG

Grosspeteranlage 5

4052 Basel

Switzerland

Tel.: +41 61 683 77 34

Materials Editorial Office

E-mail: materials@mdpi.com

www.mdpi.com/journal/materials

Disclaimer/Publisher's Note: The title and front matter of this reprint are at the discretion of the Guest Editors. The publisher is not responsible for their content or any associated concerns. The statements, opinions and data contained in all individual articles are solely those of the individual Editors and contributors and not of MDPI. MDPI disclaims responsibility for any injury to people or property resulting from any ideas, methods, instructions or products referred to in the content.

www.ingramcontent.com/pod-product-compliance
Lightning Source LLC
LaVergne TN
LVHW071940160726
843515LV00011B/2660